Next Civilization

Dirk Helbing

Next Civilization

Digital Democracy and Socio-Ecological
Finance—How to Avoid Dystopia
and Upgrade Society by Digital Means

Second Edition

 Springer

Dirk Helbing
ETH Zürich
Zürich, Switzerland

ISBN 978-3-030-62329-6   ISBN 978-3-030-62330-2   (eBook)
https://doi.org/10.1007/978-3-030-62330-2

This Springer imprint is published by the registered company Springer Nature Switzerland AG
The registered company address is: Gewerbestrasse 11, 6330 Cham, Switzerland

*I would like to dedicate this book to
Dietmar Huber
for the incredible support he has given to me
over so many years.*

Acknowledgements

I would like to thank the FuturICT community for the many inspiring discussions and also everyone, who had to be patient with me in the last couple of years, including my parents.

I am also very grateful to Philip Ball, Stefano Bennati, Anna Carbone, Andreas Diekmann, Jeroen van den Hoven, Dietmar Huber, Eoin Jones, Caleb Koch, Richard Mann, Heinrich Nax, Paul Ormerod, Evangelos Pournaras, Kay-Ti Tan, and others for their valuable feedback on the manuscript and the many improvements (but don't hold them responsible for any contents of this book). Jan Fasnacht, Petr Neugebauer, Petra Parikova, and Felix Schulz have been a great help with figures, formatting, references, permissions, and cover designs.

Furthermore, I would like to thank the Wissenschaftskolleg zu Berlin—Institute for Advanced Study for the creative atmosphere and excellent opportunity to do research and prepare this book.[1]

[1] I also like to acknowledge support from the European Research Council (ERC) under the European Union's Horizon 2020 research and innovation programme (grant agreement No. 833168), see: Using the Wisdom of Crowds to Make Cities Smarter, FuturICT Blog (March 28, 2019) http://futurict.blogspot.com/2019/03/using-wisdom-of-crowds-to-make-cities.html; furthermore, see: ERC 2018 Advanced Grants Highlighted Projects, ERC (March 27, 2019) https://erc.europa.eu/news/erc-2018-advanced-grants-highlighted-projects.

Prologue

We hold these truths to be self-evident, that all men are created equal, that they are endowed by their Creator with certain unalienable Rights, that among these are Life, Liberty and the pursuit of Happiness.

The United States Declaration of Independence after Thomas Jefferson[2]

Those who surrender freedom for security will not have, nor do they deserve, either one.

Benjamin Franklin[3]

Many people know me as the scientist who has worked, among other things, on traffic and pedestrian flows and crowd disasters. As some may still remember, I have also been the initiator and scientific coordinator of the FuturICT project.[4] This project attracted serious interest of *Google, NASA*, the US government,[5] Russia[6] and China,[7] for example.

FuturICT was a visionary project for the digital age that was in the pole position for Europe's one billion dollar flagship funding.[8] You will find more information about it in the Epilogue. In short: hundreds of scientists were ready to bring a new

---

[2]https://en.wikipedia.org/wiki/United_States_Declaration_of_Independence.

[3]http://www.goodreads.com/quotes/140634-those-who-surrender-freedom-for-security-will-not-have-nor.

[4]See https://web.archive.org/web/*/www.futurict.eu.

[5]The U.S. needs a FuturICT program to confront the challenges of the 21st century government, GovLoop (January 2, 2013) https://www.govloop.com/community/blog/the-u-s-needs-a-futurict-program-to-confront-the-challenges-of-21st-century-government/.

[6]The FuturICT project on Russian TV: https://youtu.be/H3aKfswfGag.

[7]The FuturICT project on Chinese National TV: https://youtu.be/a4rxb2r9SdI.

[8]See http://www.futurict.eu and Participatory Science and Computing for Our Complex World, EPJ Special Topics 214 (2012) https://link.springer.com/journal/11734/214/1/page/1.

paradigm for the world on the way. But, then, to everyone's—including our competitors'—surprise, another flagship candidate was chosen.[9] With the 1 billion Euro Human Brain Project,[10] Europe wanted to build a realistic brain in silico, and Barack Obama announced a multi-billion dollar Brain Initiative.[11] With this, the transhumanist era was launched.[12] In the course of this book, we will understand the worrying implications of this.

In principle, it is a normal thing that one project wins and another one loses, and people move on. However, here, things were pretty different for various reasons. Whatever part of the project I wanted to follow up on—Nervousnet,[13] for example —was seriously obstructed. This applied also to our work on "Digital Democracy",[14] which was repeatedly not even accessible on the Internet. Furthermore, I was permanently put under pressure for years, and so I wondered what was going on…

My discoveries were highly concerning.[15] In the meantime, a digital mirror world[16]*has* become reality[17]—but without privacy, ethics, participation and democracy, it seems. Instead of a visionary, almost utopian project, it appears that dystopia is on its way, as I will show in the course of this book (besides the great potentials of the digital revolution, too).

---

[9]Neelie Kroes & Prof. Henry Markram: Human Brain Project (January 29, 2013) https://www.youtube.com/watch?v=DsZ_LBdthC0.

[10]See https://en.wikipedia.org/wiki/Human_Brain_Project and https://www.humanbrainproject.eu.

[11]Obama launches multi-billion dollar brain-map project, Nature NewsBlog (April 2, 2013) http://blogs.nature.com/news/2013/04/obama-launches-ambitious-brain-map-project-with-100-million.html.

[12]The Age of Transhuman Politics Has Begun, Telepolis (April 12, 2015) https://www.heise.de/tp/features/The-Age-of-Transhumanist-Politics-Has-Begun-3371228.html?seite=all.

[13]See https://web.archive.org/web/*/nervousnet.info; Creating a Planetary Nervous System Together, TEDx Groningen (February 5, 2015) https://www.youtube.com/watch?v=BKcWPdSUJVA; A digital Nervousnet for everyone and the golden age of complexity science (talk given at the ICCSS2015) https://www.youtube.com/watch?v=pN2hAcr6ujk.

[14]D. Helbing and E. Pournaras, Build Digital Democracy, Nature 527, 33-34 (2015) https://www.nature.com/news/society-build-digital-democracy-1.18690.

[15]Some of them have found their way into the Science Fiction „iGod" by Willemijn Dicke and myself, published at CreateSpace (2017).

[16]Virtual 'mirror' of world to help predict future, The Sunday Times (January 6, 2013) https://www.thetimes.co.uk/article/virtual-mirror-of-world-to-help-predict-future-rcflmwx0dss.

[17]Sentient world: war games on the grandest scale, The Register (June 23, 2007) https://www.theregister.com/2007/06/23/sentient_worlds/; The era of digital twins and the mirror world, Medium (May 5, 2020) https://medium.com/@nomoko/the-era-of-digital-twins-and-the-mirror-world-82b33e3e3d46.

I have been warning the world about the dangers of dual use[18] of digital technologies since 2011.[19] After the FuturICT flagship was turned down, I got concerned and started to alert the public.[20] In the years since 2013, hundreds of newspaper articles appeared (these can be found with *Google* News). I have given several hundred talks, which were uploaded to the FuturICT youtube channel[21] (but not all of them are publicly accessible). I have also published dozens of blogs (see http://futurict.blogspot.com), many of which appeared in two books entitled "Thinking Ahead"[22] and "Towards Digital Enlightenment".[23]

Furthermore, I have worked on the self-published book "The Automation of Society Is Next: How to Survive the Digital Revolution",[24] which is written for a scientifically interested readership that is open to philosophical, ethical and programmatic considerations. The 10 chapters from 2015 (Chaps. 1–5 and 7–11) have become the core of this book with the title *Next Civilization*, but I have added 4 new chapters that were written in 2019 (Chap. 14) and 2020 (Chaps. 6, 12 and 13). For the sake of historical authenticity, I left the previous chapters largely unchanged.

From today's perspective, it is hard to imagine that the 2015 version of the book was highly controversial. At that time, a global digital control system was in the making. My alternative vision of the future, based on distributed organization, coordination, self-organization, and self-governance, fundamentally questioned the data-driven paradigm controlled by a powerful Artificial Intelligence (AI).

At that time, many were looking forward to the "singularity", after which we would see a superintelligent system or "digital God,"[25] which may be allowed to reign the world like a "benevolent dictator",[26] based on mass surveillance data. It could also be tasked to make the world sustainable. Unfortunately, the Artificial Intelligence system may figure out that a "depopulation strategy" might "fix the

---

[18]See https://en.wikipedia.org/wiki/Dual-use_technology.

[19]See the second part of D. Helbing and S. Balietti, From social data mining to forecasting socio-economic crises, EPJ Special Topics 195, 3–68 (2011) https://link.springer.com/article/10.1140/epjst/e2011-01401-8.

[20]Probably the first contribution of this kind was "Google als Gott?", published in the NZZ on March 20, 2013, https://www.nzz.ch/startseite/google-als-gott-1.18049950; English translation: Google as God?, in D. Helbing (ed.) Thinking Ahead (Springer, 2015) Chap. 8, https://www.springerprofessional.de/en/google-as-god-opportunities-and-risks-of-the-information-age/2369816, https://arxiv.org/pdf/1304.3271.

[21]See https://www.youtube.com/futurict.

[22]D. Helbing (ed.) Thinking Ahead: Essays on Big Data, Digital Revolution, and Participatory Market Society (Springer, 2015).

[23]D. Helbing (ed.) Towards Digital Enlightenment: Essays on the Dark and Light Sides of the Digital Revolution (Springer, 2019).

[24]D. Helbing, The Automation of Society Is Next: How to Survive the Digital Revolution (CreateSpace, 2015).

[25]Google and the Birth of a Digital God?, The Globalist (December 25, 2017) https://www.theglobalist.com/google-artificial-intelligence-big-data-technology-future/.

[26]See https://en.wikipedia.org/wiki/Benevolent_dictatorship.

problem". Later on, you will learn that this terrible scenario is not just theoretical, but a real possibility and serious threat.

I have, therefore, started to work on concepts that could make the world more sustainable in a democratic way: "digital democracy," "City Olympics," "democratic capitalism," and a "socio-ecological finance system" ("FIN4") to boost a circular economy[27]—to mention just a few of the ideas featured in Chaps. 7–13. In other words, there are surely alternatives to a dystopian future, but we would have to bring them on the way.

It turns out, however, that this is more difficult than expected, because many stakeholders seem to have different plans in mind, such as China, for example, or the big tech companies of "surveillance capitalism". These often appear to be more interested in maximizing power or profit than in "saving the world". And even those who claim that they want to save the world (such as CERN,[28] the United Nations,[29] and the World Economic Forum[30]) seem to pursue a centralized approach based on mass surveillance, with little transparency and no democratic participation.

In the end, one of the people who probably know more about this made a worrying statement in front of the United Nations' General Assembly 2019:

> *"Can we still have confidence in politics, in business, in international organizations? These are questions to which we must find answers at our General Assembly".*[31]

In fact, the past years have been a tough time characterized by a global struggle for the path into the future. The following events give just a glimpse of what has happened in these years.[32]

---

[27]See https://en.wikipedia.org/wiki/Circular_economy.

[28]CERN: The gulf between machine learning and AI, The Inquirer (July 29, 2015) https://web.archive.org/web/20150731231358/https://www.theinquirer.net/inquirer/feature/2419669/cern-the-gulf-between-machine-learning-and-artificial-intelligence.

[29]UN and CERN celebrate science for peace and development and CERN's 60th anniversary (October 20, 2014) https://home.cern/news/press-release/cern/un-and-cern-celebrate-science-peace-and-development-and-cerns-60th.

[30]Shaping the Future of Cybersecurity and Digital Trust, https://www.weforum.org/platforms/shaping-the-future-of-cybersecurity-and-digital-trust; accordingly, the Global Centre for Cybersecurity (https://www.weforum.org/videos/global-centre-for-cybersecurity) brings together actors from America, Russia, China, Israel, Saudi Arabia and many other countries, Europol, Interpol, and companies from the banking, insurance, oil and Internet sectors, in particular Cisco and Huawei, Amazon, IBM, Mastercard, Microsoft, Palantir and PayPal.

[31]Official Speech of Federal President Ueli Maurer in front of the UN General Assembly, New York (September 24, 2019) https://www.newsd.admin.ch/newsd/message/attachments/58526.pdf.

[32]A more detailed timeline seems to be given by this Connectivist blog: https://theconnectivist.wordpress.com/2019/08/06/log-of-the-rightwing-power-grab-of-society/, https://web.archive.org/web/*/https://theconnectivist.wordpress.com/2019/08/06/log-of-the-rightwing-power-grab-of-society/.

On September 25, 2015, for example, a Snowden revelation was published about the "Karma Police" program run by British secret service GCHQ, which judges (the value of) everyone's life based on mass surveillance.[33] On the same day, Pope Francis promoted the Sustainability Development Goals (Agenda 2030) at the United Nation's general assembly.[34] I mention this here, because—believe it or not —I sent him a preprint of "The Automation of Society Is Next: How to Survive the Digital Revolution" before (it was my only mail to a pope ever). Just ahead of his speech, on September 24, 2015, and on September 11, 2015, Saudi Arabia suffered from a twin disaster. It was the biggest tragedy in the history of the Muslim pilgrimage[35] and felt like part of the apocalypse.

As a follow-up to the UN general assembly, from November 30 to December 16, 2015, the Paris Agreement was worked out, which tried to bring a binding global contract to fight climate change on the way.[36] However, before the US Congress signed the agreement, Donald Trump was elected US president, who later quit the international climate deal.[37]

Shortly before the Paris Agreement, I published the Digital Manifesto ("Digital Democracy Rather than Data Dictatorship")[38] together with an interdisciplinary team of scientists. The German online version appeared on November 12, 2015, but one day later, the world got distracted by the shocking terror attacks in Paris[39] ahead of the climate summit. The publication of the English translation was ready to be published in Scientific American the week after. However, it got delayed for more than a year (in fact, until the end of the Obama administration).

---

[33]British 'Karma Police' program carries out mass surveillance of the web, The Verge (September 25, 2015) https://www.theverge.com/2015/9/25/9397119/gchq-karma-police-web-surveillance; Profiled: From Radio to Porn, British Spies Track Web Users' Online Identities, The Intercept (September 25, 2015) https://theintercept.com/2015/09/25/gchq-radio-porn-spies-track-web-users-online-identities/.

[34]Pope Francis Addresses U.N., Calling for Peace and Environmental Justice, The Guardian (September 25, 2015) https://www.nytimes.com/2015/09/26/world/europe/pope-francis-united-nations.html; United Nations Sustainable Development Summit 2015, https://www.unaids.org/en/resources/presscentre/featurestories/2015/september/20150925_UN_Summit_opening.

[35]See https://en.wikipedia.org/wiki/2015_Mina_stampede and K. Haase, M. Kasper, M. Koch, and S. Müller, A Pilgrim Scheduling Approach to Increase Safety During the Hajj, Operations Research 67(2), 376–406 (2019) https://pubsonline.informs.org/doi/abs/10.1287/opre.2018.1798; I.Ö. Verbas et al. Integrated Optimization and Simulation Framework for Large-Scale Crowd Management Application, Transportation Research Record 2560, 57–66 (2016) https://journals.sagepub.com/doi/pdf/10.3141/2560-07.

[36]See https://en.wikipedia.org/wiki/Paris_Agreement.

[37]See https://en.wikipedia.org/wiki/United_States_withdrawal_from_the_Paris_Agreement.

[38]Das Digital-Manifest: Digitale Demokratie statt Datendiktatur, Spektrum der Wissenschaft (November 12, 2015) https://www.spektrum.de/thema/das-digital-manifest/1375924; Eine Strategie für das digitale Zeitalter, https://www.spektrum.de/kolumne/eine-strategie-fuer-das-digitale-zeitalter/1376083.

[39]See https://en.wikipedia.org/wiki/November_2015_Paris_attacks.

Already before Obama left his office, it seems that China was trying to take over the role of the world's leading superpower. At the G20 summit in Hangzhou in September 2016, Obama was denied the usual red carpet treatment.[40] Before this happened, on April 30, 2016, Obama had warned the world[41]:

> "... this is also a time around the world when some of the fundamental ideals of liberal democracies are under attack, and when notions of objectivity, and of a free press, and of facts, and of evidence are trying to be undermined. Or, in some cases, ignored entirely.
> And in such a climate, it's not enough just to give people a megaphone. And that's why your power and your responsibility to dig and to question and to counter distortions and untruths is more important than ever".

This hinted at the upcoming post truth era,[42] which apparently took over after the control of the Internet was given to ICANN at the end of September 2016.[43] It seems this cleared the way for a highly personalized Internet (using surveillance data about all of us). The development was complemented by two laws allowing for (counter-)propaganda.[44] I believe, we have been in a kind of information war[45] ever since. After the Cambridge Analytica scandal,[46] we know how much this development has challenged democracies all over the world.

---

[40]Barack Obama 'deliberately snubbed' by Chinese in chaotic arrival at G20, The Guardian (September 4, 2016) https://www.theguardian.com/world/2016/sep/04/barack-obama-deliberately-snubbed-by-chinese-in-chaotic-arrival-at-g20.

[41]"Obama out": President Barack Obama's hilarious final White House Correspondents' Dinner speech (April 30, 2016) https://www.youtube.com/watch?v=NxFkEj7KPC0; see also https://obamawhitehouse.archives.gov/the-press-office/2016/05/01/remarks-president-white-house-correspondents-dinner.

[42]See https://en.wikipedia.org/wiki/Post-truth_politics.

[43]An Internet Giveaway to the U.N., Wall Street Journal (August 28 2016) https://www.wsj.com/articles/an-internet-giveaway-to-the-u-n-1472421165.

[44]See https://en.wikipedia.org/wiki/Smith–Mundt_Act, https://en.wikipedia.org/wiki/Countering_Foreign_Propaganda_and_Disinformation_Act.

[45]See https://en.wikipedia.org/wiki/Information_warfare; 'I made Steve Bannon's psychological warfare tool': Meet the data war whistleblower, The Guardian (March 18, 2018) https://www.theguardian.com/news/2018/mar/17/data-war-whistleblower-christopher-wylie-faceook-nix-bannon-trump; Before Trump, Cambridge Analytica quietly built "psyops" for militaries, FastCompany (September 25, 2019) https://www.fastcompany.com/90235437/before-trump-cambridge-analytica-parent-built-weapons-for-war.

[46]The Cambridge Analytica Files, The Guardian, https://www.theguardian.com/news/series/cambridge-analytica-files.

You are certainly aware that the struggle for the future of this planet has further intensified since the world started suffering from COVID-19.[47] So, if you want to understand what might come next and what are the alternatives, it is about time to read this book, because these developments will likely affect your life much more than you ever have imagined.

Zürich, Switzerland
August, 2020[48]

---

[47]See https://en.wikipedia.org/wiki/COVID-19_pandemic, https://en.wikipedia.org/wiki/List_of_COVID-19_pandemic_legislation.

[48]Where no date for the access of an Internet URL is given: all links were checked and accessed around August 8, 2020, if not specified otherwise. If an URL is not anymore accessible, please try to find it via the Internet Archive/Wayback Machine at https://archive.org/web/. Many scientific references may, for example, be found via https://scholar.google.com or https://www.researchgate.net

Contents

Chapter 1
Introduction: The Digital Society

A Better Future or Worse?

Smartphones, tablets and app stores with almost unlimited possibilities have become symbols of the digital revolution. However, while these innovations make our lives more comfortable and interesting, they herald a much more fundamental transformation. Advances in digital technology now affect the way we learn, decide, and interact. By harnessing „Big Data", the „Internet of Things", and Artificial Intelligence (AI), we can create smart homes and smart cities. But this is only the tip of the iceberg – our entire economy and society will also dramatically change. What are the opportunities and risks related to this? Are we heading towards digital slavery or freedom? What forces are at work and how can we use them to create a smarter society? This book offers a guided tour through the new, digital age ahead.

After the automation of factories and the creation of self-driving cars, the automation of society is next. While we were busy with our smartphones, the world has secretly changed behind our backs. In fact, our world is changing with increasing speed, and much of that change is being driven by developments in Information and Communications Technology (ICT). These technologies, such as laptop computers, mobile phones, tablets and smart watches, seemed to be about convenience. They came along and enabled us to calculate, communicate and archive with greater speed and efficiency than ever before. However, there was very little recognition that, one day, they would not only facilitate our cultural discourse and institutions, but also reshape our entire world. Large-scale mass surveillance, the global spread of *Uber* taxis and the *BitCoin* crypto-currency are just a few of the irritating symptoms of the digital era to come.

1.1 Living in the Age of "Big Data"

Suddenly, there is also a great hype about "Big Data". No wonder Dan Ariely compared the frenzy about Big Data with teenage sex:

© Springer Nature Switzerland AG 2021
D. Helbing, *Next Civilization*,
https://doi.org/10.1007/978-3-030-62330-2_1

"everyone talks about it, nobody really knows how to do it, everyone thinks everyone else is doing it, so everyone claims they are doing it..."

But some are actually doing it. In fact, "Big Data" has already given rise to many interesting applications, such as real-time language translation. So, what is "Big Data"? The term refers to massive amounts of data, which have been collected about technological, social, economic and environmental systems and activities. To get an idea of "Big Data," imagine the digital traces that almost all our activities leave, including the data created by our consumption and movement patterns. Every single minute, we produce about 700,000 *Google* queries and 500,000 *Facebook* comments. If you add all of the location data of people using smartphones, the consumption data of people who buy things, and cookies which track every click and tap of our online activity, you will begin to comprehend the enormity of "Big Data".

All the contents collected in the history of humankind until the year 2003 are estimated to amount to five billion gigabytes—the data volume that around 2015 was produced approximately every day. While we have been speaking of an "information age" since the middle of last century, the digital era started only in 2002. Since then, the digital storage capacity has exceeded the analog one. Today, more than 95% of all data are available in digital form. Even by avoiding credit card transactions, social media and digital technologies, it is no longer possible to completely avoid digital footprints on the Internet.

1.2   Data Sets Bigger Than the Largest Library

The availability of Big Data about almost every aspect of our lives, institutions and cultures has fueled the hope that we could now solve the world's problems. Every Internet purchase we make generates data about our preferences, finances and location that will be stored on a server somewhere and used for various purposes, possibly without our consent. Cell phones disclose where we are, and private messages and conversations are being analyzed. It will probably not be long before every newborn baby is genome-sequenced at birth. Books are being digitized and collated in immense, searchable databases of words that are being data-mined to enable "culturomics", a field which puts history, society, art and cultural trends under the lens. Aggregated data can be used to reveal unexpected facts in a way that would never have been possible before the digital age. For example, an analysis of *Google* searches can reveal an impending flu epidemic.

This avalanche of data continues to grow. The introduction of technologies such as *Google Glass* encourages people to document and archive almost every aspect of their lives. Further data sets include credit-card transactions, communication data, *Google Earth* imagery, public news, comments and blogs. These data sources have been termed "Big Data" and are creating an increasingly accurate digital picture of our physical and social world, as well as the global economy.

"Big Data" will certainly change our world. The term was coined more than 15 years ago to describe data sets so big that they can no longer be analyzed using standard computational methods. If we are to benefit from Big Data, we must learn to "drill" and "refine" it into useful information and knowledge. This is a significant challenge.

The tremendous increase in the volume of data is attributable to four important technological innovations. First, the *Internet* enables global communication between electronic devices. Second, the *World Wide Web* (WWW) has created a network of globally accessible websites, which emerged as a result of the invention of the Hypertext Transfer Protocol (HTTP). Third, the emergence of *social media* platforms such as *Facebook*, *Google+*, *WhatsApp* and *Twitter* has created social communication networks. Finally, a wide range of previously offline devices such as TV sets, fridges, coffee machines, cameras as well as sensors, smart wearable devices (such as activity trackers) and machines are now connected to the Internet, creating the "Internet of Things" (IoT) or "Internet of Everything" (IoE). Meanwhile, the data sets collected by companies such as *eBay*, *Walmart* or *Facebook*, must be measured in petabytes— 1 million billion bytes. This amounts to more than 100 times the information stored in the US Library of Congress, which is the largest physical library in the world.

Mining Big Data offers the potential to create new ways to optimize processes, identify interdependencies and make informed decisions. However, Big Data also produces at least four major new challenges (the "four V's"). First, the unprecedented *volume* of data means that we need immense processing power and storage capacity to deal with the huge amounts of data. Second, the *velocity* at which data must be processed has increased: now, continuous data streams must often be analyzed in real-time. Third, Big Data is mostly unstructured, and the resulting *variety* of data is difficult to organize and analyze. Finally, the *veracity* of the data may be difficult to handle because Big Data tends to contain errors and is usually neither representative nor complete.

1.3 Will a Digital Revolution Solve Our Problems?

Let us see what an evidence-based approach building on the wealth of today's data can do for us. In the past, whenever a problem had to be solved, the best course of action was to "ask the experts". These experts would go to the library, collect up-to-date knowledge, and supervise Ph.D. students who would help to fill gaps in existing knowledge. But this was a slow process. Nowadays, whenever people have a question, they ask *Google* or consult *Wikipedia*, for example. This might not always give the definitive or best answer, but it delivers quick answers. On average, decisions taken in this way may even be better than many decisions made in the past. It is no wonder, therefore, that policymakers love the Big Data approach, which seems to provide immediate answers. Business people sensing the immense commercial opportunities are getting excited too.

1.4   Big Data Gold Rush for the Twenty-First Century's Oil

The fact that we have much more information about our world than ever before is both a blessing and a curse. The accumulation of socio-economic data often implies a long-term intrusion into personal privacy and raises a number of important issues. It cannot be denied that Big Data is a powerful resource that supports evidence-based decision-making and that it holds unprecedented potential for business, politics, science and citizens. Recently, the social media portal *WhatsApp* was sold to *Facebook* for $19 billion, when it had 450 million users. This sale price implies that each employee generated almost half a billion dollars in share value.

There is no doubt that Big Data creates tremendous opportunities, not just because of its application in process optimization and marketing, but also because the information itself is becoming monetized. As demonstrated by the virtual currency *BitCoin*, it is now even possible to turn bits into monetary value. It can be literally said that data can be mined into money in a way that would previously have been considered a fairy tale. For a time, *BitCoins* were even more valuable than gold.

Therefore, it is no surprise that many experts and technology gurus claim that Big Data is the "oil of the twenty-first century", a new way of making money—big money. Although many Big Data sets are proprietary, the consultancy company *McKinsey* recently estimated that the potential value of Open Data alone is $3–5 trillions per year.[1] If the worth of this publicly available information were to be evenly distributed among the world's population, every person on Earth would receive an additional $700 per year. Therefore, the potential of Open Data significantly exceeds the value of the international free trade and service agreements that are currently under secret negotiation.[2] Given these numbers, are we currently setting the right political and economic priorities? This is a question we must pay attention to, because it will determine our future.

The potential of Big Data spans every area of social activity, from processing human language and managing financial assets, to empowering cities to balance energy consumption and production. Big Data also holds the promise of enabling us to better protect the environment, to detect and reduce risks, and to discover opportunities that would otherwise have been missed. In the area of personalized medicine, Big Data will probably make it possible to tailor medications to patients in order to increase their effectiveness and reduce their side effects. Big Data will

---

[1] https://www.mckinsey.com/business-functions/mckinsey-digital/our-insights/open-data-unlock ing-innovation-and-performance-with-liquid-information.

[2] I recommend the readers to look up Wikipedia to inform themselves about the impending international agreements coming under the abbreviations TTIP, CETA, TPP, and TISA. It seems that these would dramatically increase the power of multi-national corporations. Would this be good or bad?

also accelerate the research and development of new drugs and focus resources on the areas of greatest need.

It is clear, therefore, that the potential applications of Big Data are various and rapidly spreading. While it will enable personalized services and products, optimized production and distribution processes, as well as "smart cities", it will reveal also unexpected links between our activities. But beyond this, where are we heading?

1.5 Will Artificial Intelligence Overtake Us?

Today, an average mobile phone is more powerful than the computers used to send the Apollo rocket to the moon and even the *Cray-2* supercomputer thirty years ago, which weighted several tons and had the size of a building. This amazing progress is a result of "Moore's law", which posits that computer processing power increases exponentially. But thanks to powerful "machine learning" methods, information systems are becoming more intelligent, too. They do calculations faster than us, they play chess better than us, they remember information longer than us, and they perform more and more tasks that only humans could do in the past. Will they soon be smarter than us? Are the days counted when humans were the "crown of creation"? The famous futurist Ray Kurzweil (*1948), now a director of engineering at *Google*, was the first to claim that this critical moment (the so-called "singularity") is near.[3]

A few years ago, when I read that Artificial Intelligence (AI) might pose a serious threat to humanity, I found this hard to imagine, even ridiculous. However, experts now predict that computers will be able to perform most tasks better than humans in 5–10 years, and reach brain-like functionality within 10–25 years. The AI systems of today are no longer expert systems programmed by computer scientists—they are learning and evolving. To understand the implications, I recommend you to watch some eye-opening videos on deep learning and artificial intelligence.[4] These videos demonstrate that most of the activities we earn our money with today (such as reading and listening to language, distinguishing different patterns, and performing routines) can now be done by computers almost as well as by humans, if not better. Jim Spohrer's perspective on *IBM*'s cognitive computing products is as follows:[5] The first Artificial Intelligence applications will be our tools. As they get smarter, they will become our "partners", and when they overtake us, they will be our "coaches".

Will algorithms, computers, or robots be our bosses in a few decades from now? The Massachusetts Institute of Technology has started to study such scenarios.[6] It is extremely important therefore to realize that the digital revolution is not just

---

[3] Kurzweil [1, 2].

[4] For example, you may watch this TEDx video of Jeremy Howard to get an idea of what machine learning is currently capable of: https://www.youtube.com/watch?v=xx310zM3tLs.

[5] See https://www.youtube.com/watch?v=E7PVBGtEYyg.

[6] MIT Study Shows People Would Rather Take Orders From A Robot Than Their Boss, http://www.businessinsider.com/robots-as-bosses-2014-8?IR=T.

about more powerful computers, better smartphones or fancier gadgets. The digital revolution will change all our personal lives, and it will transform entire economies and societies. In fact, in the coming two or three decades we will see some dramatic changes. A lot of production and services will become automated, and this will fundamentally change the way we work in the future.

Quite soon, within the next two decades or so, less than 50% of people will have jobs for which they have been trained (i.e. agriculture, industry or services).[7] Even highly skilled jobs will be at risk. How will the masses of personal data collected about each of us then be used?

1.6   When Big Data Starts to Steer Our Lives

It may sound far-fetched at first, but we must ask this question: "Will we be remotely controlled by personalized information, or is this happening already?" It is clear that *Google* and *Facebook* know very well what we are interested in when they place individually tailored ads that often match our interests and tastes. *Google Now* is an example of a smart app that tells you what to do, if you have signed up for it. For instance, if there is a traffic jam on the way to your next appointment, *Google Now* may suggest you to leave 15 min earlier in order to be on time. Similarly, *Amazon* suggests what we might want to buy, and *Trip Advisor* suggests what destinations to visit and what hotels to book. *Twitter* tells us what others think—and what we should perhaps think, too. *Facebook* suggests whom to be friend with. Apps like *OkCupid* even suggest whom we might date.

While all these services can certainly be helpful we might ask: what will be the consequences? Will we end up living in a digital "golden cage"—a "filter bubble" as Eli Pariser calls it?[8] Will we just execute what our smart devices tell us to do? Modern learning software already corrects us when we make mistakes. Smart wristbands tell us how many more steps we should make today. Eye trackers can discover if we are tired or stressed, and computers can predict when our performance will decrease. In other words, we are increasingly patronized in our decision making by computer programs. Will we soon be incompetent to live on our own? And, are we sliding into a "nanny state", where we don't have a say? Has our decision-making, has democracy been "hacked"?

Why should we care? Isn't it just great that computers do calculations for us more quickly than we can do them ourselves? Isn't it fantastic that our smartphones help us manage our agendas, and that *Google Maps* tells us the way to go? Why not ask *Apple's Siri* to recommend us a restaurant? I certainly don't object to any of these functions, but it is important to recognize that this is just the beginning of what is to come. Little by little, our role as self-determined decision-makers is being eroded. The next logical step will be the automation of society. How might this look like?

---

[7]Frey and Osborne [3].

[8]Pariser [4].

1.7 The Cybernetic Society

This question brings us to an old concept that goes back to Norbert Wiener (1894–1964)[9] who was known as the father of control theory ("cybernetics"). Wiener imagined that our society could be controlled like a huge clockwork, where every company's and every individual's activity would be coordinated by a giant plan of how to run a society in an optimal way.

Many decades ago, Russia and other communist countries ran command economies. However, they failed to be competitive, while the capitalist approach based on free entrepreneurship thrived. At that time information systems were much more limited in power and scope than today. This has changed. Now there is a third approach besides communism and capitalism: socio-economic systems that are managed in a data-driven way. In the early 70ies Chile was the first country to attempt a "cybernetic society".[10] It established a control center, which collected the latest production data of major companies every week. This was a truly revolutionary approach, but despite its obvious advantages, the government was unable to stay in power, and Salvador Allende (1908–1973), the president of the country, had a tragic end.[11] Nevertheless, the dream of a cybernetic society has not ceased to exist.

Today, both Singapore and China are trying to plan social and economic activities in a top-down way using lots of data, and they enjoy larger growth rates than Western democracies. Therefore, many economic and political leaders raise the question: "Is democracy outdated?" Should we run our societies in a cybernetic way according to a grand plan? Will Big Data allow us to optimize our future?

1.8 Wise Kings and Benevolent Dictators, Fueled by Big Data

Given all the data one can now accumulate, is it conceivable that governments or big companies might try to build "God-like", almost "omniscient" information systems? Could these systems then make decisions like a "benevolent dictator" or "wise king"? Will they be able to avoid coordination failures and irrationality? Would it even become possible to create the best of all worlds by collecting all data globally and building a digital "crystal ball" to predict the future, as some people have suggested? If this were possible, and given that "knowledge is power", could a sort of digital "magic wand" be created by a government or company to ensure that the benevolent dictator's master plan remains on course?

What would it take to build such powerful tools? It would require information systems that knew us so well that they could manipulate our decisions by stimulating

---

[9] Wiener [5, 6].

[10] Medina [7].

[11] He committed suicide.

us with the right kind of personalized information. As I will show in this book, such systems are actually on their way, or they exist already.

1.9 Do We Need to Sacrifice Our Personal Freedom?

Establishing a cybernetic society has a number of important implications. For example, we would need a lot of personal data. In order to be able to control an entire society, it seems important to understand how we think, what we feel, and what we plan to do. Large amounts of personal data are essential to allow artificially intelligent machines to learn what determines our actions and how to influence them. In fact, while mass surveillance is surprisingly ineffective in fighting terrorism[12] and child abuse,[13] it seems to be very useful to establish a cybernetic society.

But as with every technology, there are serious drawbacks. We would probably lose some of the most important rights and values that have formed the bedrock of democracies and their judicial systems since the Age of Enlightenment. Secrecy and privacy would be eroded by information technologies, and with this, we would lose our security and human values such as mercy and forgiveness. With the advent of predictive policing and other proactive enforcement measures, we could see a deviation from the "presumption of innocence" principle towards the implementation of an ominous "public interest" policy at the cost of individual rights. Do we therefore need to worry about the fact that the leading Big Data nation has more people per thousand inhabitants in prison than any other country, including Russia and China?[14]

With the help of mass surveillance, it is now possible to punish even the smallest mistakes that everyone makes in an overregulated society.[15] The displayed speeding ticket for going 1 km/h too fast with my car (see Fig. 1.1) should be a wake-up call regarding what will soon possible on a much larger scale. In addition to sanctions by public authorities, will insurance companies punish us in future for eating unhealthy food? Will banks offer us punitive interest rates on loans simply because we live in the "wrong neighborhood"? Will we get restricted offers of products or services if we don't fulfill certain expectations, or will we have to pay higher prices? While

---

[12]The Washington Post (January 12, 2014): NSA phone record collection does little to prevent terrorist attacks, group says, https://www.washingtonpost.com/world/national-security/nsa-phone-record-collection-does-little-to-prevent-terrorist-attacks-group-says/2014/01/12/8aa860aa-77dd-11e3-8963-b4b654bcc9b2_story.html; http://securitydata.newamerica.net/nsa/analysis; Gill [8]; see also BBC News (August 24, 2009): 1000 cameras 'solve one crime'.

[13]The biggest sexual child abuse scandals revealed in the past years have actually not been discovered by digital surveillance techniques.

[14]In fact, in recent years there were about 45 million arrests in the USA (see the National Geographic of January 22, 2013: The war on drugs is a "miserable failure", https://blog.nationalgeographic.org/2013/01/22/the-war-on-drugs-is-a-miserable-failure/). About 2% of the overall population is in prison. This is about 10 times more than in many European countries (see the list of countries by incarceration rate, http://en.wikipedia.org/wiki/List_of_countries_by_incarceration_rate#Incarceration_rates. Despite this, it didn't really make the USA a safer place than Europe.

[15]In this connection I recommend to read J. Schmieder [9].

Stadt Zürich
Stadtpolizei

Verzeigungs-Nr.: 7136306 080 0

Zürich, 29.08.07 / UAUOBD 8

Stadt Zürich
Stadtpolizei
Abteilung Sonderleistungen
Kommissariat Zentralstelle für
Verkehrs- und Ordnungsbussen
Bahnhofquai 5
Postfach 1067
8021 Zürich

Tel 044 411 76 76
Fax 044 212 63 86

Dr.
Helbing Dirk

█████████████

Übertretungsanzeige

Wir haben festgestellt, dass die Lenkerin oder der Lenker des unten genannten Fahrzeuges folgende Verkehrsregelverletzung(en) begangen hat:

Ueberschreiten allgemeiner Höchstgeschwindigkeit innerorts um 1-5 km/h

Bussenbetrag : CHF 40.00	Devisenkurs : 1.60	**Bussenbetrag : EUR 25.00**

Kontrollschild-Nr. : ████████   Fahrzeugart : PW
Fahrzeughalterln : Helbing Dirk
Übertretungsort : Zürich 6, Wehntalerstrasse 200   Datum : So-29.07.2007
Fahrtrichtung :   stadteinwärts   Zeit : 13:03 Uhr

Gemessene Geschwindigkeit	54	km/h	Massgebende Geschwindigkeit	51	km/h
Abzug der Sicherheitsmarge	-3	km/h	Abzug der Geschwindigkeitsbegrenzung	50	km/h
Massgebende Geschwindigkeit	51	km/h	**Geschwindigkeitsüberschreitung**	**1**	**km/h**

Fig. 1.1 Illustration of how citizens could be punished for any minor transgression of law, even if it is entirely harmless to society. Note that the traffic authority figured out my foreign address to send me this ticket for going 1 km/h too fast with my car, while I was actually not even the driver (which they didn't check)....

this might sound like a dystopian science fiction fantasy, much of this is already happening. China is now even planning to rate the behavior of all its citizens on a one-dimensional scale, including what they do online.[16] Opinions that match the thinking of the communist political party will be rewarded. The resulting score will be used to determine whether or not a person gets a particular job or loan.

Can such surveillance-based technology- and data-driven approaches turn a country into a "perfect clockwork"? And given that every country is exposed to global competition, is it just a matter of time until all democracies adopt such approaches? If you think this is far-fetched, it is probably good to recall that several influential decision-makers have recently praised China and Singapore as models for the

---

[16]China rates its own citizens—including online behaviour, see http://www.volkskrant.nl/buiten
land/china-rates-its-own-citizens-including-online-behaviour~a3979668/; China: Kontrolle über alles, see http://www.zeit.de/politik/ausland/2015-07/china-plangesellschaft-xi-jinping.

world.[17] Such thinking could soon end freedom and democracy as we know it.[18] That is why we must pay attention to this now. Of course, some people might ask why we shouldn't do it, if this increases the efficiency of our lives and our society? Doesn't history teach us that society evolves over time? Why should we worry, if companies and governments take care of us?

The crucial question is whether they are doing a good job in satisfying our needs and interests. In view of financial and economic crises, cybercrime, climate change and many other problems, it seems, however, that governments have great difficulties to fulfill this promise. Similarly, if we consider the Silicon Valley as a business-driven vision of society, it also seems far-fetched to claim that everyone there is well taken care of.

1.10 Who Will Rule the World?

There is no doubt that Big Data has the potential to be totalitarian. But, in principle, there is nothing wrong with Big Data, Artificial Intelligence, and cybernetics. The question is only, how to use it? For example, who will rule the world in future: will it be big business, government elites, or Artificial Intelligence? Will citizens and experts be no longer relevant to decision-making processes? If powerful information systems knew the world and each of us, would they vote or take decisions for us? Would they tell us what to do or steer our behavior through personalized information? Or will we instead live in a free and democratic society, in which everyone makes decisions in an autonomous but well-coordinated way, empowered by personal digital assistants?

1.11 Two Scenarios: Coercion or Freedom

Will we need to sacrifice our privacy, freedom, dignity, and informational self-determination for a more efficient governance of our world? We must think about these possibilities now. Due to the important societal, economic, legal and ethical implications, we must take some crucial decisions.[19] In the aftermath of September 11, 2001, we have certainly witnessed increasing attempts to control citizen activities, including a massive surveillance of our on-line activities. Does the digital revolution imply that we will lose our human rights? Will we lose our autonomy and merely obey what powerful information systems tell us to do? Will we end up with censorship?

---

[17]For example, on March 24, 2015, The Economist featured a comment entitled "China Model. This house believes China offers a better development model than the West".

[18]If you don't like the concepts of freedom and democracy, replace them by the related ones of innovation and collective intelligence, which are important to find good solutions to complex problems (see Appendix 4.1).

[19]Helbing [10].

The digital revolution implies great opportunities and risks. Digital technologies enable different ways of running future economies and societies. If we don't want to lose our jobs, personal freedom, and democracy, we must carefully consider how to make digital technologies work for us, rather than against us. There are at least two possibilities to automate society: we may either run it in a top-down way, trying to controll the citizens' decisions and actions, using powerful information technologies, or we may instead support bottom-up self-organization based on distributed control. The latter would be compatible with individual freedom, creativity and innovation (see Fig. 1.2). However, the digital revolution enables both. As I will explain further on, to support largely autonomous decisions and processes and the coordination of them, we would have to

1. create participatory opportunities,
2. support informational self-control,
3. increase distributed design and control elements,
4. add transparency for the sake of trust,
5. reduce information biases and noise,
6. enable user-controlled information filters,
7. support socio-economic diversity,
8. increase interoperability and innovation,
9. build coordination tools,
10. create digital assistants,
11. support collective ("swarm") intelligence,
12. measure and consider external effects ("externalities"),
13. enable favorable feedback loops,
14. support a fair and multi-dimensional value exchange,
15. increase digital literacy and awareness.

In this book, I am trying to offer concepts and ideas that can contribute to a smarter and more resilient digital society. Such a framework is needed, because in many important respects, the world has become quite unpredictable and unstable. This is partially due to the increasing level of interdependency of our systems, often driven by advances in Information and Communication Technology. Therefore, which approach will be superior in a world characterized by too much data, too much speed, and too much connectivity? Will it be top-down governance or bottom-up participation? Or would it be better to combine both approaches? And how would we do this?[20] We will see that this mainly depends on the complexity of the systems surrounding us.

---

[20]In his book on The Third Industrial Revolution, Jeremy Rifkin gives interesting answers to this for the case of smart (energy) grids and some other systems.

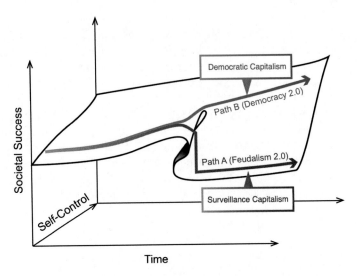

Fig. 1.2 Schematic illustration of two evolutionary paths, leading to two different types of a digital society. Path A would undermine individual freedoms, democracy, and jobs for most of us. Path B corresponds to a society based on self-control and participatory information systems supporting creative and innovative activities of everyone. Which one will we choose?

1.12 A Better Future Ahead of Us

Despite the many challenges, I am still optimistic about our *long-term* future on the whole. We have already managed societal transitions several times in human history and I am sure we can manage this one, too.

In the following chapters I hope to make a contribution to a necessary public debate, by proposing two main possible societal frameworks for the coming digital age. One of these is based upon the concept of a "big government", which takes decisions like a "benevolent dictator" or a "wise king". This framework would be empowered by huge masses of data, something like a digital "crystal ball". This might be seen as a futuristic version of Thomas Hobbes (1588–1679) "Leviathan". His belief was that social order can not exist without a powerful state, otherwise we would all behave like wild beasts.[21]

The alternative framework for the digital society is based on the concept of self-organizing systems. This vision relates to the concept of the "invisible hand", for which Adam Smith (1723–1790) is known. He assumed that the best societal outcomes are reached through self-organization based on market forces. However,

---

[21] This has been expressed by the Latin phrase "homo hominis lupus", i.e. humans treat each other like wolves.

financial meltdowns and "tragedies of the commons" such as environmental pollution or harmful climate change[22] suggest that the "invisible hand" cannot always be
relied on.

But, what if future information and communication technologies would allow us to
reach desirable systemic outcomes through decentralized decision-making and self-
organization? Can distributed control and coordination mechanisms, empowered
by real-time measurements and feedback, make the "invisible hand" work? The
feasibility of this exciting vision will be explored in the later chapters of this book, in
which I will describe a new paradigm to achieve success and socio-economic order
in the twenty-first century. Will this lead us into a new era of creativity, participation,
collective intelligence, and well-being?

In fact, examples from the spheres of traffic management and production demonstrate that it is possible to manage complex systems from the bottom-up—and to
efficiently produce desirable outcomes in this way. In the following chapters, I will
explain the general principle behind such "magic self-organization" and how it could
help us to navigate our way through a complex future. I will further explain the role
of collective intelligence and how it can help us to cope with the complexity of our
globalized world.

Therefore, rather than trying to control or combat the self-organized dynamics
of complex systems such as our economy, financial system, global trade, and transportation systems, we could learn to harness their underlying forces to our benefit.
Of course, this would involve to locally adapt the interactions of the components
of these systems. But if we could achieve this, self-organization could be used
to produce desirable outcomes, and this would enable us to create well-ordered,
effective, efficient and resilient systems!

Critics might argue that, just because self-organization has been shown to work
in complex technological systems (such as traffic control or industrial production
lines), this does not necessarily mean that it would also work for socio-economic
systems. After all, the behaviors of people can be quite surprising. In the light of this
counterargument, we will explore whether and when self-organization can outperform conventional top-down control in managing complex dynamical systems. We
will discuss institutional settings and interaction rules that will allow self-organizing
systems to be superior. One can now use real-time data to enable adaptive feedback
mechanisms, so that systems behave favorably and in a stable way. More specifically, I propose that the "Internet of Things", with its vast underlying networks of
sensors, will make socio-economic self-organization possible in a distributed and
bottom-up way. But the crucial question is how to make a data-oriented approach
based on distributed control work? The solution, as we will see, requires "complexity
science".

---

[22]Even the Pentagon considers climate change to be a threat to U.S. security, see https://www.nyt
imes.com/2009/08/09/science/earth/09climate.html.

1.13   On the Way to a Smarter Digital Society

In the long run, I am confident about our future, mainly because I believe in the power of ideas in the digital age. However, we must remain alert to the possibility that we could make serious mistakes along the way. The financial system may fail, democracies may—intentionally or accidentally—turn into surveillance societies, or we may end up fighting wars. Therefore, with this book, I am trying to explain the opportunities and risks ahead of us.

Signs of change are everywhere. Information technologies are transforming the global economy at a rapid pace. In essence, we are experiencing nothing less than a "third economic revolution",[23] leading to an "Economy 4.0". Its effects will be at least as profound as those of the first (agrarian to industrial) and second (industrial to service) revolutions. The ubiquity of digital technologies—such as social media, smart devices, the Internet of Things and Artificial Intelligence—is giving rise to a digital society. We can no longer afford to be passive bystanders of this seismic societal transition. We must prepare for it. But we should not regard these changes merely as a threat to social and global stability. In fact, we are faced with a once-in-a-century opportunity!

For example, the digital revolution is not only changing the way we learn, behave, make decisions and live. It is also altering the way we produce and consume, and our conception of ownership. Information is a very interesting resource in that it is the basis of culture and can be shared as often as we like. To get more of it, we do not have to take it away from others. We don't have to fight for it. This of course will depend on how our economy is organized in the future. In particular, on how we reward effort for the production of data, information, knowledge and creative digital products. We can either perpetuate the outdated principles of the twentieth century or open the door to a smarter twenty-first century society. Why don't we do the latter?

Many people are now talking about "smart homes", "smart factories", "smart grids", and "smart cities". It is logical that we will soon have a "smart economy" and a "smart society". Networked information systems will enable entirely new solutions to the world's problems. One thing is therefore clear: the world in the digital age will be very different. But even if we can't exactly predict what the future will hold, can we at least get a glimpse of it? I believe we can, at least to some extent, and some trends are already emerging. Clearly, the characteristics of the future world will result from the technological, social and evolutionary forces shaping it. The technological drivers include Big Data, the Internet of Things and Artificial Intelligence. The social drivers include the increase of information volumes and networking. In addition, there are evolutionary forces that will lead to new kinds of incentive systems and decision-making. I will attempt to evaluate the implications of these forces, and debate the opportunities and risks they present.

These forces, in turn, are generated by the interactions occurring within our "anthropogenic systems", i.e. our man-made or human-influenced techno-socio-economic-environmental systems. In order for interventions to be beneficial, we

---

[23] See, for example, Rifkin [11].

must understand how these interactions—and the forces they are creating—can be harnessed to our advantage in the same way as we have learned to harness the forces of nature.

How will the digital revolution reshape our socio-economic institutions? And what preparations can we make? While addressing these and other questions, I will pursue a non-ideological approach, oriented neither politically left nor right, but carefully exploring novel opportunities. I will try to explain how we can use the digital revolution to make our society more innovative, successful and resilient, through understanding the new logic of the digital era to come. I will discuss how we can adapt our systems in real-time with novel technologies. Furthermore, I will outline how we can build an information and innovation ecosystem that can create new jobs and opportunities for everyone.

We are now ready to dive into the details of why our world is troubled and how we can fix it—by using advanced "information and communication systems" in entirely new ways. The following chapters will focus on subjects such as prediction and control, complexity, self-organization, awareness and coordination, responsible decision-making, real-time measurement and feedback, systems design, innovation, reward systems, co-creation, and collective intelligence. I hope this journey through the opportunities and risks presented by the emerging digital age will be as exciting for you as it is for me!

References

1. R. Kurzweil, The Age of Spiritual Machines: When Computers Exceed Human Intelligence (Penguin, 2000).
2. Ray Kurzweil, The Singularity Is Near (Penguin, 2005).
3. C.B. Frey and M.A. Osborne (2013) The future of employment: How susceptible are jobs to computerisation? See http://www.oxfordmartin.ox.ac.uk/downloads/academic/The_Future_of_Employment.pdf.
4. E. Pariser, The Filter Bubble: What the Internet Is Hiding from You (Viking/Penguin, 2011).
5. N. Wiener, Cybernetics, Or the Control and Communication in the Animal and the Machine (MIT Press, 1965).
6. N. Wiener, The Human Use of Human Beings: Cybernetics and Society (Da Capo Press, 1988).
7. E. Medina, Cybernetic Revolutionaries (MIT Press, 2011); see also http://www.spiegel.de/ein estages/projekt-cybersyn-stafford-beers-internet-vorlaeufer-in-chile-a-1035559.html.
8. M. Gill, Spriggs: Assessing the impact of CCTV. Home Office Research, Development and Statistics Directorate (2005) https://www.newscientist.com/article/dn26801-mass-survei llance-not-effective-for-finding-terrorists/.
9. J. Schmieder (2013) Mit einem Bein im Knast: Mein Versuch, ein Jahr lang gesetzestreu zu leben (Bertelsmann).
10. D. Helbing (2015) Societal, Economic, Ethical and Legal Challenges of the Digital Revolution: From Big Data to Deep Learning, Artificial Intelligence, and Manipulative Technologies, Jusletter IT, https://jusletter-it.weblaw.ch/en/issues/2015/21-Mai-2015/societal%2c-eco nomic%2c-_588206025c.html, preprint available at http://arxiv.org/abs/1504.03751.
11. Jeremy Rifkin, The Third Industrial Revolution: How Lateral Power is Transforming Energy, the Economy, and the World (Palgrave Macmillan, 2013).

Chapter 2
Complexity Time Bomb

When Systems Get Out of Control

The digital revolution produces more data, more speed, more connectivity, and more complexity. Besides creating new opportunities, how will this change our economy and our societies? Will it make our increasingly interdependent systems easier to control? Or are we heading towards a systemic collapse? In order to figure out what needs to be done to fix the world's ills, we must explore why things, as they currently stand, go wrong. The question is, why haven't we learned how to deal with them yet?

These days, we seem to be surrounded by economic and socio-political crises, by terrorism, conflict and crime. More and more often, the conventional "medicines" to tackle these sorts of global problems turn out to be inefficient or even counter-productive. It is increasingly evident that we approach these problems with an outdated understanding of our world. While it may still look more or less as it looked for a long time, the world has changed inconspicuously but fundamentally.

We are used to the idea that societies must be protected from *external* threats such as earthquakes, volcanic eruptions, hurricanes and military attacks by enemies. Increasingly, however, we are threatened by different kinds of problems that come from *within* the system, such as financial instabilities, economic crises, social and political unrest, organized crime and cybercrime, environmental change, and the spread of diseases. These problems have become some of the greatest threats to humanity. According to the "risk interconnection map" published by the World Economic Forum, the greatest risks faced by our societies today are of socio-economic and political nature.[1] These risks, including factors such as economic inequality and governance failure, are twenty-first century problems which cannot be solved with twentieth century wisdom. They are larger in scale than ever before and result from the complex interdependencies in today's anthropogenic systems. As a result, it is of paramount importance that we develop a better understanding of the characteristics of complex dynamical systems. To this end, I will discuss the main reasons why things go wrong, such as unstable dynamics, cascading failures

---

[1] World Economic Forum, The Global Risks Report 2013 and 2014, see http://www3.weforum.org/docs/WEF_GlobalRisks_Report_2013.pdf; http://www3.weforum.org/docs/WEF_GlobalRisks_Report_2014.pdf.

© Springer Nature Switzerland AG 2021
D. Helbing, *Next Civilization*,
https://doi.org/10.1007/978-3-030-62330-2_2

in networks, and systemic interdependencies. And I will illustrate these problems through a large variety of examples such as traffic jams, electrical blackouts, financial crises, crime, wars and revolutions.

2.1 Phantom Traffic Jams

Complex systems can be found all around us and include phenomena such as turbulent flows in our global weather system, decision-making processes, opinion formation in groups, financial and economic markets, and the evolution and spread of languages. However, we must carefully distinguish complex systems from complicated ones. While a car, which consists of thousands of parts, is complicated, it is easy to control nevertheless (when it works properly). Traffic flow, on the other hand, depends on the dynamical interactions of many cars, and forms a complex dynamical system. These interactions produce counter-intuitive phenomena such as "phantom traffic jams" which appear to have no cause. Such "emergent" phenomena cannot be understood from the properties of the single parts of the system in isolation, here the driver-vehicle units. While many traffic jams occur for specific, identifiable reasons, such as accidents or road works, almost everyone has also encountered situations where a queue of vehicles seems to form "out of nothing" and where there is no visible cause.[2]

To explore the true reasons for these "phantom traffic jams", Yuki Sugiyama and his colleagues at Nagoya University in Japan carried out an experiment in which they asked many people to drive around a circular track.[3] The task sounds simple, and the vehicles did in fact flow smoothly for some time. Then, however, one of the cars caused a minor variation in the traffic flow, which triggered stop-and-go traffic—a traffic jam that moved backwards around the track.

While we often blame poor driving skills of others for such "phantom traffic jams", studies in complexity science have shown that they are actually an emergent collective phenomenon, which is the inevitable result of the interaction between vehicles. A detailed analysis demonstrates that, if the traffic density exceeds a certain "critical" threshold—that is, if the average separation of the vehicles is smaller than a certain value—then even the slightest variation in the speed of cars can eventually cause a disruption of the traffic flow through an amplification effect. As the next driver in line needs some time to adjust to a change in the speed of the vehicle ahead, he or she will have to brake a bit harder to compensate for the delay. The then following driver will have to break even harder, and so on. The resulting chain reaction amplifies the initially small variation in a vehicle's speed and this eventually produces a traffic jam which, of course, every single driver tried to avoid.

---

[2] See https://traffic-simulation.de/ring.html.
[3] See https://www.youtube.com/watch?v=7wm-pZp_mi0 and Sugiyama et al. [1].

2.2 Recessions—Traffic Jams in the World Economy?

Economic supply chains may exhibit a similar kind of behavior, as illustrated by John Sterman's "beer distribution game".[4] The game simulates some of the challenges of supply chain management. When playing it, even experienced managers will end up ordering too much stock, or will run out of it.[5] This situation is as difficult to avoid as stop-and-go traffic. In fact, our scientific research suggests that economic recessions can be regarded as a kind of traffic jam in the global flow of goods, i.e. the world economy. This insight is actually somewhat heartening, since it implies that we may be able to engineer solutions to mitigate economic recessions in a similar way as one can reduce traffic jams by driver assistant systems. The underlying principle will be discussed later, in the chapter on Digitally Assisted Self-Organization. In order to do this, however, one would need have real-time data detailing the global flow and supply of materials.

2.3 Systemic Instability

Crowd disasters are another, tragic example of systemic instability. Even when every individual within a crowd is peacefully minded and tries to avoid harming others, many people may die nevertheless. Appendix 2.1 outlines why such extreme systemic outcomes can result from normal, non-aggressive behavior.

What do all these examples tell us? They illustrate that the natural (re)actions of individuals will often be counterproductive.[6] Our experience and intuition often fail to account for the complexity of highly interactive systems, which tend to behave in unexpected ways. Such complex dynamical systems typically consist of many interacting components which respond to each other's behaviors. As a consequence of these interactions, complex dynamical systems tend to self-organize. That is, some collective dynamics may develop (such as stop-and-go traffic), which is different from the natural behavior of the system components when they are separated from each other (such as drivers who don't like to stop). In other words, the overall system may show new characteristics that are distinct from those of its components. This can result, for example, in "chaotic" or "turbulent" dynamics.

Group dynamics and mass psychology may be viewed as typical examples of spontaneously emerging collective dynamics occurring in crowds. What is it that makes a crowd turn "mad", violent, or cruel? For example, after the London riots of 2011, people have asked how it was possible that teachers and the daughters of millionaires—people you would not expect to be criminals—were participating in the looting? Did they suddenly develop criminal minds when the demonstrations against police violence turned into riots? Possibly, but not necessarily so.

---

[4]Sterman and Sterman [2], Helbing and Lämmer [3].

[5]Nienhaus et al. [4].

[6]Dorner [5].

Understanding the emergence of new properties requires an interaction-oriented perspective.

When many components of a complex dynamical system interact, they frequently create new kinds of structures, properties or functions in a self-organized way. To describe newly resulting characteristics of the system, the term "emergent phenomena" is often used. For example, water appears to be wet, extinguishes fire, and freezes at a particular temperature, but we would not expect this based on the examination of single water molecules.

Therefore, complex dynamical systems may show surprising behaviors. They cannot be steered like a car. In the above traffic flow experiment, although the aim of participants was to drive continuously at a reasonably high speed, a phantom traffic jam occurred due to the interactions between cars. While it is straightforward to control a car, it may be impossible for individual drivers to control the collective dynamics of traffic flow, which is the result of the interactions of many cars.

2.4 Beware of Strongly Coupled Systems!

Instability is just one possible problem of complex dynamical systems. It occurs when the characteristic parameters of a system cross certain critical thresholds. If a system becomes unstable, small deviations from the normal behavior are amplified. Such amplification is often based on feedback loops, which cause a mutual reinforcement. If one amplification effect triggers others, a chain reaction may occur and a minor, random variation may be enough to trigger an unstoppable domino effect. In case of systemic instability, as I have demonstrated for the example of phantom traffic jams, the system will inevitably get out of control sooner or later, no matter how hard we try to prevent this. Consequently, we should identify and avoid conditions under which systems behave in an unstable way.

In many cases, strongly coupled interactions are a recipe for disaster or other undesirable outcomes.[7] While our intuition usually works well for problems that are related to weakly coupled systems (in which the overall system can be understood as being the sum of its parts and their properties), the behaviors of complex dynamical systems can change dramatically, if the interactions among their components are strong. In other words, these systems often behave in counter-intuitive ways, so that conventional wisdom tends to be ineffective for managing them. Unintended consequences or side effects are common.

What further differences do strong interactions make? First, they may cause larger variability and faster changes, particularly if there are "positive feedbacks" that lead to reinforcement and acceleration. Second, the behavior of the complex system can be hard to predict, making it difficult to plan for the future. Third, strongly connected systems tend to show strong correlations between the behaviors of (some of) its components. Fourth, the possibilities to control the system from the outside or through

---

[7]Helbing [6].

the behavior of single system components are limited, as the system-immanent inter-actions may have a stronger influence. Fifth, extreme events occur more often than expected and they may affect the entire system.[8]

In spite of all this, most people still have a component-oriented worldview, centered on individuals rather than groups and interdependent events. This often leads to "obvious"[9] but wrong conclusions. For example, we praise heroes when things go well and search for scapegoats when something goes wrong. Yet, the discussion above has clearly shown how difficult it is for individuals to control the outcome of a complex dynamical system, if the interactions between its components are strong. This fact can also be illustrated by politics.

Why do politicians, besides managers, have (on average) the worst reputation of all professions? This is probably because we think they are hypocritical; we elect them on the basis of the ideas and policies they publicly voice, but then they often do something else. This apparent hypocrisy is a consequence of the fact that politicians are subject to many strong interactions with lobbyists and interest groups, who have diverse points of view. All of these groups push the politicians into various directions. In many cases, this forces them to make decisions that are not compatible with their own point of view—a fact, which is hard to accept for voters. However, if we believe that democracy is not just about elections every few years, but also about a continuous consolidation between citizens and their elected representatives, it could be argued that it would be undemocratic for politicians to unreservedly place their own personal convictions above the concerns of citizens and businesses in the systems they are supposed to represent. Managers of companies find themselves in similar situations. They are exposed to many different factors they must consider. This kind of interaction-based decision-making extends far beyond boardrooms and parliaments. Think of the decision-dynamics in families: if it were easy to control, there would be probably less divorces...

Crime is another example of social systems getting out of control.[10] This can happen both on an individual and collective level. Many crimes, including murders, are committed by average people, rather than career criminals (and even in countries

---

[8]In social systems, this may lead to an erosion of trust in private and public institutions, which can create social, political or economic instability.

[9]Watts [7].

[10]The traditional explanation is that a crime is committed if the expected reward is larger than the likely punishment, multiplied by the probability of being caught and convicted. Theoretically, therefore, it should be possible to eliminate all crime simply by raising the punishment. If all criminals were rational egoists, who constantly calculated their expected payoff from crime, this idea should even work. Assuming sufficiently high conviction rates, the strong punishment would make crime "unattractive" because of the expectation of making a "loss". However, empirical evidence questions these simple assumptions. On the one hand, people don't behave entirely rational. For example, they don't usually pick pockets, even though they could often escape punishment. On the other hand, deterrence strategies are surprisingly ineffective in most countries and high crime rates are often recurrent. For example, even though the USA have 10 times more prisoners than most European countries, the rates of various crimes are still much higher. Therefore, the conventional understanding of crime is wrong.

with death penalty).[11] A closer inspection shows that many crimes correlate with the circumstances individuals find themselves in. For example, group dynamics often plays an important role. Many scientific studies also show that the socio-economic conditions are a strong determining factor of crime. Therefore, in order to counter crime, it might be more effective to change these socio-economic conditions rather than sending more people to jail. I say this with one eye on the price we have to pay to maintain a large prison population—a single prisoner costs more than the salary of a postdoctoral researcher with a Ph.D. degree! It worries even more that prisons and containment camps have apparently become places where criminal organizations are formed and terrorist plots are planned.[12] In fact, our own studies suggest that many crimes spread by imitation.[13]

2.5   Cascading Effects in Complex Networks

To make matters worse, besides the dynamic instability caused by amplification effects, complex dynamical systems may produce even bigger problems. Worldwide trade, air traffic, the Internet, mobile phones, and social media have made everything much more convenient—and connected. This has created many new opportunities, but everything now depends on a lot more things. What are the implications of this increased level of interdependency? Today, a single tweet can make stock markets spin. A controversial *Youtube* video can trigger a riot that kills dozens of people.[14] Increasingly, our decisions can have consequences on the other side of the globe, many of which may be unintended. For example, the rapid spread of emerging epidemics is largely a result of the scale of global air traffic today, and this has serious repercussions for global health, social welfare and economic systems.

It is being said that "the road to hell is paved with good intentions". By networking our world, have we inadvertently created conditions in which disasters are more likely to emerge and spread? In 2011 alone, at least three cascading failures with global impact were happening, thereby changing the face of the world and the global balance of power: The world economic crisis, the Arab spring and the combination of an earthquake, a tsunami and a nuclear disaster in Japan. In 2014, the world was threatened by the spread of Ebola, the crisis in Ukraine, and the conflict with the Islamic State (IS). In the following subsections, I will therefore discuss some examples of cascade effects in more detail.

---

[11] Wilfling [8].

[12] Tagesanzeiger (December 11, 2014) «Boxershorts halfen uns, den Krieg zu gewinnen», http://www.tagesanzeiger.ch/ausland/naher-osten-und-afrika/Boxershorts-halfen-uns-den-Krieg-zu-gewinnen/story/20444791.

[13] Perc et al. [9].

[14] https://en.wikipedia.org/wiki/Reactions_to_Innocence_of_Muslims.

2.6 Large-Scale Power Blackouts

On November 4, 2006, a power line was temporarily turned off in Ems, Germany, to facilitate the transfer of a Norwegian ship. Within minutes, this caused a blackout in many regions all over Europe, from Germany to Portugal! Nobody expected this to happen. Before the line was switched off, a computer simulation predicted that the power grid would still operate well without the line. However, the scenario analysis did not account for the possibility that another line would spontaneously fail. In the end, a local overload in Northwest Germany triggered emergency switch-offs all over Europe, creating a cascade effect with pretty astonishing results. Blackouts occurred in regions thousands of kilometers away.[15]

Can we ever hope to understand such strange behavior? In fact, a computer-based simulation study of the European power grid recently managed to reproduce quite similar effects.[16] It demonstrated that the failure of a few network nodes in Spain could create an unexpected blackout several thousand kilometers away in Eastern Europe, while the electricity network in Spain would still work.[17] Furthermore, increasing the capacity of certain parts of the power grid could worsen the situation and cause an even greater blackout. Therefore, weak elements in the system may serve an important function: they can be "circuit breakers" interrupting the failure cascade. This is an important fact to remember.

2.7 From Bankruptcy Cascades to Financial Crisis

The sudden financial meltdown in 2008 is another example of a crisis that hit many companies and people by surprise. In a presidential address to the American Economic Association in 2003, Robert Lucas (*1937) said:

> *"[The] central problem of depression-prevention has been solved."*

Similarly, Ben Bernanke (*1953), the former chairman of the Federal Reserve Board, held a longstanding belief that the economy was both well understood and in sound financial shape. In September 2007, Ric Mishkin (*1951), a professor at Columbia Business School and then a member of the Board of Governors of the US Federal Reserve System, made another interesting statement, reflecting widespread beliefs at this time:

---

[15]While other areas close to the region of initial overload were not affected at all.

[16]Asztalos et al. [10].

[17]See http://www.plosone.org/article/fetchSingleRepresentation.action?uri=info:doi/10.1371/journal.pone.0084563.s005.

> *"Fortunately, the overall financial system appears to be in good health, and the U.S. banking system is well positioned to withstand stressful market conditions."*

As we all know with the benefit of hindsight, things turned out very differently. A banking crisis occurred only shortly later. It started locally, when a real estate bubble burst, which had formed in the West of the USA. As it was a regional problem then, most people thought it could be easily contained. But the mortgage crises had spillover effects on stock markets. Certain financial derivatives could hardly be sold and became "toxic assets". Eventually, hundreds of banks all over the US went bankrupt. How could this happen? A video produced by professor Frank Schweitzer and others presents an impressive visualization of the chronology of bankruptcies in the USA after Lehman Brothers collapsed.[18] Apparently, the default of a single bank triggered a massive cascading failure in the financial sector. In the end, hundreds of billions of dollars were lost.

The video mentioned above looks surprisingly similar to another one, which I often use to illustrate cascading effects.[19] It shows an experiment in which many table tennis balls are placed on top of mousetraps. The experiment impressively demonstrates that a single local disruption can mess up an entire system. The video illustrates chain reactions, which are the basis of atomic bombs or nuclear fission reactors. As we know, such cascading effects can be technologically controlled in principle, if a certain critical mass (or "critical interaction strength") is not exceeded. Nevertheless, these processes can sometimes get out of control, mostly in unexpected ways. The nuclear disasters in Chernobyl and Fukushima are well-known examples of this. We must, therefore, be extremely careful with large-scale systems potentially featuring cascading effects.

2.8 A World Economic Crisis Results

As we know, the cascading failure of banks mentioned above was just the beginning of an even bigger problem. It subsequently triggered a global economic and public spending crisis. Eventually, the financial crisis caused a worldwide damage of more than $15 trillion[20]—an amount a hundred times as big as the initial real estate problem. The events even threatened the stability of the Euro currency and the EU. Several countries including Greece, Ireland, Portugal, Spain, Italy and the US were on the verge of bankruptcy. As a consequence, many countries are suffering

---

[18]See http://www.youtube.com/watch?v=o2Budc5N4Eo.

[19]See https://www.youtube.com/watch?v=vjqIJW_Qr3c.

[20]See http://blogs.wsj.com/economics/2012/10/01/total-global-losses-from-financial-crisis-15-trillion/; http://www.bernerzeitung.ch/wirtschaft/standard/200-000000000000DollarLast/story/24865034.

from unprecedented unemployment rates. In some countries, more than 50% of young people did not have a job. In many regions, this caused social unrest, political extremism and increased rates of suicide, crime and violence.

Unfortunately, the failure cascade hasn't been stopped, yet. There is a long way to go until we fully recover from the financial crisis and from the public and private debts accumulated in the past years. If we can't overcome this problem soon, it even has the potential to endanger peace, democratic principles and cultural values, as I pointed out in a letter in 2010.[21] Looking at the situation in Ukraine, we are perhaps seeing this scenario already.

While all of this is now plausible with the benefit of hindsight, the failure of conventional wisdom to provide an advanced understanding of events is reflected by the following quote, made by the former president of the European Central Bank, Jean-Claude Trichet (*1942) in November 2010:

> *"When the crisis came, the serious limitations of existing economic and financial models immediately became apparent. Arbitrage broke down in many market segments, as markets froze and market participants were gripped by panic. Macro models failed to predict the crisis and seemed incapable of explaining what was happening to the economy in a convincing manner. As a policy-maker during the crisis, I found the available models of limited help. In fact, I would go further: in the face of the crisis, we felt abandoned by conventional tools."*

Similarly, Ben Bernanke summarized in May 2010:

> *"The brief market plunge was just an example of how complex and chaotic, in a formal sense, these systems have become… What happened in the stock market is just a little example of how things can cascade, or how technology can interact with market panic."*

Even leading scientists found it difficult to make sense of the crisis. In a letter on July 22, 2009, to the Queen of England, the British Academy came to the conclusion[22]:

> *"When Your Majesty visited the London School of Economics last November, you quite rightly asked: why had nobody noticed that the credit crunch was on its way? … So where was the problem? Everyone seemed to be doing their own job properly on its own merit. And according to standard measures of*

---

[21] See https://www3.unifr.ch/econophysics/?q=content/open-letter-george-soros.
[22] See http://wwwf.imperial.ac.uk/~bin06/M3A22/queen-lse.pdf.

success, they were often doing it well. The failure was to see how collectively this added up to a series of interconnected imbalances over which no single authority had jurisdiction. ... Individual risks may rightly have been viewed as small, but the risk to the system as a whole was vast. ... So in summary ... the failure to foresee the timing, extent and severity of the crisis ... was principally the failure of the collective imagination of many bright people to understand the risks to the systems as a whole."

Thus, was nobody responsible for the financial crisis in the end? Or do we all have to accept some responsibility, given that these problems are collective outcomes of a huge number of individual (inter)actions? And how can we differentiate the degree of responsibility of different individuals or firms? This is certainly an important question worth thinking about.

It is also interesting to ask, whether complexity science could have forecasted the financial crisis? In fact, I followed the stock markets closely before the crash and noticed strong price fluctuations, which I interpreted as advanced warning signals of an impending financial crash. For this reason, I sold my stocks in late 2007, while I was sitting in an airport lounge, waiting for my connection flight. In spring 2008, about half a year before the collapse of Lehman brothers, James Breiding, Markus Christen and I wrote an article taking a complexity science view on the financial system. We came to the conclusion that the financial system was in the process of destabilization. We believed that the increased level of complexity in the financial system was a major problem and that it made the financial system more vulnerable to cascading effects, as was later also stressed by Andrew Haldane (*1967), the Chief Economist and Executive Director at the Bank of England.

In spring 2008, we were so worried about these trends that we felt we had to alert the public. At that time, however, none of the newspapers we contacted were ready to publish our essay. "It's too complicated for our readers" was the response. We responded that "nothing can prevent a financial crisis, if you cannot make this understandable to your readers". With depressing inevitability, the financial crisis came. Although it gave us no pleasure to be proven right, a manager from McKinsey's UK office commented six months later that our analysis was the best he had seen.

Of course, some far more prominent public figures also saw the financial crisis coming. The legendary investor Warren Buffet (*1930), for example, warned of the catastrophic risks created by large-scale investments in financial derivatives. Back in 2002 he wrote:

Many people argue that derivatives reduce systemic problems, in that participants who can't bear certain risks are able to transfer them to stronger hands. These people believe that derivatives act to stabilize the economy, facilitate trade, and eliminate bumps for individual participants. On a micro level, what they say is often true. I believe, however, that the macro picture is dangerous and getting more so. ... The derivatives genie is now well out of the bottle, and these instruments will almost certainly multiply in variety and number until some event makes their toxicity clear. Central banks and governments have so far found no effective way to control, or even monitor, the risks posed by these contracts. In my view, derivatives are financial weapons of mass destruction, carrying dangers that, while now latent, are potentially lethal.

As we know, it still took five years until the "investment time bomb" exploded, but then it caused trillions of dollars of losses to our economy.

2.9 Fundamental ("Radical") Uncertainty

In liquid financial markets and many other systems, which are difficult to predict, such as our weather, we can still determine the probability of close enough events, at least approximately. Thus, we can make probabilistic forecasts such as "there is a 5% chance to lose more than half of my money when selling my stocks in 6 months, but a 60% chance that I will make a profit...". It is then possible to determine the expected loss (or gain) implied by likely actions and events. For this, the damage or gain of each possible event is multiplied by its probability, and the numbers are added together to predict the expected damage or gain. In principle, we could do this for all the actions we might take, in order to determine the one that minimizes damage or maximizes gain. The problem is that it is often impractical to determine these probabilities. With the increasing availability of data, this problem may recede, but it will often remain difficult or impossible to obtain the probabilities of "extreme events". By their very nature, there are only a few data points for rare events, which means that the empirical basis is too small to determine their probabilities.

In addition, it may be completely impossible to calculate the expected damage incurred by a problem in a large (e.g. global) system. Such "fundamental" or "radical" uncertainty can result from cascading effects, where one problem is likely to trigger other problems, leading to a progressive increase in the overall damage. In principle, the overall losses may not be quantifiable in such situations at all. This means in practice that the actual damage might be either insignificant (in the best case) or practically unbounded (in the worst case), or anything in between. In an extreme case, this might lead to a failure of the entire system, as we know it from the collapse of historical empires and civilizations.[23]

---

[23]Diamond [11], Tainter [12].

2.10   Explosive Epidemics

When studying the spread of diseases, the outcome is highly dependent on the degree of physical interactions between people who may infect each other. A few additional airline routes might make the difference between a case in which a disease is contained, and a case which develops into a devastating global pandemic.[24] The threat of epidemic cascading effects might be even worse if earlier damage reduces the ability of the system to withstand problems later on. For example, assume a health system in which the financial or medical resources are limited by the number of healthy individuals who produce them. In such a case, it can happen that the resources needed to heal the disease are increasingly used up, such that the epidemic finally spreads explosively. A computer-based study, which Lucas Böttcher, Olivia Woolley-Meza, Nuno Araujo, Hans Hermann and I performed, shows that the dynamics in such a system can change dramatically and unexpectedly.[25] Thus, have we perhaps built global networks which we can neither predict nor control?

2.11   Systemic Interdependence

Recently, Shlomo Havlin (*1942) and others made a further important discovery. They revealed that networks of networks can be particularly vulnerable to disruptions.[26] A typical example of this is the interdependence between electrical and communication networks. Another example, which illustrates the global interdependence between natural, energy, climate, financial and political systems is provided by the Tohoku earthquake in Japan in 2011. The earthquake caused a tsunami which triggered a chain reaction and a nuclear disaster in several reactors at Fukushima. Soon after this, Germany and Switzerland decided to exit nuclear power production over the next decade(s). However, alternative energy sources are also problematic, as European gas supply depends on geopolitical regions which might not be fully reliable.[27]

Likewise, Europe's DESERTEC project—a planned €1000 billion investment in solar energy infrastructure—has been practically given up due to another unexpected event, the Arab Spring. This uprising was triggered by high food prices, which in turn was partially caused by biofuel production. While biofuels were intended to improve the global CO_2 balance, their production decreased the production of food, making it more expensive. The increased food prices were further amplified by financial speculation. Hence, the energy system, the political system, the social system, the food system and the financial system have all become closely interdependent, making our world increasingly vulnerable to disruptions.

---

[24] See http://www.youtube.com/watch?v=tzTe1j5paLY.

[25] Böttcher et al. [13].

[26] Buldyrev et al. [14].

[27] Carvalho [15].

2.12 Have Humans Created a "Complexity Time Bomb"?

For a long time, problems such as crowd disasters and financial crashes have puzzled humanity. Sometimes, they have even been regarded as "acts of God" or "black swans"[28] that we had to endure. But problems like these should not be simply put down to "bad luck". They are often the consequence of a flawed understanding of the counter-intuitive way in which complex systems behave. Fatal errors and the repetition of previous mistakes are frequently the result of an outdated way of thinking. However, complexity science allows us to understand how and when complex dynamical systems get out of control.

If a system is unstable, we will see amplification effects, such that a local problem can lead to a cascading failure, which creates many further problems down the line. For this reason, the degree of interaction between the system components is crucial. Overall, complex dynamical systems become unstable, if the interactions between their components get stronger than frictional effects, or if the damage resulting from the degradation of system components occurs faster than they can recover. As a result, the timing of processes can play a key role in determining whether the overall system will remain stable. This means that delays in adaptation processes can often lead to systemic instabilities and loss of control (see Appendix 2.2).

We have further seen that an unstable complex system will sooner or later get out of control, even if everyone is well-informed and well-trained, uses advanced technology and has the best intentions. And finally, we have learned that complex dynamical systems with strong internal interactions or a high level of connectivity tend to be unstable. As our increasingly interdependent world is characterized by a myriad of global links, it is necessary to discuss the potential consequences. We must raise a fundamental question which has mammoth implications for the viability of our current economic and political systems: have humans inadvertently produced a "complexity time bomb", i.e. a global system which will inevitably get out of control? [6]

In fact, for certain kinds of networks, the potential chain reaction of cascading failures bears a disturbing resemblance to that of nuclear fission. Such processes are difficult to control. Catastrophic damage is a realistic scenario. Given the similarity with explosive processes, is it possible that our global anthropogenic systems will similarly get out of control at some point? When considering this possibility, we need to bear in mind that the speed of a destructive cascading effect might be slow, so that the process may not remind of an explosion. Nevertheless, the process may be hard to stop, and it may ultimately lead to systemic failure.[29]

What kinds of global catastrophes might today's complex societies face? A collapse of the global information and communication system or of the world economy? Global pandemics? Unsustainable growth, demographic or environmental change? A global food or energy crisis? A clash of cultures? Another world war? A

---

[28]Taleb [16].

[29]For example, the underlying processes which cause crowd disasters are slow, but deadly nonetheless.

societal shift, triggered by technological innovation? In the most likely scenario, we will witness a combination of several of these contagious phenomena. The World Economic Forum calls this the "perfect storm" (see Footnote 1), and the OECD has expressed similar concerns.

2.13 Unintended Wars and Revolutions

It is important to realize that large-scale conflicts, revolutions and wars can also be unintended outcomes of systemic interdependencies and instabilities. Remember that phantom traffic jams were unintended consequences of interactions. Similarly, wars and revolutions can happen even if nobody wants them. While there is a tendency to characterize these events as the deeds of particular historical figures, this trivializes and personalizes such phenomena in a way which distracts from their true, systemic nature.

It is essential to recognize that complex dynamical systems usually resist change if they are close to a stable equilibrium. This effect is known as Goodhart's law (1975), Le Chatelier's principle (1850–1936) or "illusion of control". Individual factors and randomness only affect complex dynamical systems if they are driven to a "tipping point",[30] where they become unstable.

In other words, the much-heralded individuals to whom so much attention is afforded in our history books were only able to influence history because much bigger systems beyond their control had already become critically unstable.[31] For example, historians now increasingly recognize that World War I was a largely unintended consequence of a chain of events. Moreover, World War II was preceded by a financial crisis and recession, which destabilized the German economic, social and political system. Ultimately, this made it possible for an individual to become influential enough to drive the world to the brink of extinction. Unfortunately, civilization is still vulnerable today and a large-scale war may happen again. This is even likely, if we don't quickly change the way we manage our world.

Typically, the unintended path towards war is as follows: Initially, resources become scarce due to a disruption such as a serious economic crisis. Then, the resulting competition for limited resources leads to an increase in conflict, violence, crime and corruption. Human solidarity and mutual tolerance are eroded, creating a polarized society. This causes further dissatisfaction and social turmoil. People get frustrated with the system, calling for leadership and order. Political extremists emerge, who scapegoat minorities for social and economic problems. This decreases socio-economic diversity, which reduces innovation and further hinders the economy. Eventually, the well-balanced "socio-economic ecosystem" collapses, such that an orderly system of resource allocation becomes impossible. As resources become scarcer, this creates an increasing "need" for nationalism or even for an external

---

[30]Such tipping points are also often called "critical points".

[31]Klein [17].

enemy to unify the fractured society. In the end, as a result of further escalation, war seems to be the only viable "solution" to overcome the crisis, but instead, it mostly leads to large-scale destruction.

2.14 Revolutionary Systemic Shifts

A revolution, too, can be the result of systemic instability. Hence, a revolution is not necessarily set in motion by a "revolutionary leader", who challenges the political establishment. The breakdown of the former German Democratic Republic (GDR) and some Arab Spring revolutions have shown that uprisings may even start in the absence of a clearly identifiable political opponent. On the one hand, this is the reason why such revolutions cannot be stopped by killing or imprisoning a few individuals. On the other hand, the Arab Spring took secret services throughout the world by surprise precisely because there were no revolutionary leaders. This created complications for countries which wished to assist these uprisings, as they did not know whom to interact with for international support.

It is more instructive to imagine such revolutions as a result of situations in which the interests of government representatives and those of the people (or particular societal groups) have drifted away from each other. Similarly to the tensions created by a drift of the Earth's tectonic plates, such an unstable situation is sooner or later followed by an "earthquake-like" release of tension (the "revolution"), resulting in a re-balancing of forces. To reiterate, contrary to the conventional wisdom which assumes that revolutionary leaders are responsible for political instability, it is due to an existing systemic instability that these individuals can become influential. To put it succinctly, in most cases, revolutionaries don't create revolutions, but systemic instabilities do. These instabilities are typically created by the politics of the old regime. Therefore, we must ask ourselves how well our society balances the interests of different groups today and how well it manages to adapt to a world, which is rapidly changing due to demographic, environmental and technological change?

2.15 Conclusion

It is obvious that there are many problems ahead of us. Most of them result from the complexity of the systems humans have created. But how can we master all these problems? Is it a lost battle against complexity, or do we have to pursue a new, entirely different strategy? Do we perhaps even need to change our way of thinking? And how can we innovate, before it is too late? The next chapters will try to answer these questions…

2.16   Appendix 1: How Harmless Behavior Can Become Critical

In the case of traffic flow, we have seen that a system can get out of control when the interaction strength (e.g. the density) is too high. Why can a change in the density make normal and "harmless" behavior become uncontrollable? To understand this better, Roman Mani, Lucas Böttcher, Hans J. Herrmann, and I studied collisions in a system of equally sized particles moving in one dimension,[32] which is similar to Newton's Cradle.[33] We assumed that the particles tended to oscillate elastically around equally spaced equilibrium points, while being exposed to random forces generated by the environment.

The following summarizes the main observations made: If the distance between the equilibrium points of neighboring particles is large enough, each particle oscillates around its equilibrium point with normally distributed velocities, and all particles have the same small variance in speed. However, when the separation between the equilibrium points reaches the diameter of the particles, we find a cascade-like transmission of momentum between particles.[34] Surprisingly, the variance of particle speeds rapidly increases towards the boundaries—it could even go to infinity with increasing system size. Due to cascading interactions of particles, this makes their speeds unpredictable and uncontrollable. While every particle in separation performs a normal dynamics, which is not excessive at all, their interactions can cause an extreme behavior of the system.

2.17   Appendix 2: Loss of Synchronization in Hierarchical Systems

When many socio-economic processes are happening simultaneously while having feedbacks on each other, a puzzling kind of systemic instability can occur, which is highly relevant for our complex societies, since many socio-economic processes happen at an increasing pace.

For the sake of illustration, let us first discuss hierarchically organized systems in physics. There, elementary particles form atoms, atoms form chemical compounds, these form solid bodies, and together they may form a planet, which is part of a planetary system, and a galaxy. Similarly, we know from biology and the social sciences that cells make up organs, which collectively form a human body. Humans, in turn, tend to organize themselves in groups, cities, organizations and nations.

Importantly, the stability of such hierarchies is based on two important principles. First, the forces are strongest at the bottom, and second, the changes are slowest at

---

[32]Mani et al. [18]: https://www.youtube.com/watch?v=RbWSal3aay8.

[33]See http://www.youtube.com/watch?v=0LnbyjOyEQ8.

[34]See https://www.youtube.com/watch?v=RbWSal3aay8.

the top. In other words, adjustment processes in these systems are faster at lower hierarchical levels (such as the atoms) as compared to higher ones (such as planetary systems). This means that lower level variables can adjust quickly to the constraints set by the higher level variables. As a result, the higher levels basically control the lower levels and the system remains stable. Similarly, social groups tend to take decisions more slowly than the individuals who form them. Likewise, organizations and states tend to change more slowly than the individuals who form them (at least this is how it used to be in the past).

Such "time-scale separation" implies that the dynamics of a system is determined only by relatively few variables, which are typically located at the higher levels of the hierarchy. Monarchies and oligarchies are good examples for this. In current-day socio-political and economic systems, however, the higher hierarchical levels sometimes change so fast that the lower levels have difficulties to keep pace. Laws are now often enacted more quickly than companies and people can adapt. In the long run, this is likely to cause systemic instability, as time-scale separation is destroyed, so that many more variables begin to influence the dynamics of the system. Such attempts to make mutual adjustments on different hierarchical levels could potentially lead to turbulence, "chaos", breakdown of synchronization, or fragmentation of the system. In fact, while it is known that delays in adaptation can destabilize a system, we are putting many of our problems on the long finger (e.g. public debts, implications of demographic change, nuclear waste, or climate change). This creates a concrete danger that our society will eventually lose control and become unstable.

References

1. Y. Sugiyama et al., Traffic jams without bottlenecks? Experimental evidence for the physical mechanisms of the formation of a jam. New Journal of Physics 10, 033001 (2008), http://iop science.iop.org/1367-2630/10/3/033001.
2. J. Sterman and John D. Sterman, Business Dynamics: Systems Thinking and Modeling for a Complex World (McGraw-Hill, 2000).
3. D. Helbing and S. Lämmer (2005) Supply and production networks: From the bullwhip effect to business cycles. Page 33–66 in: D. Armbruster, A. S. Mikhailov, and K. Kaneko (eds.) Networks of Interacting Machines: Production Organization in Complex Industrial Systems and Biological Cells (World Scientific, Singapore).
4. J. Nienhaus, A. Ziegenbein and P. Schoensleben, How human behaviour amplifies the bullwhip effect. A study based on the beer distribution game. Production Planning & Control: The Management of Operations 17(6), 547–557 (2006).
5. D. Dorner, The Logic of Failure: Recognizing and Avoiding Error in Complex Situations (Basic Books, 1997).
6. D. Helbing (2013): Globally networked risks and how to respond. Nature 497, 51–59.
7. Duncan J. Watts, Everything Is Obvious: How Common Sense Fails Us (Crown Business, 2012).
8. J. Wilfling, Unheil: Warum jeder zum Mörder werden kann (Heyne, 2010).
9. M. Perc, K. Donnay and D. Helbing (2013) Understanding recurrent crime as system-immanent collective behavior. PLOS ONE 8(10), e76063.

10. A. Asztalos, S. Sreenivasan, B. K. Szymanski, G. Korniss, PLoS ONE 9(1): e84563 (2014), http://journals.plos.org/plosone/article?id=10.1371/journal.pone.0084563; see also the interesting videos at the end of the article.
11. Jared Diamond, Collapse: How Societies Choose to Fail or Succeed (Penguin 2011).
12. J.A. Tainter, The Collapse of Complex Societies (Cambridge University, 1990).
13. L. Böttcher et al. (2015) Disease-induced resource constraints can trigger explosive epidemics, Scientific Reports 5, 16571, https://www.nature.com/articles/srep16571.
14. S.V. Buldyrev, R. Parshani, G. Paul, H.E. Stanley, and S. Havlin, Catastrophic cascade of failures in interdependent networks, Nature 464, 1025–1028 (2010).
15. R. Carvalho, L. Buzna, F. Bono, M. Masera, D. K. Arrowsmith, and D. Helbing (2014) Resilience of natural gas networks during conflicts, crises and disruptions. PLOS ONE 9(3), e90265.
16. N.N. Taleb, The Black Swan: The Impact of the Highly Improbable (Random House, 2nd ed., 2010).
17. N. Klein, The Shock Doctrine (Picador, 2008).
18. R. Mani, L. Böttcher, H.J. Herrmann, and D. Helbing, Extreme power law in a driven many-particle system without threshold dynamics, Phys. Rev. E 90, 042201.

Chapter 3
Social Forces

Revealing the Causes of Success and Disaster

Complex systems can behave in unpredictable ways and cause a lot of trouble. But it doesn't have to be like this. Their behavior depends on the interactions between the system components, the strength of these interactions, and the institutional settings. Consequently, for a complex system to work well, it is important to understand the factors that drive its dynamics. In physics, many phenomena have been understood in terms of forces, which can be measured by suitable procedures. In a similar way, the success or failure of socio-economic systems depends on hidden forces, too. Thanks to new data about our world we can now measure the forces driving socio-economic change. This will allow us to act more successfully in future.

Societies around the world are suffering from financial crises, crime, conflicts, wars, and revolutions. These "societal ills" do not occur by chance, but for a reason. The fact that they are happening time and again proves that there are hidden causes, which we haven't understood sufficiently well. This is why we keep failing to cope with these problems. In future, however, we will be able to understand societal problems and cure them. This will be akin to the discovery of the X-ray by Wilhelm Conrad Röntgen (1845–1923), which helped to reveal the causes of many diseases and to cure billions of people.[1]

Given that we are now living in a Big Data age, will we soon be able to answer all our questions and find the best possible course of action in every situation? Of course, it's far from clear that this dream will ever come true. However, the growing amount of data about our world will certainly allow us to measure the hidden forces behind our global technological, social, economic and environmental systems, just as microscopes and telescopes enabled us to discover and understand the micro- and macro-cosmos—from cells to stars—in the past. Similarly to how we have built elementary particle accelerators to discover the forces that keep our world together,

---

[1]Nevertheless, it is important to avoid an overdose of radiation, which can be very harmful. In a similar way, mass surveillance can be harmful to society, as I will show later. When measuring what's going on in society, it is therefore, important to respect privacy and the fundamental right of informational self-determination. The Nervousnet platform featured in this chapter is taking this into account.

© Springer Nature Switzerland AG 2021
D. Helbing, *Next Civilization*,
https://doi.org/10.1007/978-3-030-62330-2_3

we can now create "socioscopes" to reveal the principles that make our society succeed or fail.

3.1   Measuring the World 2.0

It's a sad but well-known fact that the loss of control over a system often results from a lack of knowledge about the rules governing it. Therefore, it is important that we learn to measure and understand the hidden "forces" determining changes in the world around us. This will eventually put us in a position where we can harness these forces to overcome systemic instability and create complex dynamical systems with particular structures, properties and functions.

Remember that some of the greatest discoveries in human history were made by measuring the world. We have discovered new continents and cultures. We have reached out to the skies and explored our universe to discover black holes, dark matter, and new worlds. Now, the Internet is offering entirely new ways to quantify what's happening on our Earth. By analyzing the sentiment of blogs, *Facebook* posts, or tweets, we can visualize human emotions such as happiness.[2] Furthermore, it is possible to get a picture of the social, economic, and political "climate", by identifying the subjects that people publicly discuss.[3] By mining data on the Web, we can also map social and economic indicators. This includes quantities such as the gross domestic product per capita[4] or the levels of violence or crime,[5] highly resolved according to geographic location and time.[6] It is even feasible to digitally re-construct our three-dimensional world based on the photos that people upload on platforms such as *flicker*.[7]

We can now create "Financial Crisis Observatories" to detect the likelihood of financial bubbles and crashes.[8] We can map crises and risks to help first-aid teams in regions struck by disasters.[9] We can analyze and visualize the production of knowledge and the spread of scientific concepts, as I did it with Amin Mazloumian, Katy Börner, Tobias Kuhn, Christian Schulz and others[10,11] (see Fig. 3.1). Furthermore, it is fascinating to examine the way culture has spread across the world over the centuries,

---

[2]See, for example, http://www.wefeelfine.org/index.html, http://hedonometer.org/about.html, and http://sentistrength.wlv.ac.uk/.

[3]http://gdeltproject.org/#watching.

[4]https://sites.brown.edu/davidweil/.

[5]https://www.trulia.com/neighborhoods/.

[6]http://www.gapminder.org/.

[7]http://www.youtube.com/watch?v=4cEQZreQ2zQ.

[8]See https://er.ethz.ch/financial-crisis-observatory.html.

[9]See http://crisismapping.ning.com/.

[10]Mazloumian et al. [1]; see also the Living Science app at https://livingscience.inn.ac.

[11]Kuhn [2].

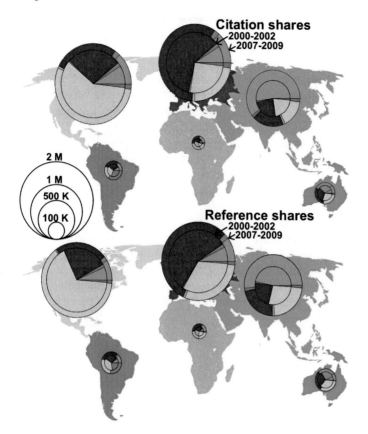

Fig. 3.1 Illustration of scientific productivity and impact. (Reproduced from Mazloumian et al. [4] with kind permission of the Springer Nature Publishing Group.)

as Maximilian Schich, Laszlo Barabasi and I did together with others.[12] Inspired by *Wikipedia* and *OpenStreetMap*, we might now create an OpenResourcesMap to visualize the resources of the world and who uses them. This could help, for example, to reduce undesirable shortages. In addition, we could produce an OpenEcosystemsMap to depict environmental change and who causes it. What else can we do? Let me elaborate a health-related example.

---

[12]Schich et al. [3]; see also the movies at https://www.youtube.com/watch?v=4gIhRkCcD4U, https://www.youtube.com/watch?v=rwmiQ75iW6Y, and http://www.youtube.com/watch?v=231 zuH3uMwc.

3.2 Monitoring the Flu and Other Diseases

Pandemics are a major threat to humanity. Some of them have killed millions of people. The Spanish flu in 1918 was a shocking example of this. In fact, such pandemics are expected to happen time and again because viruses keep mutating, such that our immune systems might be unprepared. For instance, the world was caught by surprise by the Ebola outbreak, and recently by COVID-19.

To contain the spread of epidemics, the World Health Organization (WHO) is continuously monitoring emerging diseases. It takes about two weeks to collect the data from all the hospitals in the world, meaning that each overview of the current situation is two weeks out of date. However, *Google Flu Trends* pioneered an approach called "nowcasting", which was celebrated as major success of Big Data analytics at that time. It was claimed that it is possible to estimate the number of infections in real-time, based on the search queries of *Google* users. The underlying idea was that queries such as "I have a headache" or "I don't feel well" or "I have a fever", and so on, might indicate that the user has the flu. While this makes a lot of sense, the *Google Flu* approach was recently found to be unreliable, partly because *Google* constantly changes its search algorithms and also because advertisements bias people's behavior.[13]

3.3 Flu Prediction Better Than Google

Fortunately, a model using much less data than *Google Flu* can be applied to analyze how a disease spreads, namely by augmenting data of infections with a model based on air travel data. Dirk Brockmann and I found this approach in 2012/13. About ten years back, Dirk started to investigate the spread of diseases by analyzing the time and geographic location of infections using computer simulations. He also analyzed the paths of dollar bills in his famous "Where is George?" study.[14] But when visualizing the spatio-temporal spread of epidemics, the patterns looked frustratingly chaotic and unpredictable. The relationship between the arrival time of a new disease as a function of the distance from the place where it originated was so scattered that it was hard to make much sense of the data. Eventually, however, it became clear that this problem resulted from the high volume of passenger air travel. Thus, Dirk had the idea to define an "effective distance", based on the volume of travel between airports in the world, and to study the spread of disease as a function of this alternative measure of distance.[15] In effective distance, two airports such as New York City and Frankfurt are close to each other because of the large passenger flows connecting

---

[13]Lazer et al. [5].

[14]Brockmann et al. [6], see also http://www.youtube.com/watch?v=kn32vavZqvg.

[15]See http://rocs.hu-berlin.de/ and http://rocs.hu-berlin.de/corona/docs/model/visual_analytics/, https://www.youtube.com/watch?v=zEO8yZoNBsk.

them, while two nearby cities without any direct flights between them might be largely separated.[16]

Dirk Brockmann and I started to collaborate in 2011, when Germany was witnessing the spread of the deadly, food-borne EHEC epidemic. I got in touch with Dirk and suggested that we could combine a model of the spread of epidemics with a model of food supply chains. In this way, we wanted to identify the location where the disease originated, which was unknown at that time. Unfortunately, we could not obtain proper supply chain data then. But our discussion triggered a number of important ideas. In particular, the research activities shifted from predicting the spread of diseases toward detecting the locations where they originate.

In fact, when analyzing empirical data of infections as a function of effective distance from the perspective of all airports worldwide, we found that the most circular spreading pattern identifies the most likely origin of the disease. More importantly, however, once the location of origin of a disease is known, one can use the circular spreading dynamics as a function of effective distance to predict the order in which cities will be hit by a pandemic.[17] This helps to put medical drugs (such as immunization shots) and doctors in place where they are most effective in countering the impact and spread of the disease.

When Ebola broke out, Dirk furthermore used the method discussed above to make early predictions about possible cases in other countries. This helped to inform international preparations to contain the virus.[18] However, I would also like to highlight here the fantastic research teams of Alessandro Vespignani and Vittoria Colizza, both partners of the FuturICT initiative. To predict the spread of diseases, they have built a very detailed and sophisticated simulator. Whenever a disease breaks out, this simulator can be used to test the effectiveness of countermeasures and inform policy-makers around the world.[19] It was found, for example, that closing down some airline connections can only *delay* the spread of the disease, while the best way for industrialized countries to protect themselves from diseases such as Ebola is to

---

[16]Independently of Dirk Brockmann's activitiess, I became interested in the modeling of epidemic spread back in 2002. In the wake of the September 11 attacks the year before, there were fears that terrorists could use anthrax or other deadly germs to threaten the USA and the rest of the world. At this time, I proposed to Otto Schily, the then German Minister of Internal Affairs, to build a self-calibrating epidemic simulator to predict the spread of pandemics. Directly after the outbreak of a disease, accurate data about infection and recovery rates is often not available. Thus, the idea was that a self-adaptive calibration model could produce increasingly accurate predictions, as more data became available. At that time, I received a letter stating that such an approach was not feasible. But of course, it was!

[17]See the movie http://www.youtube.com/watch?v=ECJ2DdPhMxI. It turns out that this technique can be successfully applied even in cases where certain key information (such as the infectiousness of the disease and recovery rate) is not well-known, which is typical after the outbreak of a new disease. The only data besides the outbreak location which is important for our analysis is the volume of passenger air traffic between all airports. This is needed to specify the effective distance.

[18]See http://rocs.hu-berlin.de/projects/ebola/.

[19]See https://www.youtube.com/watch?v=YstB9VWDUqE and http://www.gleamviz.org/.

spend their money on fighting the disease in those countries that are suffering from the disease first.[20]

3.4   Creating a Planetary Nervous System as a Citizen Web

Very soon, we will not only have maps, which aggregate data from the past and represent them as a function of space, time or network interdependencies. We will also have systems, which deliver real-time answers. We will be able to ask questions, which trigger tailored measurements to answer them. "How is the traffic on Oxford Street in London?" "How is the weather in Moscow?" "How could investment decisions and consumer choices be affected?" "How happy are people in Sydney today, and how much money will they spend in shops?" "What worries people in Paris at the moment?" "How many people are up between 3 am and 4 am on Sunday nights around Manhattan's Central Square, and is it worth selling pizza at that time?" "How noisy is it in the part of town I am considering to move to?" "What's the rate of flu infections in the region where I wish to spend my holidays?" "Where are the road holes in my city located?" "When did we have the last significant earthquake within a range of 500 km?" Answers to questions like these would help us to be more aware of the world around us, to make better decisions, and act more effectively. But how will we get all this real-time information?

The sensor networks, on which the "Internet of Things" is based, will enable us to perform real-time measurements of almost everything. They can be used to build a "Planetary Nervous System" (PNS), an intelligent information platform proposed by the FuturICT project (http://www.futurict.eu).[21] The In fact, my team has started to develop such an information platform, called *Nervousnet*.[22] *Nervousnet* would harness the power of the Internet of Things for everyone's benefit and would be built and managed in a participatory way, as a "Citizen Web".[23] Similar to *OpenStreetMap,* we wanted to develop this system together with an emerging network of volunteers, who are committed to developing the project further.

This collaborative approach would give citizens control over their personal data, in accordance with their right of informational self-determination, and create new opportunities for everyone. *Nervousnet* would not only offer the possibility to contribute to the measurement of our world, in order to jointly create something like a real-time data *Wikipedia. Nervousnet* would also establish a social mining paradigm, where users are given freedom and incentives to collect, share and use

---

[20]Colizza et al. [7].

[21]Helbing and Carbone [8].

[22]See http://nervousnet.ethz.ch/, https://web.archive.org/web/2020*/nervousnet.info and https://youtu.be/pN2hAcr6ujk as well as https://ieeexplore.ieee.org/abstract/document/7097988.

[23]D. Helbing, Creating ("Making") a Planetary Nervous System as Citizen Web, see http://futurict. blogspot.ch/2014/09/creating-making-planetary-nervous.html; see also the videos at https://www. youtube.com/watch?v=BKcWPdSUJVA and https://www.youtube.com/watch?v=kvnkoT4CLNk.

data in ways that do not aim to undermine privacy. Appendix 3.1 provides further information on the platform. With your help, it may very well become a cornerstone of the public information infrastructure of the emergent digital society. So why don't you join us in building the *Nervousnet* platform or in measuring the world around us?[24]

3.5 Sociophysics: Revealing the Hidden Forces Governing Our Society

But it takes more than data to understand the world and its problems. Measuring, analyzing and visualizing data is just the first step, because data mining alone may not lead to a good understanding of unstable system dynamics, which produces most of the good and bad surprises in the world. In order to help us when we really need it, we must find explanatory models, which can predict situations that haven't occurred before and cannot be understood by extrapolation.

In fact, some "social mechanisms" influence human behavior in a similar way to how the gravitational force determines planetary motion[25]. Just think of the way social norms determine our roles and behaviors.[26] The scientific approach of "agent-based simulations" aims to formalize these rules and turn them into computer codes. Complementary, the research field of sociophysics tries to express the corresponding interactions and their outcomes using mathematical formulas.[27] In this chapter, I will particularly discuss the powerful concept of "social forces", which enables researchers to understand the link between micro-level interactions of individuals and the often unexpected macro-level outcomes in socio-economic systems.

The concept of forces is one of the main pillars of physics. In order to discover it, the old geocentric worldview, in which the Earth was assumed to be the center of the universe, first had to be replaced by the heliocentric worldview, which recognized that all the planets in our solar system revolve around the sun. Later, the new understanding of the planetary system allowed Isaac Newton (1642–1727) to interpret measurement data about the planets in a new way, which led him to formulate a simple and plausible model of planetary motion based on the concept of gravitational forces. Now, most concepts used in modern physics are formulated in terms of forces and the way they influence the world. The predictive power of these models is striking and has been impressively demonstrated by the Apollo moon shot and every single satellite launch.

Another foundation of the success of physics is the tradition of building instruments to measure the forces that would otherwise be imperceptible to our senses. This has enabled physicists to explore a vast array of questions, spanning from

---

[24]You can get in touch with us through nervousnet@ethz.ch.

[25]Even though there is more diversity and randomness to human interaction (which mostly benefits our economy and society).

[26]Hechter and Opp [9].

[27]Pentland [10], Chakrabarti et al. [11].

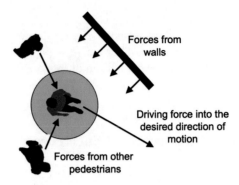

Fig. 3.2 Illustration of the various "social forces" acting upon a pedestrian (I would like to thank Mehdi Moussaid for providing this graphic.)

the early stages of our universe to the exploration of elementary particles and the study of fundamental processes in biological cells. Therefore, the next logical frontier of science is to build "Socioscopes" that can reveal the hidden forces behind the dynamics of socio-economic systems. In this way, we will eventually learn to understand the counterintuitive behaviors of complex dynamical systems.[28] I believe that we will soon be able to diagnose emergent societal problems such as financial crashes, crime, or wars before they happen.[29] This will empower us to avoid or mitigate these problems similarly to the way medical diagnostic instruments have helped us to prevent or cure diseases. Isn't that an exciting prospect?

3.6 Social Forces Between Pedestrians

To demonstrate the feasibility of this vision, let me first discuss the example of pedestrian and crowd dynamics to underline the usefulness of force models in the social sciences. Starting in 1990, when I wrote my diploma thesis,[30] I noticed that pedestrian paths around obstacles looked similar to the streamlines of fluids. So, I decided to formulate a fluid-dynamic theory of pedestrian flows. I derived it from a model for the motion of individual pedestrians, which was inspired by Newton's force model (Fig. 3.2).

This "social force model" assumes that the acceleration, deceleration an directional changes of pedestrians can be approximated by the sum of a number of different forces, each of which captures a specific desire or "interaction effect". For example, each pedestrian likes to move with a certain desired speed into a preferred direction

---

[28] Ball [12].

[29] Helbing and Balietti [13].

[30] D. Helbing, Physikalische Modellierung des dynamischen Verhaltens von Fußgängern (Physical Modeling of the Dynamic Behavior of Pedestrians), http://papers.ssrn.com/sol3/papers.cfm?abstract_id=2413177.

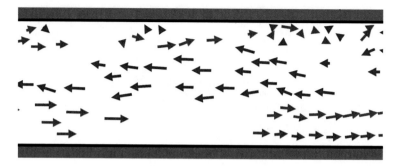

Fig. 3.3 Illustration of the formation of lanes of uniform walking direction in pedestrian counterflows (Reproduced from Helbing [18], with kind permission of Springer Publishers.)

of motion. This can be represented by a simple "driving force", which captures how the person's velocity is gradually adapted. Moreover, each pedestrian seeks to avoid collisions and to respect a certain personal "territory" of others. This is reflected by a "repulsive interaction force" between pedestrians which increases with proximity. Repulsive interactions with walls or streets can be described by similar forces. The attraction of tourist sites and the tendency for friends and family members to stay together can be represented by "attractive forces".[31] Finally, a random force may be used to reflect the individual behavioral variability.

Despite its simplicity, computer simulations of this model match many empirically observed phenomena surprisingly well. For example, it is possible to understand the emergence of river-like flow patterns through a standing crowd of people, the wave-like progression of individuals waiting in queues, or the lower density of people on a dance floor compared to the surrounding spectators watching them.[32]

3.7 Self-Organization of Unidirectional Lanes in Pedestrian Counter-Flows

There are also various self-organization phenomena that lead to fascinating collective patterns of motion. For example, when people enter a corridor on two sides, we observe the formation of lanes of unidirectional flow[33] (see Fig. 3.3). That is, people walking in opposite directions automatically coordinate each other so that they are hardly obstructed by the respective counter-flow. This makes transit more efficient for everyone.

---

[31]Moussaïd et al. [14].

[32]Helbing [15, 16].

[33]Helbing and Molnár [17]; see also the video at https://www.youtube.com/watch?v=e2WfvJ XB_8.

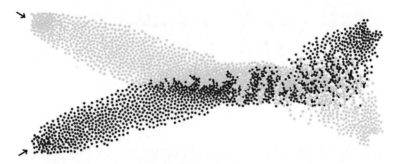

Fig. 3.4 Illustration of the phenomenon of "stripe formation" in two crossing pedestrian flows (adapted from Helbing et al. [21]. Reproduced with kind permission of INFORMS.)

While it seems as if the "invisible hand" is at work here, we can actually explain how social order is created and how collectively desirable outcomes occur from local interactions: whenever an encounter between two pedestrians occurs, the repulsive interaction force between them pushes the pedestrians a bit to the side. These interactions are more frequent between opposite directions of motion, due to the higher relative velocity. This is the main reason why people walking in opposing directions tend to separate into lanes of unidirectional flow. To explain the phenomenon, we don't need to assume that pedestrians prefer to walk on a certain side of the street.[34] In conclusion, complexity science can explain the formation of lanes of pedestrians as a so called "symmetry breaking" phenomenon, which occurs when a mixture of different directions of motion becomes unstable.

3.8   Walking Through a "Wall" of People Without Stopping

Surprisingly, the very same force model also reproduces a number of other interesting findings in pedestrian crowds, such as oscillatory changes in the flow direction of pedestrians at bottlenecks. Therefore, when a crowd of people builds up at a junction, the pressure tends to be relieved alternately on either side of the bottleneck. Another example of self-organization is the amazing phenomenon of "stripe formation", which allows pedestrians to cross another pedestrian stream without having to stop (see Fig. 3.4). This is almost akin to walking through a wall! Using the social force model, it's possible to understand how this is possible. The formation of stripes—which occur for similar reasons as the lanes discussed before—allows pedestrians to move forward with the stripes and sideways within stripes that are forming in the intersecting pedestrian flow. In combination, this enables a continuous collective motion through a pedestrian stream moving in another direction.[35]

---

[34]Note, however, that pedestrians additionally have a side preference, which is related to the emergence of a behavioral convention resulting from lane formation, see Helbing [19, 20].

[35]See the video http://www.youtube.com/watch?v=yW33pPius8E.

3.9 Measuring Forces

In physics, forces are experimentally determined by measuring the trajectories of particles, especially changes in their speed and direction of motion. It would be natural to do this for pedestrians, too. At the time when we developed the social force model for pedestrians, I could not imagine that it would ever be possible to measure social forces experimentally. But a few years later, we actually managed to do this. In around 2006, the advent of powerful video camera and processing technologies put my former Ph.D. student, Anders Johansson, into the position to detect and analyze the trajectories of pedestrians from filmed footage. Using this data, we adapted the parameters of the social force model in such a way that it optimally reproduced the trajectories of the observed pedestrians.[36] In 2006/07, similar tracking methods became essential for the analysis of dense pedestrian flows and the avoidance of crowd disasters.[37]

Later, in 2008, Mehdi Moussaid and Guy Theraulaz set up a pedestrian experiment in Toulouse, France, under well-controlled lab conditions.[38] This finally allowed us to perform data-driven modeling. While before, we had to make assumptions about the functional form of pedestrian interactions, it then became possible to determine the functional dependencies directly from the wealth of tracking data generated by the pedestrian experiment. After fitting the social force model to individual pedestrian data, it was finally used to simulate the flows of many pedestrians. To our excitement, the computer simulations yielded a surprisingly accurate prediction of the pedestrian flows observed in a wide pedestrian walkway.

So, pedestrian modeling can be considered a great success of sociophysics. One can say that, over time, pedestrian studies have turned from a social to a natural science, bringing theoretical, computational, experimental and data-driven approaches together. This has even led to practical and surprising lessons for the design of pedestrian facilities and for the planning of large-scale public events such as the annual pilgrimage in and around Mecca, as we will discuss below.

3.10 Most Pedestrian Facilities Are Inefficient

Back in 1994/95, Peter Molnar and I compared a range of different designs of pedestrian facilities. Surprisingly, we found that obstacles, if properly placed, can make pedestrian counter-flows more efficient (see Fig. 3.5). In fact, all of the conventional design elements of pedestrian facilities such as corridors, bottlenecks, and intersections turn out to be ill-designed and can be considerably improved! In many cases, "less is more" in the sense that providing less space for pedestrians can produce a

---

[36]Johansson et al. [22].

[37]Johansson et al. [23].

[38]Moussaïd et al. [24].

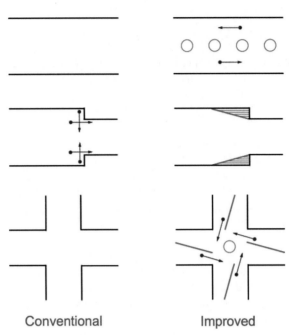

Fig. 3.5 Illustration of conventional and improved elements of pedestrian facilities (adapted from Helbing [25]. Reproduction with kind permission of Springer Publishers.)

better flow. This surprising discovery can be best understood for bottlenecks such as doors. Here, a funnel-shaped design can reduce disturbances in the pedestrian flow, which otherwise result when the directions of motion are not well enough aligned (e.g. when some people approach the door from the front and others from the side).

In the case of busy bi-directional pedestrian flows, the efficiency of motion can be improved by a series of pillars in the middle. These pillars help to stabilize the interface between the opposite flow directions, thereby reducing disturbances. The effectiveness of the design becomes particularly clear in subway tunnels, where pedestrians move both ways and pillars exist for static reasons.

Finally, an obstacle in the middle of a pedestrian intersection may also improve the flow. When Peter Molnar and I discovered this, it took us some time to understand this unexpected finding. Eventually we noticed that, at intersections, many different collective patterns of motion can emerge, for example, clockwise or counterclockwise rotary flows, or oscillatory patterns of the crossing flows. The problem is that the different collective patterns of motion conflict with each other, so that none of them are stable. Putting a column in the center increases the likelihood of rotary flows and thereby increases the overall efficiency of pedestrian traffic. The flow can be improved even more by replacing a four-way intersection by four intersections of bidirectional flows, which can be achieved by placing railings in suitable locations. This encourages a rotary flow pattern, which greatly reduces disturbances.

3.11 Crowd Disasters

Unfortunately, pedestrian flow doesn't always self-organize in an efficient way. Sometimes, terrible crowd disasters happen, and hundreds of people may be injured or killed, even when everyone has peaceful intentions and does not behave aggressively. How is this possible?

When I got interested in the problem in 1999, crowd disasters were often regarded as "acts of God" similar to natural disasters that are beyond human control. However, the root cause of the breakdown of social order in crowds is similar to the reason behind "phantom traffic jams". If the density of people gets too high, the flow of pedestrians becomes unstable. The resulting crowd dynamics can be uncontrollable for individual people and even for hundreds of security guards. Nevertheless, crowd disasters can be avoided, if their causes are well enough understood and proper preparations made.

Crowd disasters have happened since at least Roman times. That's why building codes were developed for stadiums, as exemplified by the Coliseum in Rome. The Coliseum had 76 numbered entrances and could accommodate between 50,000 and 73,000 visitors, who would exit through the same gate through which they had entered. These rules and the generous provision of exits meant that the Coliseum could be evacuated within five minutes. Modern stadiums, which generally have a smaller number of exits, can rarely match this performance.

Despite the frequent and tragic occurrences of crowd disasters in the past, they continue to happen due to common misperceptions. Media reports often suggest that crowd disasters occur when a crowd panics, causing a stampede in which people are crushed or trampled. Therefore, crowd disasters are claimed to be the result of unreasonable or aggressive behavior, with some individuals pushing others relentlessly as they try to escape. But why would people panic? My colleague Keith Still put it like this:

> *"People don't die because they panic, they panic because they die."*

In fact, studies that I conducted with Illes Farkas, Tamas Vicsek, Mehdi Moussaid, Guy Theraulaz and others revealed that many crowd disasters have physical rather than psychological causes.[39] They may occur even if everybody behaves reasonably and tries not to harm anyone else. Therefore, the common view that crowd disasters are mostly a result of panic is outdated. I don't negate that people are in a danger to be crushed when the inflow of people into a spatially constrained area exceeds the outflow for an extended period of time. Certainly, a high density crowd can become life-threatening under such conditions, as more and more people accumulate in too little space. However, most of the time crowd disasters occur for a different reason.

---

[39]Helbing et al. [26], Moussaïd et al. [27].

Fig. 3.6 Time-lapse photograph of stop-and-go flows in dense pedestrian crowds. (Reproduced from Helbing and Johansson [29], with kind permission of Springer Publishers.)

For example, during the annual Muslim pilgrimage around Mecca in 2006, a crowd disaster occurred on a large plaza. Anders Johansson and I were asked to evaluate video footage of this accident. Initially, due to the high density of pilgrims, we could not see anything else than very slow motion, a few centimeters per second. However, when I asked Anders to play the videos 10 or even 100 times faster, we made some surprising discoveries![40]

The accelerated videos showed some striking phenomena. First, we observed an unexpected, sudden transition from smooth pedestrian flows to stop-and-go flows (see the long-term photograph in Fig. 3.6).[41] In contrast to highway traffic, however, these stop-and-go waves were previously unknown and unlikely to result from delayed adaptation. Eventually, we discovered that these waves were caused by a competition of too many pedestrians for too few gaps in the crowd, i.e. by a coordination problem.[42] The stop-and-go movement emerged when the overall flow suddenly dropped to lower values, similarly to the capacity drop phenomenon in vehicle traffic. As a consequence, the outflow from the area drastically decreased, while the inflow stayed the same. Thus, the density increased quickly, but while this certainly created a dangerous situation, it was not the ultimate cause of the tragedy!

To our further surprise, some minutes later we witnessed another unexpected transition—from stop-and-go flows to a phenomenon that we call "crowd turbulence" (see Fig. 3.7).[43] In this situation, people were pushed around in random ways. So, Anders Johansson and I discovered that it was not the density, but the density multiplied by the variability of velocities—the so-called "crowd pressure"—which

---

[40]Helbing et al. [28].

[41]Also watch the video at https://www.youtube.com/watch?v=muKC5bZezlo.

[42]Helbing et al. [30].

[43]A related video can be found at https://www.youtube.com/watch?v=F6EJnMbyM-M.

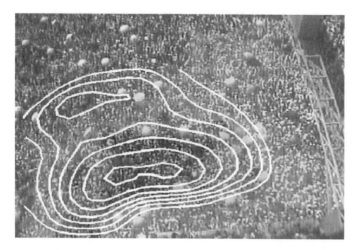

Fig. 3.7 Illustration of the phenomenon of crowd turbulence under extremely crowded conditions. (Reproduced from Helbing and Johansson [29], with kind permission of Springer Publishers.)

triggered the disaster at a certain time and location. Moreover, we found that even unintentional body movements could exert a force on the bodies of pedestrians in the immediate neighborhood, when the crowd density exceeded a certain critical threshold. These forces could add up from one body to the next, meaning that the resulting force quickly changed in strength and direction. As a consequence, people were pushed around in unpredictable and uncontrollable ways.

It was just a matter of time until someone lost balance, stumbled, and fell to the ground. This produced a "hole" in the crowd, which unbalanced the forces acting on the surrounding people, because the counter-force from where the person stood before was missing. Therefore, the surrounding people tended to fall on top of those who had previously fallen or they were forced to step on them. The situation ended with many people piled up on top of each other, suffocating the people on the ground. Similar observations were made in other crowd incidents, such as the Love Parade disaster in Duisburg, Germany, for example.[44]

3.12 Countering Crowd Disasters

Can we use the above knowledge to avoid crowd disasters in the future? The answer is yes! Some years back, together with a team of various colleagues, I got temporarily involved in a project aiming to improve the flow of pedestrians during the annual Muslim pilgrimage to Mecca. We were asked to find a better way of organizing the crowd movements around the Jamarat Bridge, a focal point of the

---

[44]Helbing and Mukerji [31].

pilgrimage, where thousands of pilgrims had died in the past due to a number of tragic crowd disasters. How could one avoid them?

This was a challenge that was not simply related to technical matters such as crowd densities. We also needed to take dozens of religious, political, historical, cultural, financial and ethical factors into account. Our previous experience of modeling crowds led us to propose a range of measures including the counting of crowds through a newly developed video analysis tool, the implementation of time schedules for groups of pilgrims, re-routing strategies for crowded situations, contingency plans for possible incidents, an awareness program informing pilgrims in advance about the procedures during the Hajj, and an improved information system to guide millions of pilgrims who spoke 200 different languages.[45] As many of the proposals had been implemented, the Hajj (in 1427H) was indeed performed safely without any incident in 2007, [46] after which I turned to other projects. The main success principles were to avoid crossing and counter-flows, and to suitably adapt to real-time information gathered.

Since then, the principle of providing real-time feedback to crowds has become a trend. An interesting example is an app to improve the safety of mass events, which was developed by Paul Lukowicz, a member of the FuturICT project, and a number of other scientists such as Ulf Blanke.[47] By using this *app*, festivalgoers at a number of festivals in London, Vienna and Zurich voluntarily provided GPS data about their locations, which was used to determine their speeds and directions. This data was then returned to the festivalgoers to give them a picture of the areas which they should better avoid due to over-crowding.

3.13 Forces Describing Opinion Formation and Other Behaviors

Is the usefulness of the concept of social forces restricted to pedestrian flows (and traffic flows[48]), or can it be applied to various other kinds of social phenomena such as crime and conflict as well? The success of force models in describing pedestrian flows is related to the fact that pedestrians are moving continuously in space. Therefore, the dynamics of a pedestrian can be represented by an equation of motion, which states that the change of his/her spatial position over time is given by their velocity. An additional equation expresses that the change in velocity over time (i.e.

---

[45] Haase et al. [32]; also see Ref. 82.

[46] In the meantime, three further levels of the new Jamarat Bridge were completed, and the organization of the pilgrim flows was changed several times. However, since 2007 I haven't been involved in this anymore. To the best of my knowledge, the Hajj was safe as long as the recommendations of the international expert panel were approximately implemented. It also seems that the international expert panel, which is now in charge, has not been responsible for the sad crowd disaster in 2015.

[47] See https://ieeexplore.ieee.org/document/6827652 and Helbing et al. [33].

[48] Helbing and Tilch [34].

the acceleration) can be modeled by a sum of forces. But can we also understand how people form opinions or other behavioral changes based on social forces? Surprisingly, the answer is "yes", if the changes in opinion are more or less gradual on a continuous opinion scale or in a continuous opinion space.[49] Otherwise, generalized models would have to be used, which exist as well.[50]

After formulating the social force model for pedestrians in Göttingen, Germany, in 1990, I joined the team of Professor Wolfgang Weidlich (1931–2015) at the University of Stuttgart, Germany. He was probably the only physicists working on socioeconomic modeling at that time. So, Professor Weidlich might be seen as grandfather of sociophysics.[51] When I joined his team, my plan at this time was to learn how to model opinion formation and decision-making. Since my work on pedestrians, I had the idea that both individual and collective human behavior could be understood as a result of social forces, and I formulated a corresponding theory (see Appendix 3.2).

Interestingly, it is possible to develop social force models for migration, too, if one assumes that people relocate within a certain (not too large) geographic range. A model that I formulated in 2009 examines "success-driven migration".[52] According to this, individuals try to avoid locations where they expect bad outcomes and are attracted to locations that appear to be favorable. Bad neighborhoods (in which people were uncooperative) were found to have a repulsive effect, whereas good neighborhoods (where people were cooperative) attracted migrants. It is even possible to calculate the direction and strength of this repulsion and attraction effects, i.e. the forces which imply the average direction and speed of motion in a certain location.

In general, a great advantage of using the concept of "social forces" is that it can help us to develop a better idea of the complex processes underlying social change. Movements towards a subject or object are reflected by attractive forces, while movements away from a subject or object are reflected by repulsive forces. It is also important to recognize that such forces may not be attributable to individuals, but rather to groups of individuals, companies or institutions. In other words, social forces may be a collective effect. Group dynamics or "group think", as a result of the emergence of a particular group identity is probably a good example for this. Here, a collective "group" perspective emerges from the interactions of individuals, which in turn changes their characteristic opinions and behaviors. In fact, the theory of social milieus posits that the behavior of individuals is largely influenced by their social environment (Helbing et al. [40]). Very soon, it will be possible to quantify the underlying forces and to derive mathematical formulas for them. But what is more powerful, physical or social forces?

---

[49]Helbing [35].

[50]Helbing [36].

[51]Weidlich [37].

[52]Helbing [38], Helbing and Yu [39].

3.14   Culture: More Persistent Than Steel

It has often been claimed that civilizations were born out of war, and that the world is ultimately ruled by those with the greatest military power. However, I don't buy this. Even though war certainly played a role in establishing the modern world, I believe the main mechanisms underlying the spreading of civilization are migration and the exchange of goods and ideas. Today, the Internet can certainly advance civilizations in ways that don't need to be paid by human lives.

But what is the basis of civilizations? It's culture, and to a large extent, culture is the result of numerous sets of rules, such as social conventions, values, norms, roles, and routines.[53] These rules determine the success or failure of societies and guide their evolution. Just take religious values for instance, which can determine the behavior of millions of people for thousands of years. It is not an exaggeration, therefore, to say that culture is more persistent than steel[54] and probably also more relevant to the success of civilizations than weapons.[55] In other words, social forces can be stronger than physical forces. A good example of this is ancient Greek culture, which spread to the Roman occupants because it was more advanced.

3.15   Reducing Conflict

While we all learned about physical forces at school, very few people have an explicit understanding of the social forces, which determine the behavior of socio-economic systems. This has to change, if we want to overcome or at least mitigate the problems we are faced with. Conflicts, wars and revolutions can be understood as a result of social forces, too. Certain forces can destabilize systems and cause them to disintegrate. There are at least three types of conflict situations: (1) An encounter (say, between two countries) causes losses on both sides. This might be avoided by increasing the awareness of the likely outcomes of such an encounter in advance. (2) The encounter is beneficial to one party but unfavorable to the other, and causes damage overall. Here, the second party needs to be protected from exploitation (e.g. through solidarity from third parties, or by separating the disputing parties). (3) The encounter is advantageous for one side and undesirable for the other, but the overall outcome is positive. In such situations, the benefits can be redistributed to make the interaction beneficial for both sides, i.e. it's possible to align interests to create a win-win situation.[56]

---

[53] Helbing et al. [41], Helbing and Johansson [42].

[54] It is surprising enough that one can cut steel with light (i.e. a mighty laser beam), but the persistence of culture is even more astonishing.

[55] Even though protection from aggression and exploitation is needed, of course.

[56] I have recently proposed "Social (Information) Technologies" that aim to reduce the occurrence of conflict between people and companies.

Would it also be possible to actually measure the forces creating conflict? Yes, I think one can build a ConflictMap, which illustrates regional and international tensions and explains how they come about. In fact, when working in my team, Thomas Chadefaux mined millions of news articles over a period of more than 100 years and performed a sentiment analysis for words indicating conflict. This allowed him to quantify the level of tension between countries in the world. Moreover, he could show that the level of tension could be used to predict the likelihood of outbreaks of war within a 6–12 months time period.[57] Such advance warning signals might give politicians enough time to engage in diplomacy to peacefully resolve the tensions before it's too late. Our analyses also revealed how tension spreads from one country to the next, destabilizing a huge region, as it happened after the war in Iraq. This might also have produced fertile ground for the rise of the Islamic State (IS).

3.16 What We Can Learn from Jerusalem

Another data-driven study analyzed a problem that has worried the world for many decades: the conflict in the Middle East. Why haven't we been able to end this conflict yet? A classical Big Data approach, even if we knew the trajectories of all the bullets shot, couldn't really reveal the causes of this conflict. Nevertheless, it is possible to understand the roots of the conflict. A few years ago, I initiated a study with Ravi Bhavnani, Dan Miodownik, Maayan Mor, and Karsten Donnay, which lead to an empirically grounded, agent-based model.[58] Our model suggests that intercultural distance is the main driver of the conflict.

A further analysis reveals that violent events are correlated with each other.[59] So, there is a responsive dynamic, whereby each side retaliates for previous attacks by the other side.[60] For example, Palestinians retaliate against Israeli violence and vice versa. What does this tell us? Basically, both sides punish each other for violence that they suffered before. It seems that each party tries to send the message: "Stop being violent to us or you will have to pay a high price!" From the point of view of rational choice theory, this should stop the chain of violence. As one event triggers another, usually bigger one, or even several, the conflict becomes increasingly costly for both sides over time. However, rather than creating peace, a deadly spiral of violence sets in. An Israeli documentary film entitled "The Gatekeepers",[61] which interviewed previous secret service chiefs, came to a remarkable conclusion: "We have won every battle, but we are losing the war." In other words, it doesn't pay off to be violent, quite the opposite!

---

[57]Chadefaux [43].

[58]Bhavnani et al. [44].

[59]Jaeger and Paserman [45].

[60]See the movies http://youtu.be/JG86BPezqAU and http://youtu.be/J64kEPa2LaI.

[61]See https://en.wikipedia.org/wiki/The_Gatekeepers_(film).

So why does such retaliation cause an escalation rather than a calming of hostilities? This occurs because both sides think their actions are right. In fact, they are applying sanctioning mechanisms that are intended to create social order, but these mechanisms are only suited for a mono-cultural context.

3.17   Punishment Doesn't Always Work

To understand the problem better, we must ask the question: "Why do we punish others?" This has a simple reason: we have learned that punishment can establish and stabilize social norms. Therefore, who doesn't follow our norms is usually sanctioned. But such punishment is only effective, if it is accepted by the punished party. Otherwise this party will strike back and inflict revenge, which escalates the conflict. Therefore, it is important to recognize that punishment is only effective if people share the same values, norms and culture.

In multi-cultural settings, punishment is often not suited to create social order. Under such circumstances, however, it might be possible to reduce the level of conflict by physically separating the opposing parties so that they live in different areas.[62] Another option is to develop a culture of tolerance, understanding and respect. In fact, as we will see later, there are many social mechanisms that foster social order, such as reputation systems, for example. I am confident, therefore, that a deeper understanding of the mechanisms and forces producing conflicts will eventually help us to overcome or mitigate them. In a multi-cultural world, I would strongly recommend to move away from a punitive culture. Instead, it seems more promising to engage with each other in a differentiated, reputation-based culture in which diversity is welcomed.[63] This brings us to another important set of invisible factors, which determines the success of societies, namely "social capital".

3.18   Why "Social Capital" Is so Terribly Important

Most of us have probably heard the proverb that "money makes the world go round", but there are other, intangible factors that matter. For example, human capital (like education) can boost individual careers, and social capital can act like a catalyst of socio-economic success. But what is social capital? I define it as everything that results from interactions within a social network which could potentially (be used to) create benefits. Examples include cooperativeness, public safety, a culture of punctuality, reputation, trust, power, and respect. However, while our own actions influence our social capital, we can't fully control it. This is in contrast to money.

---

[62]Bhavnani et al. [44].

[63]We will discuss this further in Chap. 8: How Society Works.

In many cases, we can't buy social capital (or only to a limited extent), but social capital creates added value.

Moreover, we cannot automatically generate a certain amount of social capital by doing certain things. Similarly to reputation and respect, these things are given to us by others. They depend on the effects of social interactions. Note that the amount of social capital within a system also determines its resilience or failure. Social capital influences both the probability and extent of damage. This became clear to me at a seminar of ETH Zurich's Risk Center,[64] when we discussed the disproportional effect of large disasters on public opinion. Plane crashes and terror attacks, for example, matter a lot to people, while they seem to feel less threatened by everyday risks such as car accidents or fatalities caused by smoking. Therefore, it is often believed that "size matters", in the sense that large disasters make people respond irrationally or even in panic.

However, having studied the phenomenon of panic for some time, I came to a different conclusion. People realize that the damage is not just physical in nature. Social capital can be damaged, too. For example, a large-scale disaster often reduces public trust in the risk management of companies or public authorities, particularly if it was caused by unprofessional conduct or corruption. While people care about such things, no insurance company covers damage to social capital.

Hence, we must protect social capital similarly to how we protect economic capital or our environment. Social capital can be destroyed or exploited, but this should be prevented. In order to do this, we must learn to measure social capital and to quantify its value. Quantifying the value of our environment also helped to protect it.

3.19 Trust and Power

To stress the importance of social capital, it is important to acknowledge that the financial crisis resulted from a loss of trust. Banks did not trust other banks anymore and did not want to lend out money; customers did not trust their banks anymore and emptied their bank accounts; banks did not want to give loans to companies anymore; people did not want to invest in financial derivatives anymore—the list goes on. In the end, the resulting financial meltdown cost an estimated $15 trillion at least.[65] So, trust is highly valuable and when it erodes, the economic losses are tremendous. To give another example, the recent loss of trust in US cloud storage companies due

---

[64]The Risk Center brings together experts in probability theory and experts in complexity and network theory.

[65]See http://blogs.wsj.com/economics/2012/10/01/total-global-losses-from-financial-crisis-15-trillion/;   http://www.bernerzeitung.ch/wirtschaft/standard/200-000000000000DollarLast/story/24865034.

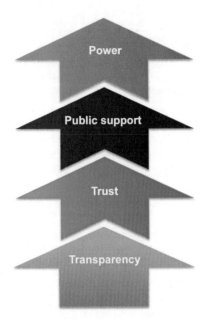

Fig. 3.8 Illustration of how power results from trust and transparency

to the revelations concerning mass surveillance by the National Security Agency (NSA) has substantially reduced their business volume.[66]

Trust is also the basis of power and legitimacy. When I studied in Göttingen, Germany, a deadly car accident caused by a mistake of the police triggered a large public outcry and massive demonstrations. This was the first time when I noticed that public institutions can easily lose public support. In other words, they can easily lose social capital. Trust is eroded whenever authorities do something that is contrary to the moral beliefs of the public. I made a similar observation in Zurich, Switzerland, when many people complained about the policies of the migration office. During this time, the windows of the migration office were repeatedly smashed in, but when the director was replaced, the problem disappeared.

It is very interesting that "soft" factors such as credibility, transparency and trust are the basis of public support and power (see Fig. 3.8). For example, the London Riots of 2011 occurred after a person was shot by the police, but no sufficient justification was given to the public. Something similar happened in 2014 in the city of Ferguson, Missouri, and 2015 in the city of Baltimore, Maryland. In fact, riots in many other countries were triggered by events where public authorities lost legitimacy in the eyes of the public. The Arab spring, for example, started in Tunisia, after Mohamed Bouazizi burnt himself to death to protest against police corruption and brutality.

---

[66] See also 4 Ways that Mass Surveillance Destroys the Economy, https://web.archive.org/web/201 41207003757/, https://www.washingtonsblog.com/2014/06/surveillance-makes-us-poor.html.

In other words, both legitimacy and power are contingent on the authorities doing what the people regard to be "the right thing". If people withdraw their support, authority and power vanish. While the state can purchase weapons and, with this, acquire destructive power, constructive power depends on the trust and support of the people. While weapons might create fear, this is not a good substitute for genuine legitimacy. As the situation becomes increasingly unacceptable, more and more people will lose their fear, and start to resist the previously respected authorities. Some may even be willing to sacrifice their own lives. Remember that many extremists and terrorists previously led normal family lives. But even the passiveness of citizens can make a country fail within just a few years. This could be observed, for example, in the former German Democratic Republic, which had a horrible surveillance system. For these reasons, I am convinced that trust is the only sustainable basis of power and social order.[67]

3.20 Appendix 1: Nervousnet: A Decentralized Digital Nervous System

The open standards of the World Wide Web (WWW) have unleashed a digital economy worth many billion dollars, and participatory projects such as *Linux*, *Wikipedia* and *OpenStreetMap* have created opportunities for everyone. Therefore, it makes a lot of sense to build a Planetary Nervous System, which uses the sensor networks behind the Internet of Things, including those in our smartphones, to measure the world around us and build a data commons together. The question is how to do this while respecting privacy and minimizing misuse. It is time to learn this now.

The Nervousnet project has started to work on such an open and participatory, distributed information platform for real-time data.[68] Nervousnet is an open source project, which believes in the importance of privacy, informational self-determination and trust. If you download the Nervousnet app to your smartphone, you can choose to turn about 10 different sensors separately on or off, such as the accelerometer, light or noise sensors. You can measure data about your environment for yourself (kept on your smartphone) or share it with others (as decided by yourself). External sensors for "smart home" and other applications can be added (for example, weather or health sensors). To maximize informational self-control, the user can also determine the recording rate, and potentially the storage time after which the data will be deleted. Shared data are anonymized. In addition, we are working on data encryption and

---

[67]The following TEDx talk by Detlef Fetchenhauer explains why trust is often better than control: https://www.youtube.com/watch?v=gZlzCc57qX4.

[68]See http://www.nervous.ethz.ch/ and http://www.futurict.eu, also https://www.youtube.com/watch?v=BKcWPdSUJVA. The Nervousnet app can be downloaded via Apple's AppStore and Google's play.

plan to add a personal data store,[69] which will allow you to determine what kind of data you want to share with whom, for what purpose, and for what period of time.

Nervousnet would be run as a Citizen Web, built and managed by its users. It would allow all developers to add measurement procedures and apps. For example, you may run games, scientific measurement projects, or business applications on top of the Nervousnet platform. So, anyone could add data-driven services and products. For security and conceptual reasons (such as scalability and fault-tolerance), Nervousnet would be based on distributed data and control. To promote responsible use, Nervousnet would integrate reputation, qualification, and community-based self-governing mechanisms, determining accessible sensors, data volume and functionality. All this is intended to catalyze a novel information, innovation and production ecosystem to create societal benefits, business opportunities, and new jobs.

Nervousnet would offer five main functionalities. First, it would empower us to collectively measure the world around us in real time. For example, it would allow us to quantify external effects of our interactions with the environment and others (such as noise and emissions, but also economic, social and immaterial value created). Such measurements could help to create a circular economy and more sustainable systems.

Second, Nervousnet would help to create awareness about the problems and opportunities around us. It would warn us of the side effects of certain decisions and actions (e.g. the amount of CO_2 emissions produced) and support us in identifying and implementing better alternatives (e.g. how to use public transport comfortably).

Third, the Nervousnet data stream would establish something like a "CERN for society". It would allow us to reveal the hidden regularities and forces underlying socio-economic change. This would create the knowledge to establish a Global System Science,[70] which is needed to master our future in a highly complex and interdependent world.

Fourth, the combination of this knowledge with real-time data would enable us to build self-organizing systems, using real-time feedbacks. With the right kinds of interactions, a complex dynamical system could create a huge variety of self-stabilizing structures, properties and functions in a way that is self-organized and enormously efficient. For example, self-controlled traffic lights, which flexibly respond to local vehicle flows, can dramatically reduce urban congestion compared to the classical, centralized control approach.[71] By using the Internet of Things or Internet of Everything, as some people say, many production processes could benefit as well (often summarized under the label "industry 4.0").

Fifth, Nervousnet would support a network of distributed artificial and human intelligence. Digital assistants would support us not only in everyday situations, but also in bringing knowledge, ideas, and resources efficiently together. By creating such "collective intelligence", we would be able to master the combinatorial complexity of our increasingly interdependent world much better.

---

[69] de Montjoye et al. [46].

[70] Helbing [47].

[71] Lämmer and Helbing [48], Helbing [49]; see also http://www.stefanlaemmer.de/.

3.21 Appendix 2: Social Fields and Social Forces

When I worked on the social force model, I soon discovered a book by Kurt Levin (1890–1947) on the concept of the "social field".[72] I immediately liked his idea, even though a behavioral and theoretical foundation of the concept was missing. So I decided to develop this foundation in my Ph.D. thesis in 1992.[73] This involved deriving Boltzmann-like and Boltzmann-Fokker-Planck equations using behavioral assumptions.[74]

These equations contain a quantity describing a systematic motion in behavioral space, which can be interpreted as a "social force" and is often expressed as the slope of a "social field". This social field can be imagined like a mountain range in behavioral space, where the steepest slope in a given location determines the social force that a person is subject to. This social force indicates the expected size and direction of the behavioral change. "Valleys" in the social field correspond to social norms. If someone complies with a norm, the social force is zero, but a deviation from the social norm will cause a social force (just as in real-life situations).

Note, however, that the "mountain range" discussed above (and its corresponding social field) is variable in time. It changes as a result of the behavioral changes of others. Therefore, while the social field influences the behavior of a person, at the same time, it is modified by that person's behavior and the behavior of others. In other words, social norms may change over time as a result of social interactions.

References

1. A. Mazloumian, D. Helbing, S. Lozano, R. P. Light and K. Börner (2013) Global multi-level analysis of the 'Scientific Food Web'. Scientific Reports 3, 1167.
2. T. Kuhn, M. Perc, and D. Helbing, Inheritance patterns in citation networks reveal scientific memes, Phys. Rev. X 4, 041036.
3. M. Schich, C. Song, Y. Y. Ahn, A. Mirsky, M. Martino, A.L. Barabási and D. Helbing, A network framework of cultural history. Science 345, 558–562 (2014).
4. A. Mazloumian et al. (2012) Global multi-level analysis of the 'scientific food web', Scientific Reports 3: 1167.
5. D. Lazer, R. Kennedy, G. King, and A. Vespignani, The parable of Google Flu: Traps in Big Data analytics. Science 343, 1203–1205 (2014).
6. D. Brockmann, L. Hufnagel, and T. Geisel, The scaling laws of human travel, Nature 439, 462–465 (2006).
7. V. Colizza, A. Barrat, M. Barthelemy, A.-J. Valleron, and A. Vespignani, Modeling the world-wide spread of panemic influenza: Baseline case and containment interventions. PLoS Medicine 4(1): e13.
8. D. Helbing and A. Carbone (eds.) Participatory Science and Computing for Our Complex World. EPJ Special Topics 214, 1–666 (2012).
9. M. Hechter and K.-D. Opp, Social Norms (Russell Sage Foundation, 2009).

---

[72]Lewin [50].

[73]Helbing [51]; for an English translation see Ref. 96.

[74]Helbing [52].

10. A. Pentland, Social Physics: How Good Ideas Spread (Penguin, 2014).
11. B.K. Chakrabarti, A. Chakraborti, and A. Chatterjee, Econophysics and Sociophysics (Wiley-VCH, 2006).
12. P. Ball, Why Society Is a Complex Matter: Meeting Twenty-First Century Challenges with a New Kind of Science (Springer, 2012).
13. D. Helbing and S. Balietti, From social data mining to forecasting socio-economic crises, EPJ Special Topics 195, 3–68 (2011).
14. M. Moussaïd, N. Perozo, S. Garnier, D. Helbing, and G. Theraulaz (2010) The walking behaviour of pedestrian social groups and its impact on crowd dynamics. PLoS One 5(4), e10047.
15. D. Helbing, Verkehrsdynamik: Neue physikalische Modellierungskonzepte (Springer, 1997).
16. D. Helbing (2001) Traffic and related self-driven many-particle systems. Reviews of Modern Physics 73, 1067–1141.
17. D. Helbing and P. Molnár (1995) Social force model for pedestrian dynamics. Physical Review E 51, 4282–4286.
18. D. Helbing, Verkehrsdynamik (Springer, Berlin, 1997), pp. 38.
19. D. Helbing (1992) A mathematical model for behavioral changes by pair interactions. Pages 330–348 in: G. Haag, U. Mueller, and K. G. Troitzsch (eds.) Economic Evolution and Demographic Change. Formal Models in Social Sciences (Springer, Berlin).
20. D. Helbing (1996) A stochastic behavioral model and a 'microscopic' foundation of evolutionary game theory. Theory and Decision 40, 149–179.
21. D. Helbing et al. (2005) Self-organized pedestrian crowd dynamics: Experiments, simulations, and design solutions, Transportation Science 39(1), 1–24.
22. A. Johansson, D. Helbing, and P. S. Shukla (2007) Specification of the social force pedestrian model by evolutionary adjustment to video tracking data. Advances in Complex Systems 10, 271–288.
23. A. Johansson, D. Helbing, H. Z. A-Abideen, and S. Al-Bosta (2008) From crowd dynamics to crowd safety: A video-based analysis. Advances in Complex Systems 11(4), 497–527.
24. M. Moussaïd, D. Helbing, S. Garnier, A. Johansson, M. Combe, and G. Theraulaz (2009) Experimental study of the behavioural mechanisms underlying self-organization in human crowds. Proceedings of the Royal Society B 276, 2755–2762.
25. D. Helbing, Verkehrsdynamik (Springer, Berlin, 1997), pp. 50.
26. D. Helbing, I. Farkas, and T. Vicsek (2000) Simulating dynamical features of escape panic. Nature 407, 487–490.
27. M. Moussaïd, D. Helbing, and G. Theraulaz (2011) How simple rules determine pedestrian behavior and crowd disasters. PNAS 108 (17) 6884–6888.
28. D. Helbing, A. Johansson, and H. Z. Al-Abideen (2007) The dynamics of crowd disasters: An empirical study. Physical Review E 75, 046109.
29. D. Helbing and A. Johansson (2010) Pedestrian, crowd and evacuation dynamics, in: Encyclopedia of Complexity and Systems Science 16, 6476–6495.
30. D. Helbing, A. Johansson, J. Mathiesen, M.H. Jensen, and A. Hansen (2006) Analytical approach to continuous and intermittent bottleneck flows. Physical Review Letters 97, 168001.
31. D. Helbing and P. Mukerji (2012) Crowd disasters as systemic failures: Analysis of the Love Parade Disaster. EPJ Data Science 2012, 1:7.
32. K. Haase, M. Kasper, M. Koch, S. Müller, and D. Helbing, OR Practice: A pilgrim scheduling approach to increase public safety during the Hajj, Operations Research, in print (2015).
33. D. Helbing, D. Brockmann, T. Chadefaux, K. Donnay, U. Blanke, O. Woolley-Meza, M. Moussaïd, A. Johansson, J. Krause, S. Schutte, and M. Perc (2014) Saving Human Lives: What Complexity Science and Information Systems can Contribute. Journal of Statistical Physics, 1–47.
34. D. Helbing and B. Tilch (1998) Generalized force model of traffic dynamics. Physical Review E 58, 133–138.
35. D. Helbing (1994) A mathematical model for the behavior of individuals in a social field. Journal of Mathematical Sociology 19(3), 189–219.

36. D. Helbing (2010) Quantitative Sociodynamics. Stochastic Methods and Models of Social Interaction Processes (Springer, Berlin).
37. W. Weidlich, Sociodynamics: A Systematic Approach to Mathematical Modelling in the Social Sciences (Dover, 2006).
38. D. Helbing, Pattern formation, social forces, and diffusion instability in games with success-driven motion, Eur. Phys. B 67, 345–356 (2009).
39. D. Helbing and W. Yu (2009) The outbreak of cooperation among success-driven individuals under noisy conditions. Proceedings of the National Academy of Sciences USA (PNAS) 106(8), 3680–3685.
40. D. Helbing, W. Yu, and H. Rauhut (2011) Self-organization and emergence in social systems: Modeling the coevolution of social environments and cooperative behavior. The Journal of Mathematical Sociology 35, 177–208.
41. D. Helbing, W. Yu, K. D. Opp, and H. Rauhut (2014) Conditions for the Emergence of Shared Norms in Populations with Incompatible Preferences. PLOS one, 9(8), e104207.
42. D. Helbing and A. Johansson (2010) Cooperation, norms, and revolutions: A unified game-theoretical approach. PLoS ONE 5(10), e12530.
43. T. Chadefaux, Early warning signals for war in the news, Journal of Peace Research 51(1), 5–18 (2014).
44. R. Bhavnani, K. Donnay, D. Miodownik, M. Mor and D. Helbing (2014) Group segregation and urban violence. American Journal of Political Science 58(1), 226–245.
45. D.A. Jaeger and M.D. Paserman (2008) The cycle of violence? An empirical analysis in the Palestinian-Israeli conflict, American Economic Review 98:4, 1591–1604, https://pubs.aea web.org/doi/pdfplus/10.1257/aer.98.4.1591.
46. Y.-A. de Montjoye, E. Shmueli, S.S. Wang, and A.S. Pentland, openPDS: Protecting the privacy of metadata through SafeAnswers, PLoS ONE 9(7): e98790, https://journals.plos.org/plosone/ article?id=10.1371/journal.pone.0084563.
47. D. Helbing (2013): Globally networked risks and how to respond. *Nature* **497**, 51–59.
48. S. Lämmer and D. Helbing (2008) Self-control of traffic lights and vehicle flows in urban road networks. JSTAT P04019.
49. D. Helbing (2013) Economics 2.0: The natural step towards a self-regulating, participatory market society. Evolutionary and Institutional Economics Review 10, 3–41.
50. K. Lewin Resolving Social Conflicts: Field Theory in Social Science (American Psychological Association, Washington D.C., 1948).
51. D. Helbing (1992) Stochastische Methoden, nichtlineare Dynamik und quantitative Modelle sozialer Prozesse. (Ph.D. thesis, University of Stuttgart. Published by Shaker, Aachen, 1993, 2nd ed. 1996).
52. D. Helbing (1993) Boltzmann-like and Boltzmann-Fokker-Planck equations as a foundation of behavioral models. Physica A 196, 546–573.

Chapter 4
Google as God?

The Dangerous Promise of Big Data

> *As for the future, your task is not to foresee it, but to enable it.*
> —*Antoine de Saint-Exupery*

Almost every activity in the world now leaves numerous digital traces. What if a „wise king" or „benevolent dictator" had real-time access to all the data in the world? Could he take perfect decisions to benefit society? Could he predict the future? Could he control the world's path? What would be the limits and side effects of such an approach? Or is the attempt to create a digital „crystal ball" to predict the future and a digital „magic wand" to control the world a dangerous dream?

Historically, human civilization developed by establishing mechanisms to promote cooperation and social order. One such mechanism is based on the idea that everything we do is seen and judged by God. Bad deeds will be punished, while good ones will be rewarded. This might be seen as one of the mechanisms through which religion has established social order. More recently, the information age has inspired the dream of some keen strategists that we might be able to know everything about the world ourselves, and to shape it as desired. Would it be possible to acquire God-like omniscience and omnipotence fueled by Big Data? There are now hopes and fears that such power lies within the reach of technology and data giants such as *Google* or *Facebook*, or secret services such as the CIA or the National Security Agency (NSA). In 2013, CIA Chief Technology Officer Ira "Gus" Hunt[1] explained how easy it is for such institutions to gather a great deal of information about each of us:

> *"You're already a walking sensor platform... You are aware of the fact that somebody can know where you are at all times because you carry a mobile*

---

[1] See: http://www.businessinsider.com/cia-presentation-on-big-data-2013-3?op=1 and https://gig aom.com/2013/03/20/even-the-cia-is-struggling-to-deal-with-the-volume-of-real-time-social-dat a/2/. For similar recent FBI priorities see http://www.slate.com/blogs/future_tense/2013/03/26/and rew_weissmann_fbi_wants_real_time_gmail_dropbox_spying_power.html.

© Springer Nature Switzerland AG 2021
D. Helbing, *Next Civilization*,
https://doi.org/10.1007/978-3-030-62330-2_4

device, even if that mobile device is turned off. You know this, I hope? Yes? Well, you should... Since you can't connect dots you don't have, it drives us into a mode of, we fundamentally try to collect everything and hang on to it forever... It is really very nearly within our grasp to be able to compute on all human generated information."

Will this massive data-collection process be good for the world, helping us to eliminate terrorism, crime, and other societal ills? Will it allow us to turn our society into a perfect clockwork and overcome the mistakes that people make? Or will so much information create even bigger problems?

4.1   Technology to Empower a "Wise King"?

Imagine you are the president of a country and aim to maximize the welfare of its people. What would you do? You would probably want to prevent financial crashes, economic crises, social problems and wars. You would like to ensure a safe and reliable supply of food, water and energy. You would perhaps try to prevent environmental degradation. You may wish people to be rich, happy and healthy. You may further like to avoid problems such as traffic jams, corruption, and drug abuse. In sum, you would probably like to create a prosperous, sustainable and resilient society.

What would it take to achieve all of this? You would certainly need to take the right decisions and avoid those that would imply harmful, unintended side effects. For each impending decision you would need to know alternative courses of action that could be taken and their associated opportunities and risks. For your country to thrive, you would need to avoid ideological or impulsive decision-making and let data-based evidence guide your actions. In order to have enough information for this type of decision-making, you would need a lot of data about all quantifiable aspects of life and society, and excellent data analysts to interpret it. You might even decide to collect all the data in the world (as much as this can be done), just in case it might be useful one day to counteract threats and crises, or to exploit opportunities that might arise.

In the past, rulers and governments haven't had this possibility: they often lacked the quality or quantity of data needed to take well-informed decisions. But this is now changing. In recent decades, the processing power of computers has exploded, and the volume of stored data has dramatically increased. Each year, we are now generating as much data in the world as in the entire history of humankind.[2] What could we do with all this data?

---

[2]Knowledge doubling every 12 months, soon to be every 12 hours, see http://www.industrytap. com/knowledge-doubling-every-12-months-soon-to-be-every-12-hours/3950.

4.2 A Digital "Crystal Ball"?

Until recently, we have successfully used supercomputers for almost everything, except for understanding our economy, society and politics. Every new car or airplane is designed, simulated and tested on a computer before they are built. Increasingly, the same is true for the research and development of medical drugs. Thus, why shouldn't we use computers to understand and guide our economy and society, too? In fact, we are moving in this direction. As a minor (yet illustrative) example, computers have been used for traffic control since their early days. In the meantime, also modern production and supply chain management would be inconceivable without computerized control. Already since some time, airplanes are controlled by a majority decision among a few computers, and driverless, computer-controlled cars will soon traverse our streets. In all of these cases, computers do a better job than humans. If they are better pilots, drivers, and chess players, why shouldn't computers eventually make better policemen, administrators, lawyers, and politicians, too?

Given today's data volume and computational power, it seems no longer unreasonable to imagine a gigantic computer system that tries to simulate the actions and interactions of all humans globally. In such a digital mirror world our virtual doubles might be even equipped with cognitive abilities and decision-making capacities. If we fed these virtual humans with our own personal data, how similar to us would they behave?[3] Would it eventually be possible to create a digital copy of our world, a virtual reality as realistic as life itself?

Recent studies using smartphone data and GPS records suggest that the activity of a person at a particular time can often be forecast with an accuracy of more than 90%.[4] Our lives are surprisingly predictable due to our repetitive daily and weekly schedules and routines. It is also possible to determine our personality traits and attributes.[5] There are even companies such as *Palantir* and *Recorded Future*, which have developed sophisticated tools for predictive analytics and are trying to build a digital "crystal ball". The military in the USA is engaged in similar projects,[6] and it is likely that the same applies to other countries.

Although the prospect of building a digital crystal ball might sound ominous to some, we should carefully discuss it. The potential benefits are obvious, as there are many huge problems that might be solved with such predictive capabilities. The financial crisis has created global losses of at least $15 trillion. Crime and corruption consume about 2–5% of the gross domestic product (GDP) of all nations on earth—about $2 trillion each year. A major influenza pandemic infecting 1% of the world's population might cause losses of $1–2 trillion per year. The wars post September 11

---

[3] We certainly could not expect real humans and their virtual doubles to be more similar than "identical" twins.

[4] Song et al. [1].

[5] Kosinski et al. [2].

[6] Can the military make a prediction machine?, see http://www.defenseone.com/technology/2015/04/can-military-make-prediction-machine/109561/.

have cost many trillion dollars. And cybercrime costs more than €750 billion a year in Europe alone.

If a computer simulation of the entire global socio-economic system could produce just a 1% improvement in dealing with these problems, the benefits to society would be already immense. But if the management of smaller complex social systems provides any guide, even an improvement of 10–30% seems possible. Overall, this would amount to savings of more than $1 trillion annually. Even if we had to invest billions in creating such a system, we could see a hundred-fold return on investment nevertheless. Even if the success rates were significantly smaller, this would represent a substantial gain. Thus, a digital crystal ball seems to be a worthwhile investment. But how could one ensure that the world stays on the predicted path, and individual decision-makers don't mess it up?

4.3   The Digital "Magic Wand": A Remote Control for Humans?

Whenever we spend time on the Internet, we leave digital traces that are collected by electronic "cookies" and other means, often without our consent. But the more data is generated about us, stored and interpreted, the easier it becomes to find out things about us, which were not intended to be shared. Our computers and smart devices leave unique digital fingerprints—such as their configuration, our wireless network logins, behavioral patterns, and movement records. As a result, it is now possible for data analysts to infer our interests, passions, ways of thinking, and even our feelings. Some companies analyze "consumer genes" to offer personalized products and services. 3000–5000 different kinds of metadata derived from personal data have already been collected from almost a billion people worldwide, including names, contact data, incomes, consumer habits, medical information and more.[7] Therefore, everyone with a certain level of income and an Internet connection has been mapped to some extent.

This raises a controversial, but unavoidable question. Would it be good if a company or well-intentioned government had access to all this data? Would it help politicians and administrations to make decisions that reduce terrorism, crime, energy consumption, environmental degradation, traffic congestion, financial meltdowns and recessions? Furthermore, could the data be used to optimize our health and education systems and to provide public services which are tailored to the needs of the citizens?

The question is to what extent can this idea be realized? If one had enough data about every aspect of life, could one be omniscient? Would it even become possible to predict the world's future? For this, one would certainly have to determine people's choices to avoid deviations from what was predicted before. How could this be done? This is easier than one might think! One could use information specifically tailored

---

[7]See https://www.democraticmedia.org/acxiom-every-consumer-we-have-more-5000-attributes-customer-data.

to us in a way that is made to manipulate our choices. Personalized ads do this every day. They let us buy certain products or services that we wouldn't have cared about otherwise. The more personal data is collected about us, the more efficient do personalized ads get.

It would be surprising if such possibilities to manipulate our choices weren't attractive to politicians, too. Given the overwhelming amount of data available today, information needs to be filtered to be useful for us. Such filtering will inevitably be done in the interests of those who do the filtering or pay for it. A recent *Facebook* experiment involving almost 700,000 users showed that it is, in fact, possible to manipulate people's feelings and moods.[8]

Therefore, will it soon be possible to turn "omniscience" into "omnipotence"? In other words, could those who have access to all the data finally control everything? Let's call the hypothetical tool creating such power a digital "magic wand". Suppose again you were the president of a country and had such a magic wand. Could you be "omnibenevolent", i.e. take the best decisions for society and every single one of us? Many people might say that forecasting societal trends is different from forecasting the weather. While the weather will not react to a forecast, people will. This seems to suggest that societal predictions will be unreliable (or even "self-defeating prophecies"), if the predictions are published. But what if a secret institution evaluated our data and advised the government about the right decisions to be made? The magic wand would then be used to manipulate people's behavior based on the evidence provided by the crystal ball. Could such a scheme work?

4.4   A New World Order Based on Information?

A wishful "benevolent dictator" or "wise king" would probably see the crystal ball and the magic wand as perfect tools to create the desired societal outcomes. Of course, the "wise king" could not fulfill our wishes all the time, but he might be able to create better outcomes on average for everyone as long as we obey him. So, a "wise king" would sometimes interfere with our individual freedom, if the autonomous decisions of people would otherwise obstruct his plan. This, however, would result in a form of totalitarian technocracy, where all of us would have to follow personalized instructions, as if they were commands from God. One might distinguish two kinds of such "big government" approaches. The "Big Brother" society would build on mass surveillance and the punishment of people, who don't obey the commands or are unlikely to obey them as determined by "predictive policing" algorithms (even though the effectiveness is questionable[9]). The "Big Manipulator" society would

---

[8] Kramer et al. [3]. This experiment created a big «shit storm»: http://www.wsj.com/articles/furor-erupts-over-facebook-experiment-on-users-1404085840.

[9] Journalist's Resource (6 November 2014), The effectiveness of predictive policing: Lessons from a randomized controlled trial, https://journalistsresource.org/studies/government/criminal-justice/predictive-policing-randomized-controlledtrial/. ZEIT Online (29 March 2015), Predictive Policing—Noch hat niemand bewiesen, dass Data Mining der Polizei hilft, http://www.zeit.de/digital/datenschutz/2015-03/predictive-policing-software-polizei-precobs.

build on mass surveillance and mass manipulation using personalized information. Of course, these approaches could also be combined.

It turns out that both approaches are already on their way. Predictive policing is now trying to anticipate the hotspots of criminal activities, but it also puts people in jail who are assumed to potentially commit a crime in future.[10] We must be aware that, in an over-regulated society, almost everyone violates some laws or regulations within the course of a year.[11] One day, we may have to pay for all these little mistakes, because it's easy to record them now. With new, section-based speed controls, such technologies are already being tested. The principle of "assumed innocence" which has been the basis of democratic legal systems for ages, is now called into question. With the use of mass surveillance, the assumption that most people are good citizens has effectively been replaced by the assumption that everyone is violating rules or may become an obstacle to the plans of the "wise king", "big government", or "benevolent dictator". In many countries, specialized police forces have increasingly been equipped like soldiers, and it seems that new, coercive strategies are being tested during demonstrations, also when people didn't behave in a criminal, violent or improper way. So far, however, this tended to cause a public outcry or even countrywide protests.[12] But what about the "Big Manipulator" approach? Wouldn't it be just perfect to reach good societal outcomes in a way that doesn't hurt? While we still don't have sufficient data to understand the way humans decide, there are now enough metadata about everyone to manipulate people's choices.

4.5 Nudging: When the State Takes Care of Our Decisions

In times of mass collection of personal data, everyone is not only a potential suspect, but also susceptible to manipulation attempts. While Big Data analytics is far from being able to understand the complexity of human behavior, it is advanced enough to manipulate our decisions based on a few thousand metadata that have been collected about every one of us. Some companies earn many billions each year for influencing us with personalized ads. *Google*, for example, one of the most valuable companies

---

[10] See, for example, http://www.nzz.ch/zuerich/stadt-zuerich/die-akte-babini-1.18506892, but there are increasingly more cases in various countries due to new "disturbing the peace" laws, which make dissent a legal offence.

[11] Schmieder [4].

[12] Frankfurt, Germany: http://www.fr-online.de/blockupy-frankfurt/blockupy-demo-entsetzen-ueber-polizei-kessel,15402798,23144620.html; Zurich, Switzerland: http://www.tagesanzeiger.ch/zuerich/stadt/Jetzt-reden-die-Eingekesselten/story/21545861, http://www.nzz.ch/zuerich/stadtrat-wolff-und-polizeikommandant-blumer-drohen-strafverfahren-1.18574181; Ferguson and Baltimore, USA: http://www.bbc.com/news/world-us-canada-30193354, https://en.wikipedia.org/wiki/2015_Baltimore_protests.

in the world, is making more than 90% of their money with ads. One might, therefore, think that they are making money mainly with manipulation, not information.[13]

With the huge amounts of personal data that such companies are storing about us, they probably know us better than anybody else, including our closest friends. Therefore, it is easy for these companies to manipulate our choices with personalized information. Some of this information may even affect our actions subconsciously by "subliminal messaging".[14] As a consequence, we might do many things without our conscious control.[15] To some extent we are now remotely controlled by others, and with increasing amounts of personal information and powerful machine learning algorithms, attempts to manipulate our behavior will become ever more successful.[16]

Possibilities to influence peoples' decisions are not only extremely attractive for business, but also for politics. In political contexts, this method is known as "nudging" or "liberal paternalism", and it has been promoted as the perfect tool to improve peoples' choices.[17] It must be very appealing to political leaders to have an instrument allowing them to make us behave in healthier and environmentally friendly ways without having to convince us. However, the same approach may be used to promote nationalism, reservations against minorities, or pro-war sentiments. So far, one can certainly say that the use of nudging hasn't reduced the number of problems in our society—on the contrary.

Nevertheless, it may be tempting to use nudging to influence our voting behavior. In fact, experts think that companies such as *Google* and *Facebook* can affect election outcomes.[18] Most likely, this has happened several times already.[19] At least, it is publicly known that our personal data (which has been accumulated largely without our consent) is being used to win elections.[20] It is clear that such attempts to manipulate millions of people in ways that are neither transparent nor well regulated undermine our conscious and autonomous decision-making.[21] In addition, they endanger the fundament of democracies, which is based on pluralism and a fair

---

[13]This fact has recently been criticized by Apple chief Tim Cook, see http://www.malaysiandigest. com/technology/556240-tim-cook-silicon-valley-s-most-successful-companies-are-selling-you-out.html.

[14]Companies such as *Facebook* even seem to apply strategies that make people addicted to their information, using methods copied from the gaming salons in Las Vegas, see http://www.dailymail.co.uk/news/article-2174024/Facebook-creating-generation-gam bling-addicts-sites-Las-Vegas-style-games.html.

[15]In some sense, this might be best compared with acting under hypnosis.

[16]In the past, the privilege to influence our behavior was mainly restricted to our partners, friends, and bosses.

[17]Thaler and Sunstein [5].

[18]Surveillance-based manipulation: How Facebook or Google could tilt elections, see http://arstec hnica.com/security/2015/02/surveillance-based-manipulation-how-facebook-or-google-could-tilt-elections/; some experts believe that up to 20% of votes may be changed by nudging.

[19]Bond et al. [6].

[20]The real story of how big data analytics helped Obama win, see http://www.infoworld.com/art icle/2613587/big-data/the-real-story-of-how-big-data-analytics-helped-obama-win.html.

[21]Note that a lack of control over our personal data also implies that we will progressively lose the possibility to control our own life.

competition of the best ideas for our votes. Are personal freedom and democracy perhaps outdated?[22] (see Appendix 4.1).

Besides the use of nudging, many governments also pay special agents and deploy computer bots to produce fake information in order to make us believe and do certain things.[23] There are even "cybermagicians" working for secret services, who change Web and social media contents.[24] In other words, facts are being manipulated, but the outcomes can sometimes be quite unexpected. For example, while the United Kingdom tried to strengthen its central government powers, there is now a trend towards greater regional autonomy. Moreover, even though Turkey "streamlined" the reporting in the public media to support the candidacy of the president, his power is being questioned more than ever. The public reporting was so biased that nobody expected the 2015 election outcome. This caused significant disruptions of the financial markets.[25]

I am deeply convinced that the digital society can only thrive, if we learn to reduce manipulation and information pollution. The more information we must process and assess, the more we must be able to rely on it.[26] If we don't have trustable information systems, almost nobody will finally be able to judge the situations we are faced with. Consequently, we will probably make more mistakes even though we could use a quickly growing amount of information.[27]

One problem of today's information systems is their lack of transparency. We have usually no idea of the quality of results that search engines are delivering. These algorithms determine the ranking of results, when we make Internet searches: they decide what information appears on the top and what information is hidden. This has dramatic consequences: it can ruin peoples' lives[28] and cause the bankruptcy of

---

[22]Note that a wrong use of Big Data implies numerous dangers for democracy, including (1) a loss of freedom, autonomy, and diversity, (2) an undermined wisdom of crowds, (3) discrimination, (4) self-censorship, reduced innovation and failed adaptation, (5) security issues (such as cybercrime and cyber war).

[23]See http://www.theguardian.com/technology/2011/mar/17/us-spy-operation-social-networks.

[24]RT, Western spy agencies build 'cyber magicians' to manipulate online discourse (February 25, 2014), https://www.rt.com/news/five-eyes-online-manipulation-deception-564/; https://theinterc ept.com/2014/02/24/jtrig-manipulation/, https://firstlook.org/theintercept/2014/07/14/manipulat ing-online-polls-ways-british-spies-seek-control-internet/, http://praag.org/?p=13752.

[25]Turkey's online censorship, see https://web.archive.org/web/20201114002447/https://takingnote. blogs.nytimes.com/2015/04/08/turkeys-onlinecensorship/; Market turmoil in Turkey after crushing election, see http://www.cnbc.com/2015/06/08/turkey-election-result.html.

[26]This is particularly a precondition for collective intelligence and for a successful management of the increasing complexity of our world.

[27]See, for example, Lazer et al. [7].

[28]Lawsuits against Google autocomplete, see http://www.techtimes.com/articles/12296/20140806/ billionaire-albert-yeung-allowed-to-sue-google-over-the-autocomplete-of-his-name.htm, Persön- lichkeitsverletzende Wortkombinationen, see http://www.sueddeutsche.de/politik/persoenlichkeit sverletzende-wortkombinationen-bettina-wulff-schliesst-vergleich-mit-google-1.2306708.

companies.[29] It's no wonder that there are currently a number of lawsuits,[30] which challenge the manipulation of search results.[31] Is this manipulation just a business model, or is it more?

4.6 Google as God?

We have seen that, powered by the giant masses of information, a crystal ball to predict the future and a magic wand to shape it are already being built, used, and further improved. It's not anymore far-fetched to imagine that companies such as *Google* or the USA's National Security Agency (NSA) might try to take the role of a "wise king" or "benevolent dictator". Or do they have even bigger ambitions?

There is no doubt that many people in the Silicon Valley believe in their ability to change and re-invent the world. Companies like *Google* aren't just building powerful search engines, smart software tools, and self-driving cars. It seems they are also trying to create omnipresence (with the *Google Loon* project[32]) and omniscience (by collecting the world's knowledge with a giant search engine). Given that "knowledge is power", *Google* may also be striving for omnipotence (by manipulating peoples' choices—after all the company makes most of its money with personalized ads). It is further known that *Google* wants to overcome death[33] (e.g. with *Google Calico*) and that the company wants to create Artificial Intelligence (e.g. with their *Google Brain* project).[34] In other words, one may get the impression that *Google* wants to build or be a "digital God". What kind of God would this be? A God that watches us, a God that controls our destiny, or a God that gives us responsibility and freedom? And how well would such a plan work?

For the time being, let's make the naive assumption that a company or government having all these digital powers would like to act in a benevolent way and not selfishly at all (i.e. it would take care of us rather than trying to exploit us). Would such an approach be able to optimize the state of the world? This is actually more tricky than one might think.

---

[29]EU Antitrust Probe Threatens To Hit Google With Heavy Fines, see http://www.techtimes.com/articles/62230/20150622/eu-antitrust-probe-threatens-to-hit-google-with-heavy-fines.htm.

[30]See http://www.bloomberg.com/news/articles/2013-04-10/google-sued-by-streetmap-in-u-k-for-anti-competitive-behavior.

[31]In conclusion, in many cases manipulating our decisions might be illegal, dangerous or counterproductive.

[32]Google's giant balloons set to connect remote parts of the world to the internet, see http://www.dailymail.co.uk/sciencetech/article-3127331/Project-Loon-commercial-year-Google-s-giant-balloons-set-connect-remote-parts-world-internet.html.

[33]How Google's Calico aims to fight aging and 'solve death', see http://edition.cnn.com/2013/10/03/tech/innovation/google-calico-aging-death/.

[34]Inside the artificial brain that's remaking the Google empire, see http://www.wired.com/2014/07/google_brain/.

4.7 Errors or First and Second Kind: Doing Good Isn't Easy

First of all, what's the goal function one should optimize? Money? Happiness? Health? Security? Innovation? Peace? Or sustainability? Surprisingly, there is no scientific method that can give a definite answer, but if we chose the wrong goal, it could end in disaster.[35] Therefore, choosing and implementing a single solution everywhere is highly dangerous.[36] But even if the goal function was clear, distinguishing "good" from "bad" solutions is often difficult, for statistical reasons. For example, good and bad risks or behaviors can hardly ever be perfectly separated, as all measurements are noisy. Consequently, there is the problem of false positive classifications (false alarms, so-called type I errors) and of false negatives (type II errors, where the alarm is not triggered when it should be).

Imagine a population of 500 million people, in which there are 500 terrorists. Let's assume that we can identify terrorists with an extremely impressive 99% accuracy. In this scenario, there are 1% false negatives (type II error), which means that 5 terrorists are not detected, while 495 will be discovered. Newspapers reported that about 50 terrorist acts in Western countries were prevented over the past 12 years or so, while the attack on the Boston marathon and on the Charlie Hebdo editorial office in Paris could not be prevented, even though the perpetrators were listed in databases of terrorist suspects (in other words, they turned out to be false negatives).

How many false positives (false alarms) could we expect? If the type I error is just 1 in 10,000, there will be 50,000 innocent suspects, while if it is 1 in 1000 there will be 500,000 innocent suspects. If it is 1% (which is entirely plausible), there will be 5 million innocent suspects! Indeed, it has been reported that there are between 1 and 8 million people on the various lists of suspects in the USA.[37] If these figures are correct, this means that for every genuine terrorist, around 10,000 innocent citizens will be wrongly categorized as potential terrorists, perhaps even more. Since the 9/11 attacks, about 40,000 suspects have had to undergo special questioning and screening procedures at international airports. In 99% of these cases, it was later concluded

---

[35] For a number of years, using asbestos in buildings has been the preferred choice for fire protection and other means. However, in the end it turned out to cause lung cancer, and its widespread use was very costly. The use of Contergan, which caused the abnormal development of childrens' bodies, is another well-known disaster of this kind. The reasons for the widespread increase of allergies and obesity are not yet known for sure. However, the underlying problem is again that a huge number of people has been simultaneously exposed to new substances. The same applies to the massive deaths of bees.

[36] That's probably why social, ecological and economic systems happen to "explore" a large diversity of goal functions, where some competitive or selection mechanism helps good solutions to spread. This speaks for pluralism. In fact, the reduction of diversity is one of the greatest dangers. As we will see later on, socio-economic diversity is as important as biodiversity. Many species (including many plants) multiply themselves in a way that increases diversity. If this is not done, handicaps or low IQ might result. Therefore, laws even forbid that brothers and sisters marry.

[37] It recently got known that the police of Bavaria (a federal state of Germany with 12.7 million in habitants) stores about 1 million people in a list of suspects, see http://www.sueddeutsche.de/bay ern/datenbank-der-polizei-verdaechtig-fuers-leben-1.2549799.

that the suspect was innocent. And yet the effort needed to reach even this level of accuracy is considerable and costly: according to media reports, one million people with National Security Agency (NSA) clearance on the level of Edward Snowden are employed by the USA alone.[38]

Thus, large-scale surveillance doesn't seem to be an effective means of fighting terrorism. Indeed, this conclusion has been reached by several independent empirical studies.[39] Attempting to keep the entire population under surveillance is not sensible for the same reason why it is generally not useful to screen the entire population for a particular disease or to administer a particular preventative drug nationwide; since such mass screenings imply large numbers of false positives, millions of people would be needlessly treated, often resulting in negative side effects for their health.[40] Thus, for most diseases, patients should only be tested if they show worrying symptoms.

Apart from these type I and type II errors, a third type of error may be encountered: the application of wrong models. For example, models which did not sufficiently account for rare, extreme events were one of the reasons for the recent financial and economic crisis. The risks of many financial products turned out to be much greater than anticipated, creating immense losses. Adair Turner, head of the UK Financial Service Authority, pointed out that there is

> *"a strong belief ... that bad or rather over-simplistic and overconfident economics helped create the crisis. There was a dominant conventional wisdom that markets were always rational and self-equilibrating, that market completion by itself could ensure economic efficiency and stability, and that financial innovation and increased trading activity were therefore axiomatically beneficial."*

---

[38]Business Insider, How a GED-holder managed to get 'top secret' government clearance (June 10, 2013), http://www.businessinsider.com/edward-snowden-top-secret-clearance-nsa-whistleblower-2013-6.

[39]The Washington Post (January 12, 2014): NSA phone record collection does little to prevent terrorist attacks, group says, https://web.archive.org/web/20150815210511/https://www.washingtonpost.com/world/national-security/nsaphone-record-collection-does-little-to-prevent-terrorist-attacks-group-says/2014/01/12/8aa860aa-77dd-11e3-8963-b4b654bcc9b2_story.html; http://securitydata.newamerica.net/nsa/analysis; Gill [8]; see also BBC News (August 24, 2009): 1000 cameras 'solve one crime'.

[40]A good example are recent large-scale immunization campaigns and their unexpected side effects: Swine flu vaccine can trigger narcolepsy, UK government concedes, see http://www.theguardian.com/society/2013/sep/19/swine-flu-vaccine-narcolepsy-uk; see also New research implies potential link between European H1N1 flu vaccine and narcolepsy, http://www.extremetech.com/extreme/209324-a-vaccine-health-scare-thats-actually-real.

4.8 Limitations of the "Crystal Ball"

One might think that these three kinds of errors could be overcome if we just had enough data. But is this really true? There are a number of fundamental scientific factors that will impair the predictive power of the crystal ball. For example, the problem, which is known as "Laplace's Demon", posits that all future developments are determined by the world's history. As a result, our ability to make predictions is fundamentally constrained by our inability to measure all historical information needed to predict the future[41] (assuming that the world would change according to deterministic rules at all, [42] which is questionable).

In general, the parameters of a computer model representing certain aspects of our world can only be determined with a certain accuracy. But even a small variation of the assumed model parameters within their respective "confidence intervals" may fundamentally alter the model predictions. In complex anthropogenic systems, such "parameter sensitivity" is often expected to be large. While the confidence intervals may be narrowed down, if enough data are available, having too much data can be a problem, too: it may reduce the quality of predictions due to "over-fitting", "spurious correlations" or "herding effects".

Furthermore, many complex dynamical systems show phenomena such as "turbulence" or "(deterministic) chaos", for which even the slightest difference may fundamentally change the outcome sooner or later. This well-known property is sometimes called the "butterfly effect". It imposes a time limit beyond which no useful forecast can be made. The particular physics of weather phenomena is the reason why meteorologists cannot forecast the weather reasonably well for more than a few days,[43] and even a million times more data could not fundamentally change this.

In social systems there is the additional problem of ambiguity: the same information may have several different meanings depending on the context, and the way we interpret it may influence the future course of the system. Beyond this, since Kurt Gödel (1906–1978) it is known that some questions are fundamentally undecidable in that the correctness of certain statements can neither be proved nor disproved with formal logic. Appendix 4.2 explains the above problems in more detail.

Thus, it is safe to say that Big Data is not the universal panacea it is often claimed to be.[44] Attempts to predict the future will be mostly limited to probabilistic and short-term forecasts. This particularly applies to unstable systems. It is, therefore,

---

[41] For example, we are influenced by cultural inventions, ideas and social norms which are sometimes thousands of years old.

[42] Assuming determinism is rather questionable, given that there are probably many random influences on the world's course, as we know from quantum mechanics, for example.

[43] These limitations in predictive power are not merely a matter of having insufficient data or not enough computer power. The physical nature of the underlying processes fundamentally limits the precision of forecasts.

[44] To convince me otherwise, in an analogy to the "Turing test" which checks whether a computer can communicate in a manner which is undistinguishable from a human, a computer system would have to find all the fundamental laws of physics discovered by scientists so far by mining the experimental data accumulated in the past centuries.

dangerous to suggest that a crystal ball could be built that might reliably predict the future. As a consequence, any real-world application of Big Data requires critical assessment and particular care.

4.9 Limitations of the "Magic Wand"

If the crystal ball is "cloudy", it doesn't augur well for the magic wand that depends on it. Therefore, using the magic wand would often have unintended consequences.[45] It is questionable, whether we would ever learn to use it well. In fact, top-down control is still very ineffective, as the abundance of problems in our world shows. To control our complex world, i.e. to make it change in the desired ways, we would need to understand it much better than today. In many cases, attempts to control a complex dynamical system in a top-down way will miserably fail, as this tends to undermine the normal functionality of the system (for example, the guidance of our behavior by socio-cultural cues[46]). The result may be a "broken" system, for example, an instability or crisis. An example of the failure of top-down control is the fact that the greatest improvement of airplane flight safety was not reached by technical control mechanisms but by introducing a non-hierarchical culture of collaboration in the cockpit, where co-pilots were encouraged to question the decisions and actions of the pilot. In another example, the Fukushima nuclear disaster in Japan, the official investigation report stresses that it was not primarily the earthquake and tsunami that were responsible for the nuclear meltdowns, but

> *"our reflexive obedience; our reluctance to question authority; our devotion to 'sticking with the program'; our groupism."*

Unfortunately, there are many more examples that illustrate the failure of top-down control. In spite of 45 million arrests, the US "war on drugs" was largely ineffective. Similarly, many attempts to enforce "law and order" in the world after September 11, 2001 failed. The wars in Afghanistan and Iraq didn't reach the goal of stabilizing the geopolitical region. Instead, the world is faced with a chaotic situation in the aftermath of the Arab Spring and the war of the Islamic State. Torture that was

---

[45]Johann Wolfgang von Goethe articulates this problem in his poem "The Sorcerer's Apprentice", see https://web.archive.org/web/20130620035237/https://germanstories.vcu.edu/goethe/zauber_dual.html.

[46]In other words, nudging may "overwrite" the guiding effects of culture, while it is not able to replace it by a sufficiently sophisticated kind of top-down control. I would not be surprised if this created social chaos in the long run.

used to win the "war on terror" was found to be ineffective, and drone strikes seem to trigger new conflict rather than reducing it.[47]

Why do these control attempts fail despite all the power that some governments have? The issue is that a complex dynamical system such as our society cannot be steered like a bus. Let's compare this with our body, which is a complex dynamical system, too. Here, we know very well that taking more medicine doesn't help more, but it might poison our body. In our economy and society, therefore, power needs to be applied in the right dose, in the right place, in the right way. Typically, we don't need more power but more wisdom. In many cases, too much top-down control is a problem, not the solution. It is also quite expensive, and we find it increasingly hard to pay for it: most industrialized countries already have debt levels of at least 100 or 200% of their gross domestic product.

4.10   Complexity Is the Greatest Challenge

The complexity of today's human-influenced systems is the main reason why the concept of a "super-government", "wise king" or "benevolent dictator" can't work for a long time. There are at least four kinds of complexity that matter: dynamic complexity, structural complexity, functional complexity and algorithmic complexity. I have already addressed the problem of complex dynamics. In the following, I will focus on implications of structural, functional and algorithmic complexity. Within a centralized super-computing approach we can only solve those optimization problems, which have sufficiently low algorithmic complexity. But many problems are "NP-hard", i.e. computationally so demanding that they cannot be handled in real-time even with today's super-computing. This problem is particularly acute in systems that are characterized by a large variability. In such cases, top-down control can often not reach optimal results. In a later chapter, I will illustrate this by the example of traffic light control.

Given the quick increase in computing power, couldn't we overcome this challenge in future? The surprising answer is "no". While the processing power doubles every 18 months (blue curve in Fig. 4.1), the amount of data doubles even every 12 months (green curve), and the growth rate will further increase.[48] This implies that we are

---

[47]U.S. torture techniques unethical, ineffective (January 6, 2011), http://www.livescience.com/9209-study-torture-techniquesunethical-ineffective.html; http://en.wikipedia.org/wiki/Effectiveness_of_torture_for_interrogation and http://www.huffingtonpost.com/2014/04/11/cia-harsh-interrogations_n_5130218.html; The Guardian, US drone attacks 'counter-productive', former Obama security adviser claims (January 7, 2013), http://www.theguardian.com/world/2013/jan/07/obama-adviser-criticises-drone-policy; http://www.huffingtonpost.com/2013/05/21/us-drone-strikes-ineffective_n_3313407.html; The NSA isn't just spying on us, it's also undermining Internet security (April 30, 2014), http://www.nationaljournal.com/daily/the-nsa-isn-t-just-spying-on-us-it-s-also-undermining-internet-security-20140429; http://www.slate.com/blogs/future_tense/2014/07/31/usa_freedom_act_update_how_the_nsa_hurts_our_economy_cybersecurity_and_foreign.html.

[48]Helbing [10].

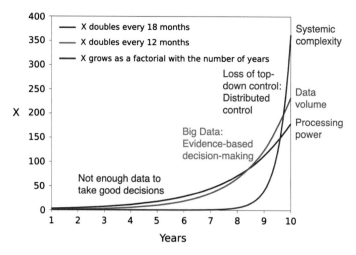

Fig. 4.1 Schematic illustration of the increase of processing power, data volume and systemic complexity with time. (Reproduced from Helbing [9] with kind permission of Springer Publishers.)

heading from a situation in which we did not have enough data to take good decisions, to a situation where we can take evidence-based decisions. But despite the rising processing power, we will be able to process only a decreasing share of all the data in the world.[49] Moreover, the gap in processing power will be quickly increasing. So we are moving towards a situation where we may see anything we want with a kind of "digital flashlight", but many things will remain unseen in the dark. In addition, some systems are irreducibly complex, i.e. every little detail can matter.[50]

The "flashlight effect" creates a new kind of problem: paying too much attention to some problems, while neglecting others. Sometimes we overlook really big problems. For example, while the world focused on fighting post-9/11 terrorism, the biggest financial crises ever emerged. While governments were trying to overcome the financial crisis, they didn't see the Arab Spring coming. While they were busy with its aftermath, they didn't see the crisis in the Ukraine emerge. Later on, they failed to anticipate the conflict with the Islamic State. Thus, keeping a well-balanced overview of everything will get progressively more difficult. Instead, politics will be increasingly driven by problems that suddenly happen to gain attention. Therefore, decisions will often be made in a reactive way rather than in a proactive way, as it should be.

---

[49]In fact, in contrast to what is sometimes suggested, we are far from being able to process all data in the world. In a talk of Robert Munro on Distributed Computing and Disaster Response, I learned that only 7% of all communication is digital, and only about 5% of communication is in English. There are 6000 more languages in the world. Consequently, many problems won't be noticed by us, while we will pay too much attention to a few of them, as if we used a telescope. Unfortunately, this produces a "tunnel view."

[50]Kondor et al. [11].

But let's now have a look at the question of how the complexity of the world is expected to grow. The possibility to network the components of our world creates ever more options. We have, in fact, a combinatorial number of possibilities to produce new systems and functionalities. If we have two kinds of objects, we can combine them to produce a third one. Connecting them in various ways, these three kinds of objects allow us to create six ones, and those already 720. This is mathematically reflected by a factorial function, which grows much faster than exponential (see the red curve in Fig. 4.1). For example, we will soon have more devices than people communicating with the Internet. In about 10 years from now, 150 billion (!) "things" will be communicating with the Internet. Thus, even if we realize just every thousandth or millionth of all combinatorial possibilities, the factorial curve will eventually overtake the exponential curves representing data volumes and computational power. It has probably overtaken both curves already for some time. Due to limited communication bandwidth, an even smaller fraction of data can be processed centrally, such that a lot of local information, which is needed to produce good solutions, is ignored by centralized optimization approaches.

In other words, attempts to optimize systems in a top-down way will become less and less effective—and cannot be executed in real time. Even revolutionary new technologies such as quantum computing are unlikely to change this (see Appendix 4.3). Paradoxically, as economic diversification and cultural evolution progress, a "big government", "super-government" or "benevolent dictator" would increasingly struggle to take good decisions, as it becomes more difficult to satisfy the diverse local expectations and demands.[51] This means that any attempt to govern our increasingly complex world in a top-down way by centralized control is probably destined to fail. Given the situation in Afghanistan and Iraq, Syria, Ukraine, and the states experiencing the Arab Spring, and given the financial, economic and public debt crisis, have we perhaps lost control already? Are we fighting a hopeless battle against complexity?[52]

4.11 Appendix 1: Democracy and Freedom—Outdated Concepts?

These days, many people seem to be asking: "Are the concepts of freedom and democracy obsolete?" Let us look into this. The "freedom of will" or "freedom of decision-making" has, in fact, been questioned many times and for various reasons. For example, some religions believe that our destiny is predetermined. If this were true, then our life would basically be like playing back a 3D movie. We would not have any possibilities to choose, but also no responsibility for what we do. Consequently,

---

[51] Communication rates will also keep imposing serious limitations on centralized top-down control.

[52] Simplifying our world by homogenizing or standardizing it would not fix the complexity problem. It would rather undermine cultural evolution and innovation, thereby causing a failure to adjust to our ever-changing world.

people would be sent to prison not for wrongdoing, but for disturbing public order. There would be also no freedom of press. However, in such a framework, business leaders and policy-makers couldn't take any decisions as well. We would just have to accept what is happening to us.

In other cultures, it is being recognized that people can make their own decisions, but everyone is expected to subordinate personal decisions to the interests of higher-level institutions or to those who are more powerful. Such societies are hierarchically organized. Compared to this, in Western democracies the power of people and institutions has been limited by law in favor of individual freedom. This has boosted entrepreneurial activities, growth, prosperity, and social well-being for a long time. Remarkably, the most developed economies are the most diverse ones (and vice versa).[53]

New critique of the "freedom of decision-making" comes from some neurobiologists.[54] Their experiments suggest that the feeling of free decision-making arises *after* the decision was actually made in the brain. However, this does not prove anything. Decision, execution and conscious recognition are separate things.[55] The feeling of having taken a free decision might be just a conscious confirmation of execution.

Of course, our decisions are often influenced by external factors. However, the old idea of behavioralism that people would be programmed like a computer (by education and other external influences such as public media) and that they would execute what they have been programmed for has failed long ago.

There is little doubt that most people can learn to take different decisions in identical situations (if they are not in a deprived state). A conscious decision resulting from a deliberation process compares possible alternatives from a variety of perspectives, before a decision is made. In a sense, it is the art of decision-making to consider many aspects that might matter, and some people are very good at simulating possible decision outcomes in their brain.

From a societal perspective, it does not matter much whether different decision outcomes in practically identical situations are the consequence of free will, randomness, or other mechanisms (such as "deterministic chaos" or "turbulence"). What really matters is to have socio-economic institutions that support innovation and the spreading of good ideas.

To solve the world's problems and adapt to environmental, social, economic and technological change, we must increase our problem solving capacity. We must get more efficient in taking the best ideas on board and combining them, in other words, to catalyze collective intelligence. It turns out that countries supporting individual freedom are better at boosting new ideas (see the green bars in Fig. 4.2), while others are better at applying them (red bars).

---

[53]Hidalgo et al. [12].

[54]Neuroscience of free will, see https://en.wikipedia.org/wiki/Neuroscience_of_free_will; also see http://www.wired.com/2014/09/belief-free-will-threatened-neuroscience/.

[55]This becomes clear when we consider the example of a boss who asks his or her staff to do a certain job and to report back after execution.

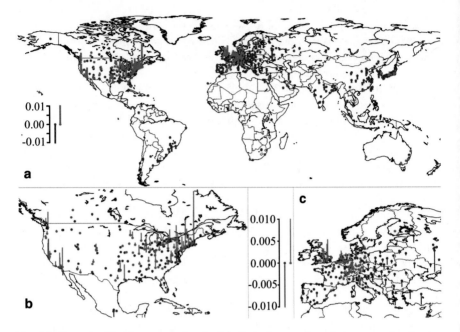

Fig. 4.2 New scientific ideas are produced primarily in democratic societies supporting freedom of thought, speech, and markets. (Reproduced from Mazloumian et al. [13] with kind permission of the Springer Nature Publishing Group.)

4.12   Appendix 2: Limitations to Building a Crystal Ball

4.12.1   Laplace's Demon and Measurement Problems

Laplace's Demon is a hypothetical being that could calculate all future states of the world, if it knew the exact positions and speeds of all particles in the universe and the physical laws governing their motion and interactions. But Laplace's Demon cannot exist in reality, not least because measurements to determine all particle speeds would be impossible due to the restriction of Einstein's theory of special relativity: all velocities must be less than the speed of light. This would prevent one from gathering all the necessary data.

4.12.2   Parameter Sensitivity

How close can computer-modeled behavior ever come to real human social behavior? I would say, a digital mirror-self wouldn't resemble us more than two "identical twins" resemble each other. To specify the parameters of a computer double (digital twin), they are varied by calibration procedures until the difference between measurement data and model predictions does not get smaller anymore. However, the best-fitting model parameters are usually not the correct parameters. These parameters are typically located within a certain "confidence interval". If the parameters are randomly picked from the confidence interval, however, the model predictions may vary a lot. This problem is known as "sensitivity". To illustrate the problem: such parameter sensitivity could make some people rich over night, while others may lose their property.[56]

4.12.3   Instability, Turbulence and Chaos: When All the Data in the World Can't Help

Two further problems of somewhat similar nature are "turbulence" and "chaos". Rapid flows of gases or liquids produce swirly patterns—the characteristic forms of turbulence. In chaotically behaving systems, too, the motion becomes unpredictable after a certain time period. Even though the way a "deterministically chaotic" system evolves can be precisely stated in mathematical terms, without any random elements, the slightest change in the starting conditions can eventually cause a completely different global state of the system. In such a case, no matter how accurately we measure the initial conditions of the system, we will effectively not be able to predict the behavior after some time. This also holds for the formation of "phantom traffic jams". Even if we would collect huge masses of information about the behavior of each driver, we could not predict who causes the traffic jam. It's a collective phenomenon produced by the interactions of drivers, which occurs whenever the vehicle density grows beyond a certain critical density. We have made a similar observation in a decision experiment under well-controlled laboratory conditions. Here, we could predict 96% of all individual decisions based on a theory called the "best response rule". Nevertheless, the rule failed to predict the macro-level (systemic) outcome, as even small deviations from this deterministic theory caused a different result.[57] Surprisingly, adding noise to the highly accurate decision model (i.e. making it less accurate) improved the macro-level predictions, because this could reproduce effects of small deviations in situations of systemic instability.

---

[56]China's richest man lost $15 billion in one hour, see http://money.cnn.com/2015/05/21/investing/china-hanergy-stock-plunge/ and also http://www.zerohedge.com/news/2015-05-21/crash-contagion-second-hk-billionaire-wiped-out-seconds-after-stock-instacrash; Twitter spikes on a fake story about a fake bid, see http://blogs.wsj.com/moneybeat/2015/07/14/twitter-spikes-and-falls-on-fake-story-about-a-bid/.

[57]Mäs and Helbing [14].

4.12.4 Ambiguity

Information can have different meanings. In many cases, the correct interpretation can be found only with some additional piece of information: the context. This contextualization is often difficult and not always available when needed. Different pieces of information may also be inconsistent, without any chance to resolve the conflict.

Another typical problem of "data mining" is that data might be plentiful, but nevertheless incomplete and non-representative. Moreover, a lot of data might be wrong, because of measurement errors, data manipulation, misinterpretations, or the application of unsuitable procedures.

4.12.5 Information Overload

Having a lot of data does not necessarily mean that we'll see the world more accurately. A typical problem is known as "over-fitting", where a model with many parameters is fitted to the fine details of a data set in ways that are actually not meaningful. In such a case, a model with less parameters might provide better predictions. Spurious correlations are a somewhat similar problem: we tend to see patterns that actually don't have a meaning.[58]

It seems that we are currently moving from a situation where we had too little data about the world to a situation where we have too much. It's like moving from darkness, where we can't see enough, to a world flooded with light, in which we are blinded. We will, therefore, need "digital sunglasses": information filters that will extract the relevant information for us. But as the gap between the data that exists and the data we can analyze increases, it might become harder to pay attention to those things that really matter. Although computer power increases exponentially, we will only be able to process an ever-decreasing fraction of data, since the storage capacity grows even faster. In other words, there will be increasing volumes of data that will never be touched (which I call "dark data"). To reap the benefits of the information age, we must reduce information biases and pollution, otherwise we will increasingly make mistakes.

4.12.6 Herding

In many situations, which are not entirely well defined, people tend to follow the decisions and actions of others. This produces undesirable herding effects. The Nobel laureates in economics, George Akerlof (*1940) and Robert Shiller (*1946), have called this problematic behavior "animal spirits", but the idea of herding goes back

---

[58] See http://www.tylervigen.com/spurious-correlations for some examples.

at least to the French mathematician Louis Bachelier (1870–1946). Bubbles and crashes in stock markets are examples of undesired consequences of such herding effects.

4.12.7 Randomness and Innovation

Randomness is a ubiquitous feature of socio-economic systems. However, even though we would often like to reduce the risks it generates, it would be unwise to try to eliminate randomness completely. Randomness is an important driver of creativity and innovation, while predictability excludes positive surprises ("serendipity") and cultural evolution. We will later see that some important social mechanisms can only evolve in the presence of randomness. While newly emerging behaviors are often costly in the beginning (when they are in a minority position), the random coincidence or accumulation of such behaviors in the same neighborhood can enable their success (such that new behaviors may eventually spread).

4.13 Appendix 3: Will New Technologies Make the World Predictable?

How can we know that the processing power, data volumes, and complexity will always grow according to the same mathematical laws, which I have assumed above (in Fig. 4.1)? In fact, we can't, but Moore's law for the computational processing power has been valid for many decades. A similar thing applies to the curve for the stored data volumes. In future, quantum computers may potentially change the game. They are based on a different computing architecture and concept. For example, classical data encryption would become easily breakable, but new encryption schemes would become available, too. Thus, could such a paradigm shift in computing make optimal top-down control possible? I doubt it, because there are also new technologies that will dramatically increase data production rates, such as the Internet of Things.[59] Moreover, the production of hardware devices, both of computers and communicating sensors, might be a limiting factor, as are data transmission rates. Finally, when computer power and data rates are increasing, this is promoting complexity, too, as new devices and functionalities can be produced.

We must also realize that a small subsystem of our universe such as a quantum computer can, by its very nature, not evaluate and simulate the entire world, including its own state and temporal evolution. Simplifications will always be needed. Unfortunately, sheer data mining and machine learning are often quite bad at predicting future changes such as paradigm shifts due to systemic instabilities or innovations, which

---

[59]See http://www.industrytap.com/knowledge-doubling-every-12-months-soon-to-be-every-12-hours/3950.

may fundamentally change the system and its properties. If we wanted to change this, we would have to forbid innovations, but then our world would become as boring as a graveyard, and could not adjust to environmental, social and technological change that is happening.

I believe, we could also not be interested in creating an information infrastructure that creates a broadband information exchange between all brains in the world, or in creating a computer having the computational capacity of all brains. Why? Because it would create strongly coupled networks of networks, which would be highly prone to showing undesirable cascading effects. Harmful mass psychology, as it occurred in many totalitarian regimes, riots, and revolutions are some examples illustrating the danger of too much connectivity between people's minds. As I pointed out before, many of our anthropogenic systems are vulnerable to cascading effects already, which often makes them uncontrollable. Privacy may be seen as a mechanism to avoid such undesirable socio-psychological cascading effects. Without it, we would probably be involved in many more conflicts.

References

1. C. Song, Z. Qu, N. Blumm, and A.-L. Barabasi, Limits of predictability in human mobility, Science 327, 1018–1021 (2010).
2. M. Kosinski, D. Stillwell, and T. Graepel, Private traits and attributes are predictable from digital records of human behavior, Proceedings of the National Academy of Sciences of the USA 110, 5802–5805 (2013).
3. A.D.I. Kramer, J.E. Guillory, and J.T. Hancock, Experimental evidence of massive-scale emotional contagion through social networks. Proceedings of the National Academy of Sciences of the United States of America 111, 8788–8790.
4. J. Schmieder, Mit einem Bein im Knast — Mein Versuch, ein Jahr lang gesetzestreu zu leben (Bertelsmann, 2013).
5. R.H. Thaler and C.R. Sunstein, Nudge: Improving Decisions About Health, Wealth, and Happiness (Penguin, New York, 2008).
6. R.M. Bond et al., A 61-million-person experiment in social influence and political mobilization, Nature 489, 295–298 (2012).
7. D. Lazer, R. Kennedy, G. King, and A. Vespignani, The parable of Google Flu: Traps in Big Data analytics. Science 343, 1203–1205 (2014).
8. M. Gill, A. Spriggs: Assessing the impact of CCTV. Home Office Research, Development and Statistics Directorate (2005), https://www.newscientist.com/article/dn26801-mass-survei llance-not-effective-for-finding-terrorists/.
9. D. Helbing (2015) Thinking Ahead (Springer, Berlin), pp. 181.
10. D. Helbing, The world after Big Data: What the digital revolution means for us, see http://fut urict.blogspot.mx/2014/05/the-world-after-big-data-what-digital.html.
11. I. Kondor et al. Strong random correlations in networks of heterogeneous agents, J. Econ. Interact. Coord. 9, 203–232 (2014)
12. C. Hidalgo et al. (2007) The product space conditions the development of nations. Science 317, 482–487.
13. A. Mazloumian et al. (2012) Global multi-level analysis of the 'scientific food web', Scientific Reports 3: 1167.
14. M. Mäs and D. Helbing (2017) Random deviations improve micro-macro predictions: An empirical test. Sociological Methods & Research 49(2), 387–417, https://journals.sagepub. com/doi/pdf/10.1177/0049124117729708. Title of preprint: "Noise in behavioral models can improve macro-predictions when micro-theories fail".

Chapter 5
Genie Out of the Bottle

Major Socio-economic Shifts Ahead

The technological forces unleashed by the „digital revolution" will fundamentally transform our economy and society within a very short time. We need to prepare ourselves for this transformation and the new era to come. In fact, we may just have 20 or 30 years to adapt. Even though we may dislike such socio-economic change, we will probably not be able to stop it. In order to be able to cope with surprises, we need to design our socio-economic systems in a resilient way. Moreover, if we learn to understand the fundamentally new logic of our digital future and harness the driving forces behind it for our purposes, we can benefit tremendously and fix a number of long-standing problems that humanity has struggled with!

It seems the world is changing at a rapidly increasing pace and is getting ever more complex. However, today's democratic institutions are slow, and they are expensive, too. In the meantime, most industrialized countries have piled up debt levels that can hardly be dealt with. This calls for a new and more efficient approach to decision-making. Two concepts have been around: (1) the "China model", as China has developed much faster than Western countries recently,[1] and (2) giving more control to multi-national corporations, as they tend to be more efficient than government institutions.

5.1 The China Model

Has the French Revolution been a tragic historical accident that is now making lives in Western democracies more difficult? Are citizens and public institutions obstacles, or have we not yet learned to use them as a valuable resource? Is democratic decision-making outdated and blocking our way into a better future? Would more top-down control be better?

Given that countries such as Singapore and China are experiencing larger economic growth rates, should other countries copy their model? Most likely, this wouldn't be a good idea. First, a political system must be culturally fitting. Second,

---

[1]Bell [1].

© Springer Nature Switzerland AG 2021
D. Helbing, *Next Civilization*,
https://doi.org/10.1007/978-3-030-62330-2_5

even though top-down control might accelerate decision-making, it is questionable, whether the resulting decisions would be better and sustainable, too.[2] If we implemented the Chinese approach in Europe or the USA, we could easily decide to build a road over here, a shopping mall over there, and even new city quarters or entire towns. Moreover, these projects could be realized in just a few years. This may sound tempting to decision-makers. However, when space and other resources are limited, improving a system becomes increasingly difficult as it evolves: Given the many interdependencies, improvements in one part of a system may come with undesired side effects in others. Over time, many decisions will turn out to be less favorable than expected, and quick decisions will often result in mistakes. For example, there are many "ghost malls" in China, which almost nobody wants to use, and empty "ghost towns", too. Moreover, environmental pollution has become a serious issue. Some cities are now suffering from smog levels that imply considerable risks to health and make them almost dysfunctional.

Even though China has quickly developed, it should also be considered that the satisfaction of its people hasn't everywhere increased. Social unrests are frequent. Due to the world economic crisis, China is now even confronted with a dangerous reduction in its growth rate and financial turmoil, so that it must reorganize its economy. To reduce the likelihood of a revolution, information flows to and from the country are increasingly controlled by the government.

In other words, China shows worrying signs of destabilization.[3] India's democracy, in contrast, is doing increasingly well.[4] Also in Europe, federally organized systems such as Germany and Switzerland are performing better than more centrally governed countries such as France. In fact, it can be said that the most advanced economies in the world are the most diverse and complex economies.[5]

---

[2] A recent data analytics study performed in my team shows the following: The transition from autocratic to democratic governance can fuel economic development (probably because free entrepreneurships allows new products, services, and business models to come up). Eventually, however, the growth rate slows down. Those countries that have then turned to top-down governance tended to have an initial economic boost (which is probably due to a streamlining of society at the cost of diversity). However, on the long run, such a transition doesn't pay off. On the contrary, the price for it is high (probably because a loss of diversity reduces innovation rates).

[3] Based on a statistical analysis of Jürgen Mimkes, China will now undergo a major transformation towards a more democratic state in the coming years. First signs of instability of the current autocratic system are visible already, such as the increased attempts to control information flows. The following recent articles support the conclusion regarding increased instability in China (please use, for example, Google Translate where needed): https://www.chinausfocus.com/society-culture/instability-and-unpredictabi lity-in-china-the-new-normal; https://www.sciencedirect.com/science/article/abs/pii/S1043951X 13000035; http://www.zeit.de/politik/ausland/2015-03/china-wachstum-fuenf-vor-acht; http://baz online.ch/wirtschaft/konjunktur/China-uebernimmt-die-rote-Laterne/story/20869017.

[4] Why India will soon outpace China, see http://www.forbes.com/fdc/welcome_mjx.shtml and http://www.economist.com/news/business-and-finance/21642656-indias-economy-grew-faster-chi nas-end-2014-catching-dragon.

[5] Hidalgo et al. [2].

5.2   Can Corporate Control Fix the World?

If today's democratic systems are facing difficult times and autocratic systems struggle, too, would more corporate control be able to solve our problems? In fact, this is often claimed, and companies keep demanding more control. This is probably what the free trade and service agreements, which have recently been negotiated under secrecy, are about: governments will give up some of their power and hand it over to corporations.

So, would it be better to let multi-national corporations run the world? In fact, companies are often more efficient than governments in accomplishing specific tasks that can be well monetized. However, if we look at a map displaying what companies control what regions of the world, it doesn't look less fragmented than the map illustrating the hypothetical "clash of cultures". So, we can't expect that more corporate control will cause more agreement in the world. We may just see more "economic wars".

There are other issues, too.[6] While large corporations have certainly a lot of power to move things ahead, they often show surprisingly low innovation rates and tend to obstruct innovations of others.[7] There are many examples where even the value of own inventions hasn't been realized. For instance, Xerox did not see the value of the Windows software they invented. The value of the mp3 music file format was totally underestimated, and nobody expected that text messaging would become important. To compensate for their innovation weakness, large corporations buy innovative small and medium-size companies. Nevertheless, they often fail to stay on top for a long time. Within a period of just 10 years, 40–50% of top 500 companies are predicted to disappear. Given such high takeover and "death" rates, societies would be extremely unstable if run by corporations. Compared to this, countries and cities persist for hundreds of years, exactly because they are governed in a more participatory way, unlike corporations.

In summary, there is little evidence that more corporate control would solve the problems of the world. I don't deny that many companies have laudable goals. The use of self-driving cars, for example, intends to eliminate accidents, which have killed a lot of people in the past. Moreover, by means of personalized medicine, genetic engineering and biological enhancements, companies try to overcome death altogether. However, none of these ambitious goals have been accomplished yet. I might start to believe in corporate control of our globe, if we had a perfect world everywhere within a 100 km radius of the Silicon Valley, but we are far from this. In the Silicon Valley, there is a lot of light, but lots of shadow, too. In the past, not a

---

[6]Monopolies often produce efficiently, but they are known to imply a "too big to fail" problem. Unfortunately, they are the form of economic production which is most vulnerable to disruptions (due to their lack of diversity). Besides this, they might misuse their power (e.g. by selling over-priced products, which leads to an inefficient resource allocation on the side of the consumers).

[7]Is Google with its approximately 50 software platforms and 100 secret Google X projects an exception to the typical lack of innovation of big companies? Probably not, as it earns most of its money on personalized ads.

single city in the USA has been in the top 10 list of most livable cities. In other words, we need entirely new medicines to heal the world's ills, in particular as we are now confronted with another unsolved challenge: the destabilization of our economy and society by the digital revolution.

5.3   The Digital Revolution on Its Way

The digital revolution arguably deserves even more attention than climate change because it will dramatically affect almost every aspect of our economy and society within our lifetimes. Just about ten years ago, most of us didn't have the slightest idea that *Facebook, Twitter*, and *iPhones* would be an integral part of modern life. Even large companies such as *Microsoft, Yahoo* and *Nokia* didn't see some important developments coming. Now, we are in the middle of a socio-economic transformation, which will create a world ruled by a different logic, and the progress of digital technologies is further accelerating. What can we expect to happen in the next ten, twenty, or fifty years? In the analysis that follows, I offer some suggestions which are not wide-eyed futurology, but considerations based on existing evidence and trends. However, before we discuss these trends, let us look at how the current situation emerged.

According to a well-known anecdote, Thomas John Watson Sr. (1874-1956), the then chairman of *IBM*, said in 1943:

> *"I think there is a world market for maybe five computers."*

Although the statement is surprisingly inaccurate with the benefit of hindsight, at that time almost nobody could imagine a mass market for computers. For decades, computers were barely useful to anyone. Even in 1968, an engineer at the Advanced Computing Systems Division of *IBM* asked of the microchip: *"what … is it good for?"* In 1981, Bill Gates is claimed to have said that a computer memory of 640 kilobytes *"ought to be enough for anybody"*. Today, our smartphones have a hundred thousand times more memory than that, and an iPhone has more processing power than the Apollo rocket, which flew to the moon and back. Nevertheless, the early innovators who pushed for the acceptance of information technology didn't have an easy life. The founder of *Apple* Computers Inc., Steve Jobs (1955–2011), remembers his failed attempts to interest Atari and Hewlett-Packard in the personal computer he co-developed with Steve Wozniak:

> *"So we went to Atari and said, 'Hey, we've got this amazing thing, even built with some of your parts, and what do you think about funding us? Or we'll*

*give it to you. We just want to do it. Pay our salary, we'll come work for you.'
And they said, 'No.' So then we went to Hewlett-Packard, and they said, 'Hey,
we don't need you. You haven't got through college yet.'"*

Later, Steve Jobs even found himself forced to leave the *Apple* board for a number
of years. But when he returned, he had the *iPod, iPhone, iPad* as well as *iTunes* and
the *AppStore* developed, i.e. products which made *Apple* the most valuable company
in the world for some time.

This is just a small part of an even larger story. There was, in fact, another important development. In the late sixties, *Arpanet* was created to allow a few military
computers to exchange information with each other. But it was then opened up
for public use and became the Internet. This eventually unleashed the power of
information. Later, to support the collaboration between Swiss and French teams
contributing to the CERN elementary particle accelerator, physicist Tim Berners-
Lee (*1955) invented a hyperlink protocol supporting the development of web pages
that can be linked with each other. This gave rise to the World Wide Web (WWW)
and eventually made computers useful for ordinary people rather than just experts.
It also made the Internet attractive for doing business and enabled a multi-billion
dollar market. However, Tim Berners-Lee's ideas did not have an easy start either.
At CERN he didn't get the support he requested, so he went to the Massachusetts
Institute of Technology (MIT), where he founded the World Wide Web Consortium
(W3C).

Now, there are about 3 billion Internet users in the world. Eventually, the creation
of *Facebook* in 2004 and *Twitter* in 2006 linked more than a billion people together
in giant social networks. We are no longer just users of this system. We are also
(co-)creators of data and services. In some sense, we are now "human processors"
and nodes in a globe-spanning information system.

5.4 Computers More Intelligent Than Humans?

For decades, the processing power of computers has doubled roughly every
18 months. If this trend continues, some information systems may surpass the power
of a human brain sooner than we think. Our brain has approximately 100 billion
neurons, a firing rate of 200 times a second, and a signal speed of 120 meters per
second. This has to compete with the current generation of supercomputers, which
have 100 times fewer transistors, but operate 20 million times faster, and have a
signal speed which is two million times faster.

Therefore, computers are outperforming humans in increasingly many ways.
Whereas a century ago, some companies maintained departments with hundreds
of employees to perform business calculations, these were later replaced by simple
calculators. Computers are far better than us at doing calculations—and even at chess.

In 1996, *IBM*'s Deep Blue computer defeated the best chess player in the world at that time, Garry Kasparov (*1963). Computers can now also beat the best backgammon players, the best scrabble players and the best players in many other strategic games. In 2011, *IBM*'s Watson, a "cognitive computer", which is able to judge the relevance of information from the Internet, even beat the best human experts in answering questions in the game show "Jeopardy!"[8] Now, Watson is taking care of customer hotlines, as the computer is better at managing all the knowledge required to answer customer queries. Watson understands natural language and comes increasingly close to what humans can do. In many respects Watson is even superior to humans. Now, by investing 1 billion dollars in cognitive computing technology, *IBM* hopes to earn 100 times that amount in the years to come.

Europe invests a comparable sum—1 billion Euros—to build a supercomputer which can simulate the human brain. The USA even decided to spend $3 billion on a brain project. The *Google* Brain project, led by the technology guru Ray Kurzweil, is another attempt to turn the Internet into an intelligent entity that can think and decide.[9]

While building computers with performance comparable to humans may still take a few years, the *Google* car—a driverless vehicle developed by the tech giant— is already there. It may soon navigate our roads more safely than human drivers. Furthermore, about 70% of all financial transactions in the world's stock markets are now performed by autonomous trading algorithms. Therefore, computers and robots are increasingly doing our work, and they will replace many jobs that can be performed according to rules and routines. We may even lose highly qualified jobs which depend on skilled judgments, including medical doctors, care workers, scientists, lawyers, managers, and politicians, and to some degree even teachers and parents. For example, South Korea recently invested in robotic childcare, and Japan is building robots to take care of elderly people. All of these developments raise the question: what role will humans play in future? To stay competitive, would we have to become more robotic, or would we better distinguish ourselves by engaging in less predictable activities and more creative jobs?

5.5 When Will We See Artificial Superintelligences and Superhumans?

Only two years ago, most people would have considered it impossible that algorithms, computers, or robots would ever challenge humans as the "crown of creation". This has changed.[10] Intelligent machines are learning by themselves. For example,

---

[8]Kelly and Hamm [3], Hurwitz et al. [4].

[9]Kurzweil [5].

[10]Barrat [6]. Edge Question 2015: What do you think about machines that think? http://edge.org/annual-question/what-do-youthink-about-machines-that-think.

Google's Deep Mind learned to play and win 49 *Atari* video games.[11] One could also recently read about a computer that passed the Turing test, i.e. it was mixed up with a human.[12] Furthermore, a robot recently mastered a university entry exam.[13] Some robots even show self-awareness,[14] and they might build other robots, which are superior. The resulting evolutionary progress is quickly accelerating, and it is just a matter of time until there are machines smarter than us.[15] Perhaps such superintelligences already exist.

Some notable people have recently commented on this new situation. For example, Silicon Valley's technology genius, Elon Musk of Tesla Motors, tweeted that Artificial Intelligence is *"potentially more dangerous than nukes"* and warned[16]:

> *"I think we should be very careful about artificial intelligence. If I had to guess at what our biggest existential threat is, it's probably that. So we need to be very careful. ... I am increasingly inclined to think that there should be some regulatory oversight, maybe at the national and international level, just to make sure that we don't do something very foolish. ..."*

Similar critique comes from Nick Bostrom at Oxford University.[17]

Stephen Hawking (1942–2018), perhaps the most famous physicist recently, voiced concerns as well[18]:

> *"Humans who are limited by slow biological evolution couldn't compete and would be superseded. ... The development of full artificial intelligence could spell the end of the human race. ... It would take off on its own, and re-design itself at an ever increasing rate."*

---

[11]Google machine learns to master video games, see http://www.bbc.com/news/science-enviro nment-31623427.

[12]A "chatbot" named Eugene Goostman passed the "Turing test", meaning that 33% of humans could not tell whether they were chatting with a computer or a person, see http://www.theguardian. com/technology/2014/jun/08/super-computer-simulates-13-year-old-boy-passes-turing-test.

[13]See http://www.dailymail.co.uk/sciencetech/article-2821132/Smarter-average-student-Japanese-researchersreveal-artificial-intelligence-software-BEAT-real-students-university-entrance-exams. html and http://www.itworld.com/article/2720680/enterprise-software/fujitsu-to-build-software-robot-to-pass-college-entranceexams.html.

[14]See http://www.dailymail.co.uk/sciencetech/article-3165282/A-polite-robot-uprising-Humano ids-glimmer-selfawareness-apologises-scientific-experiment.html#ixzz3girYdohe, https://www. youtube.com/watch?v=MceJYhVD_xY, http://m.jetzt.de/texte/593752.

[15]Armstrong [7], Barrat [8].

[16]http://www.theguardian.com/technology/2014/oct/27/elon-musk-artificial-intelligence-ai-big gest-existential-threat.

[17]Bostrom [9].

[18]http://www.bbc.com/news/technology-30290540.

Bill Gates of *Microsoft* contributed to the discussion on Artificial Intelligence, too[19]:

"I am in the camp that is concerned about super intelligence. ... I agree with Elon Musk and some others on this and don't understand why some people are not concerned."

Last but not least, Steve Wozniak, co-founder of *Apple*, formulated his worries as follows[20]:

"Computers are going to take over from humans, no question ... Like people including Stephen Hawking and Elon Musk have predicted, I agree that the future is scary and very bad for people ... If we build these devices to take care of everything for us, eventually they'll think faster than us and they'll get rid of the slow humans to run companies more efficiently ... Will we be the gods? Will we be the family pets? Or will we be ants that get stepped on? I don't know ..."

So, how will our future look like? Will superintelligent robots enslave us? Or will we upgrade our biological hardware in order to become superhumans? I openly admit, this sounds like science fiction, but such scenarios can't be any longer ignored, given the current pace of technological advances. The first cyborgs, i.e. humans that have been technologically enhanced, already exist. The most well-known of them is Neil Harbisson.[21] With the *Crisper* technology, it is even possible now to genetically engineer humans.[22] But there is an alternative: If suitable information platforms and personal digital assistants support us in creating collective intelligence, we might be able to keep up with the exponentially increasing performance of Artificial Intelligence.[23]

---

[19] http://www.cnet.com/news/bill-gates-is-worried-about-artificial-intelligence-too/.

[20] https://web.archive.org/web/20150501122305/http://www.washingtonpost.com/blogs/theswitch/wp/2015/03/24/apple-co-founder-on-artificial-intelligence-the-future-is-scary-and-very-bad-for-people/.

[21] See, for example, the TED talk "I listen to color" accessible at https://www.youtube.com/watch?v=ygRNoieAnzI.

[22] The Promise and Peril of Crispr: A cheap gene-editing method could lead to cures—and frankenbabies, see http://www.bloomberg.com/news/articles/2015-06-25/the-promise-and-peril-of-crispr.

[23] See this video by Jim Spohrer: https://www.youtube.com/watch?v=E7PVBGtEYyg. Therefore, I believe we should better engage in distributed collective intelligence rather than in creating a few extremely powerful superintelligences that may get out of hand. We might lose control due to their smartness, but also because of their attractiveness for organized criminals, terrorists, or political extremists, and because of unsolved cyber security issues. Moreover, superintelligences may start to act selfishly or lie, or they could make mistakes. This could be disastrous if their proposals

It is clear, however, that new socio-economic forces will be created when, within the next 10–30 years, computers, algorithms and robots increasingly take over our classical roles and jobs. In the past, we witnessed the transition from the "economy 1.0", characterized by an agricultural society, to the "economy 2.0", driven by industrialization, and finally to the "economy 3.0", based on the service sector. Now, we are seeing the emergence of the "economy 4.0", a "digital" sector driven by information and knowledge production. The third socio-economic transformation, which is currently on its way, is fueled by the spread of Big Data and Artificial Intelligence, and the Internet of Things.[24] Unfortunately, history tells us that the emergence of a new economic sector tends to cause social and economic disruptions, as the innovation that feeds the new sector undermines the basis of the old ones.[25] Financial and economic crises, revolutions, or war may be the results. If we want to avoid this, we must quickly adapt our perspective of the world and the way we organize it. I am totally convinced that we can master the digital revolution, but only if we manage to create a win-win-win situation for politics, business, and citizens alike.

5.6   Everything Will Change

The digital revolution is fundamentally changing many of our activities and institutions. This includes how we educate (using massive open on-line courses), how we do research (through the analysis of Big Data), how we move around (with self-driving cars), how we transport goods (with drones), how we shop (with *Amazon* and *eBay*, for example), how we produce goods (with 3D printers), and how we run our health systems (with personalized medicine). Our political institutions are likely to change as well, for example, by data-driven decision-making and demands for more citizen participation. Likewise, the basis of our economy is altered by a new wave of automation and a number of other trends. Financial transactions, which used to be performed by bankers, are being replaced by algorithmic trading, *Paypal*, *BitCoin*, *Google Wallet*, and so on. Moreover, the biggest share of the insurance business, amounting to a multiple of the global gross domestic product, is now in financial derivatives traded at stock markets. According to military experts, war may fundamentally change, too—with states preparing for cyberwar. If one succeeded to build a "digital God", even religion may change.

Let us now look at the problem of the coming economic transformation more closely. Are our societies well-prepared for the digital revolution? In some industrialized countries, the proportion of jobs in the agricultural ("primary") sector is now

---

are applied to an entire country. Therefore, we should at least be able to judge the solutions that superintelligences would come up with, and this would require collective intelligence.

[24]Rifkin [10].

[25]For example, during the transition to industrial production methods in British textile mills in the 1810s, the term "Luddite" was used to describe workers who deliberately smashed and destroyed modern machinery because they feared it would threaten their jobs.

just about 3%, whereas 300 years ago, it stood at 80% or more. How did we get there? Among other developments, the great success of James Watt's steam engine, invented in 1769, gave rise to the industrial revolution. As a result, many people lost their jobs due to automation—in the production of textiles, for example. However, the industrial revolution also gave rise to consumerism, meaning that industrial production eventually created a secondary sector with many new jobs. In fact, during its heyday, industrial production provided more than 30% of all jobs in developed countries. But later, efficiency increases in industrial production led to a loss of half of all industrial jobs. Unfortunately, the number of industrial jobs will continue to fall—probably below 10%. Very soon, therefore, only about 10 to 15% of jobs will be provided by the primary and secondary sectors of economy.

The tertiary sector—services—is the result of another big innovation: the expansion of education to all levels of society from the 1870s onwards. Reading and writing were no longer restricted to elites, and many more people could perform jobs requiring special qualifications. As a result, the service sector became more developed and brought about an increase in planning, administrative and management tasks. In modern societies, up to 70% of jobs are currently in the service sector. But again, history will repeat itself and more than half of these employees will be replaced—this time by algorithms, computers and robots.[26] What will happen, if computer power and Artificial Intelligence progress so far that humans can no longer keep up?

5.7   The Third Economic Revolution

The digital revolution is driving another dramatic wave of automation, leading to a "second machine age".[27] Not only philosophers become increasingly worried about our future, but leading experts as well. The techno-revolutionaries are losing control of their revolution. If we don't find a new way to produce jobs quickly, we will face unemployment rates far beyond the levels that our current socio-economic system can handle. Indeed, all of the previous socio-economic transitions were accompanied by mass unemployment, which is one of the major forces transforming societies. Given that computers will match the capabilities of the human brain within 10–25 years, we have very little time to adapt to the current societal transition. If we don't adapt fast enough, the transition will be sudden, harsh, and discontinuous: we should expect a bumpy road ahead!

In control theory, it's a well-established fact that delayed adaptation causes systems to become unstable. The consequences of such instabilities for our society may include financial and economic crises, social and political conflicts, and even wars. In fact, there are already signs of destabilization in many countries around the world. These might be interpreted as advance warning signals of socio-economic transitions to come. The list of countries that have recently encountered social

---

[26]Frey and Osborne [11].
[27]Brynjolfsson and McAfee [12].

unrest is surprisingly long. It includes Afghanistan, Pakistan, Iraq, Syria, Somalia, Nigeria, South-Sudan, Uganda, Congo, Central African Republic, Egypt, Bahrain, Libya, Lebanon, Yemen, Tunisia, Turkey, China, India, Malaysia, Thailand, Brazil, Mexico, Ukraine, Spain, Greece, Romania, the United Kingdom, and even Sweden, Switzerland and the USA. Note, however, that these countries are faced with different scenarios. While many African countries are now in transition from an agricultural to an industrial society, some Asian countries are in transition from an industrial to a service society. Finally, the US, Europe, Japan and a few other countries are in transition from a service to a digital society. These transitions may also interact with each other, of course. In the worst case, they might all occur around the same time, and pretty soon.

Can we avoid or suppress these transitions? I don't think so. Stopping technological innovations does not seem to be compatible with most economic systems we currently have, and it wouldn't be effective, too. Countries trying to obstruct technological progress by means of legal regulations would have to find other ways to create comparable economic efficiency gains, as all countries are exposed to international competition in the world market. And even if a global moratorium against the development of Artificial Intelligence was signed, it is not clear how one could ensure compliance.

Therefore, we may be able to delay historical developments, but we cannot stop them, and it wouldn't be reasonable to try. Delaying these transitions makes them more costly, and implies competitive disadvantages. On the long run, no country can afford to miss the opportunities resulting from these transitions, which will eventually lead to more efficient systems and a higher quality of life. This is all part of the cultural evolution which takes place as societies progress. It is driven by socio-economic forces such as increases in complexity, diversity, Big Data and computer-supported (artificial and collective) intelligence.

In the case of the digital revolution, the genie is out of the bottle. Information and communication systems have grown increasingly powerful, but a single mistake could be disastrous. We must, therefore, learn to make the genie work for us. How can we do this? We must learn to understand the new logic implied by digital technologies and the trends they are producing. And we must learn it soon.

5.8   The New Logic of Prosperity and Leadership

Most likely the twenty-first century will be governed by principles which are fundamentally different to those of the twentieth century. That is why we must change our way of thinking about socio-economic systems. We need to recognize some fundamental trends and truths.

Overall, our anthropogenic systems are becoming more connected and more complex. As a result, they are often more variable, less predictable, and less controllable. So, most of our personal knowledge about the world is outdated, and digitally literate people may gain informational advantages over today's experts. Therefore, classical hierarchies may dissolve.

It is also important to recognize that we are entering an increasingly non-material age, characterized by an abundance of information and ideas rather than a shortage of goods. Data can be replicated as often as we like. It's a virtually unlimited resource, which may help us to overcome conflicts over scarce resources. In fact, this opens up almost unlimited possibilities for new value generation.

In the digital society, ideas will spread more quickly. Information is both ubiquitous and instantly available almost everywhere, such that borders tend to dissolve. Furthermore, the more data we produce, the more difficult is it to keep secrets. Data will also become cheaper, i.e. one can make increasingly less profits on data. In addition, the rapid growth in data volumes will result in potential information overload. So, we need to learn how to convert raw data into useful information and actionable knowledge. Most value will be derived from algorithms rather than data, and from individualized, personalized, user-centric products and services. I further believe that the organization of our future socio-economic system will increasingly be based on collective intelligence.

Finally, what used to be regarded as science fiction may become reality one day, and possibly much more quickly than we think. As Sir Arthur Clarke realized:

> *"Any sufficiently advanced technology is indistinguishable from magic."*

The countries which are first to recognize these new principles and use them to their advantage will be the leaders of the digital age. Conversely, those which fail to adapt to these trends on time will be in trouble. Given that computers may soon match the capabilities of the human brain, we may only have two or three decades to adapt our societies—a very short time considering that the planning and building of a road often takes 30 years or more.

5.9 Creating a Resilient Society

The best way to prepare for a future, which is hard to predict, is to create a society capable of flexibly adapting to new kinds of situations. The digital revolution certainly has the potential to precipitate disruptive change. But are our societies well-prepared for such shocks? Our response to September 11, 2001, calls this into question. Even though it was initially a local event, it has changed the face of the entire world. In the aftermath, we have built a security architecture to protect us from terrorism. But even with mass surveillance and armed police, are we now safe? The bomb attack at the Boston marathon and the Charlie Hebdo attack in Paris, for example, question this. Moreover, less than 10 years later, an even more serious event changed the face of the world: the financial crisis. Again, this started locally, but ultimately had a major global impact.

Therefore, how can we better prepare for future challenges? A whole range of measures are at our disposal, including risk assessment, probabilistic prediction, prevention, intervention, insurance, and hedging (which basically means to choose a portfolio strategy rather than betting on one horse). Nevertheless, we must realize that problems will sometimes occur and accidents will sometimes happen in a world that is not entirely predictable. That is why we need resilient systems.

But what exactly does resilience mean? Resilience is the ability of a system to absorb shocks and to recover from them both quickly and thoroughly. If we fall and hurt us, we will recover quickly because our body is resilient to such shocks. So, how to build resilient systems that are not prone to undesired cascading effects, but recover quickly and well from disruptions? This is primarily a matter of systems design and management. Safety margins, reserves, backups, and alternatives (a "plan B", "plan C") can certainly help. Furthermore, *modularization* is a well-known principle to make the complexity of a system manageable. This basically means that the organization of a system is broken down into substructures or "units", between which there is a lower level of connectivity or interaction compared to the connectivity or interaction within the units. This allows one to reduce the complexity within substructures to a manageable level. Furthermore, it decreases interaction effects between units. Well-designed systems have "engineered breaking points", "shock absorbers", or "dynamic decoupling strategies", which can counter the amplification of problems and undesirable cascading effects. For example, think of electrical fuses in your flat or the crumple zones of a car, which are there to protect the sensitive parts of the system (such as our home or our life).

In principle, of course, the modular units of a system can be organized in a hierarchical way. This can be efficient, when the units (and the interactions between them, including information flows and chains of command) work reliably, with very few errors. However, as much as hierarchical structures help to define accountability and to generate power, control might already be lost if a single node or link in the hierarchy is dysfunctional. This problem can be mitigated by redundancies and *decentralization*. In particular, if the dynamics of a system is hard to predict, local *autonomy* can improve proper adaptation, as it is needed to produce solutions that fit local needs well. More autonomy, of course, requires the decision-makers to take more responsibility. This calls for high-level education and suitable tools supporting a greater awareness of potential problems, in particular reliable information systems.

A further important principle that can often support resilience is *diversity*. The benefits of diversity are multifold. First of all, diversity makes it more likely that some units stay functional when the system is disrupted, and that solutions for many kinds of problems already exist somewhere in the system when needed. Second, diversity supports collective intelligence, as we will see later. Third, the innovation rate typically grows with diversity, too. However, diversity also poses challenges, as we know, for example, in intercultural settings. For this reason, *interoperability* is important. I will come back to this issue, when we discuss "digital assistants" and "externalities", i.e. external effects of decisions and (inter-)actions. Finally, using the principles of assisted self-organization, it is possible to control complex dynamical system in a distributed way.

In the light of the principles discussed above, can we be confident that we are currently well prepared to master the digital revolution? Does mass surveillance, combined with armed police, create a resilient society? On the contrary (see Appendix 5.1)! Sustainable political power requires legitimacy, and this requires trust. However, undermining privacy reduces trust (see Appendix 5.2). Mass surveillance also promotes fear and self-censorship, thereby obstructing innovation and the ability of our society to adapt to a changing world.

I personally believe that we must—and can—prepare ourselves much better for the challenges ahead. To illustrate this, let us discuss the problem of traffic safety. We know that accidents keep happening and that this is bad. But how would it be to live in a world without traffic? Our economy would not work, and we would not be able to live in modern societies. Therefore, we have developed measures to reduce the number accidents and their severity. How have we done this? We have improved traffic rules, developed better technology, created emergency services and built hospitals. We have developed better brakes and driver assistance systems that empower drivers to do a better job. We provide traffic news to drivers in order to prevent accidents from messing up the entire traffic system. And we have designed better cars. Decades ago, whereas the body of a car was scarcely damaged by an accident, the driver and passengers were often killed. Nowadays, cars are constructed to absorb shocks, so that even a small accident may damage the entire car, but the risk of injuries and deaths is dramatically reduced.

This serves as a useful metaphor to guide our thinking about how to deal with societal challenges. It shows us that trying to protect institutions from changing is wholly counterproductive and would, in fact, make societies less resilient. Many institutional structures today are like old cars with a rigid body—they are constructed to stay the same rather than to flexibly adjust to a changing world. But in order to be resilient, our institutions must learn to flexibly adapt in ways that serve the citizens best.[28]

It is, therefore, alarming to see so many global crises. Within just a few months, our world was confronted with several major challenges: worrying epidemics (such as Ebola and MERS), the crisis in Ukraine endangering world peace, and the raise of the Islamic State, to mention just a few. While all these problems started locally, their implications have become globally relevant. The question is how much more has to happen until we prepare ourselves better and start building a more resilient society?

A society should be prepared for *all* eventualities. It should be hedging its risks, and its capacities to respond. A society should, therefore, have a portfolio of options

---

[28] So, rather than a predictable society we need a flexible one that can cope with surprises and benefit from them. Therefore, pluralism, diversity, and participation shouldn't be seen as concessions to citizens, but as ways to produce innovation and collective intelligence, in other words: to generate better solutions.

to (re)act. But as I will show, we have seriously neglected some of the most promising options. It's time to do something about this!

5.10 Time for a New Approach

Importantly, in contrast to the approach followed today, simplifying our world by homogenizing or standardizing it would not fix our problems. It would rather undermine cultural evolution and innovation, thereby causing a failure to adjust to our everchanging world. Thus, do we have alternatives? Actually, yes: rather than fighting the properties of complex systems, we can use them for us, if we learn to understand their nature. The fact that the complexity of our world has surpassed our capacity to grasp it, even with all the computers and information systems assisting us, does not mean that our world must end in chaos. While our current system is based on administration, planning, and optimization, our future world may be built on (co-)evolutionary principles and collective intelligence, i.e. intelligence surpassing that of the brightest people and best expert systems.

How can we get there? In the second part of this book, I will show how the choice of suitable local interaction mechanisms can create surprising and desirable outcomes by the process of self-organization. Information and communication systems will enable us to let things happen in a favorable way. This is the path we should take, because there are currently no better alternatives. The proposed approach will create more efficient socio-economic institutions and new opportunities for everyone: politics, business, science, and citizens alike. As a positive side effect, our society will become more resilient to the future challenges and shocks, which we will surely face.

5.11 Appendix 1: Side Effects of Massive Data Collection

Like any technology, Big Data has not only great potential but also harmful side effects. Of course, not all Big Data applications come with the problems below, but they are not uncommon. We must focus, in particular, on those problems that can lead to large-scale effects and major crises.

5.11.1 Crime

In the past years, cybercrime has exponentially increased, now costing the entire world 3 trillion dollars a year. Some of this has resulted from undermining security standards for the purpose of surveillance (e.g. by creating hardware and software "backdoors"). Other common problems are financial, data or identity theft, data manipulation, and the fabrication of false evidence. These crimes are often committed by means of "Trojan horses"—computer codes that can steal passwords and PIN codes. Further problems are caused by computer viruses or worms that damage software or data.

5.11.2 Military Risks

Because most of our critical infrastructures are now connected with other systems via information and communications networks, they have become quite vulnerable to cyber attacks. In principle, malicious intruders can manipulate the production of chemicals, energy (including nuclear power stations), as well as financial and communication networks. Attacks are sometimes possible even if the computers controlling such critical infrastructures are not connected to the Internet. Given our dependence on electricity, information and money flows as well as other goods and services, this makes our societies more vulnerable than ever before. Coordinated cyber-attacks might be launched just within microseconds and bring the functioning of our economy and societies to a halt.

Digital weapons (so-called "D weapons") are now considered to be as dangerous as ABC (atomic, biological, and chemical) weapons.[29] Therefore, the US government reserves the right to respond to cyber war with a nuclear counter-strike. Everywhere in the world, we are now seeing a digital arms race for the most powerful information-based surveillance and manipulation technologies. It is doubtful whether governments will be able to prevent a serious misuse of such powerful tools. Just imagine, a crystal ball or magic wand or other powerful digital tools would exist. Then, of course, everyone wanted to use them, including criminals and our enemies. It is obvious that, sooner or later, these powerful tools would get into wrong hands and finally out of control. If we don't take suitable precautions, mining massive data may (intentionally or not) create problems of any scale—including digital weapons of mass destruction. Therefore, international efforts towards confidence-building and digital disarmament are crucial and urgent.[30]

5.11.3 Economic Risks

Similarly, cybercrime harms our economy, as does the illicit access to sensitive business secrets or the theft of intellectual property. Furthermore, a loss of customer trust in products can reduce sales that would be worth billions of dollars. This has recently

---

[29]The Digital Arms Race: NSA Preps America for Future Battle, see http://www.spiegel.de/intern ational/world/newsnowden-docs-indicate-scope-of-nsa-preparations-for-cyber-battle-a-1013409. html.

[30]Recently, thousands of scientists and Artificial Intelligence experts have signed an Open Letter to ban the use of autonomous weapons, see http://futureoflife.org/AI/open_letter_autonomous_w eapons.

been experienced, for example, by companies selling cloud-based data storage. Many digital systems and services can only work with a sufficient level of trust, including electronic banking, eBusiness, eVoting, social media, and sensitive communication by email. Yet more than two thirds of all Germans say they don't trust government authorities and Big Data companies any longer that they wouldn't misuse their personal data. More than 50% even feel threatened by the Internet.[31] Similarly, trust in governments has dramatically dropped in the USA and elsewhere.[32] The success of the digital economy is further undermined by information pollution, as it results from spam and undesired ads, for example, and recently from fake news as well.

5.11.4 Social and Societal Risks

To contain "societal ills" such as terrorism and organized crime, it often seems that surveillance is needed. However, the effectiveness of mass surveillance in improving the level of security is frequently questioned: the evidence is missing or weak.[33] At the same time, mass surveillance undermines privacy, whereby it questions the government's trust in citizens. This, in turn, undermines the citizens' trust in their government, which is the basis of its legitimacy and power.

The saying that "trust is good, but control is better" is not entirely correct: control cannot fully replace trust.[34] A well-functioning and efficient society needs a suitable combination of both. In particular, the perceived loss of privacy is likely to promote conformism and to endanger diversity and useful criticism. Independent judgments and decision-making could be undermined. On the long run, this would impair a society's ability to innovate and adapt, which could finally make it fail.[35]

---

[31]Umfrage zum Datenschutz: Online misstrauen die Deutschen dem Staat (June 5, 2014), http://www.spiegel.de/netzwelt/web/umfrage-deutsche-misstrauen-dem-staat-beim-online-datenschutza-973522.html.

[32]See, for example, https://www.bloomberg.com/news/articles/2013-11-26/nsa-spying-risks-35-billion-in-u-s-technology-sales, http://www.npr.org/sections/alltechconsidered/2014/06/05/318 770896/a-year-after-snowden-u-stech-losing-trust-overseas, and https://rightedition.com/2013/08/01/nsa-scandal-shakes-americans-trust-in-government/.

[33]The Washington Post (January 12, 2014): NSA phone record collection does little to prevent terrorist attacks, group says, http://www.washingtonpost.com/world/national-security/nsa-pho nerecordcollection-does-little-to-prevent-terrorist-attacks-group-says/2014/01/12/8aa860aa-77dd-11e3-8963-b4b654bcc9b2_story.html?hpid=z4; http://securitydata.newamerica.net/nsa/analysis; Gill [14], https://www.newscientist.com/article/dn26801-mass-surveillance-not-effective-for-fin ding-terrorists/ see also BBC News (August 24, 2009): 1000 cameras 'solve one crime'.

[34]Detlef Fetchenhauer, Six reasons why you should be more trustful, TEDx Groningen, https://www.youtube.com/watch?v=gZlzCc57qX4.

[35]In fact, it's mostly the collapse of diversity that leads to the failure of systems that are centrally controlled in a top-down way, see Helbing [15].

For such reasons, the constitutions of many countries consider it of fundamental importance to protect privacy, informational self-determination, private communication, and the principle of assumed innocence without proof of guilt. These principles are also considered to be essential preconditions for human dignity and for democracies to function well.[36]

However, today the Internet lacks good mechanisms for forgetting, forgiveness, and re-integration. There are also concerns that the increasing use of Big Data could lead to greater discrimination, which in turn could promote increasing fragmentation of our society into subcultures.[37] For example, it is believed that the spread of social media and personalized information has promoted the polarization of US society and politics.[38]

5.11.5 Political Risks

It is often pointed out that leaking confidential communication can undermine the success of sensitive political negotiations. This is probably true, but there are other political problems, too, which aren't less serious. For example, if incumbent governments have better access to Big Data applications than opposition parties, this can result in unfair competition, biased election outcomes and non-representative governments. This could seriously undermine democracies. The greatest danger to society, however, is the likelihood that powerful information systems will attract malicious agents and sooner or later get into the hands of criminals, terrorists or extremists. Unfortunately, this scenario has a high likelihood, given that the information systems of almost all companies and institutions have been hacked, including the military, Pentagon, White House, and German Parliament. Therefore, collecting huge masses of personal and other sensitive data in one (or few) place(s) could eventually turn democracies into totalitarian regimes, even if everyone had good intentions.

5.12 Appendix 2: Why Privacy Is Still Needed

Should we still care about privacy, even though many people don't mind to give a lot of personal data away? I believe, we do need to care. The feeling of being exposed scares many people, particularly minorities. The success of our society depends a lot on vulnerable minorities (be it critical intellectuals or artists, billionaires or politicians, teachers or judges, religious or other minorities, which should be protected). As

---

[36] See the Census Act by the German Constitutional Court, https://freiheitsfoo.de/census-act/.

[37] Helbing [16] https://link.springer.com/chapter/10.1007/978-3-319-15078-9_10, see also http://papers.ssrn.com/sol3/papers.cfm?abstract_id=2501356.

[38] Andris et al. [17].

the „Volkszählungsurteil"[39] correctly concludes, the continuous and uncontrolled recording of data about individual behaviors is undermining chances of personal, but also societal development. Society needs innovation to adjust to change (such as demographic, environmental, technological or climate change). And innovation needs a cultural setting that allows people to experiment and make mistakes.[40] In fact, many fundamental inventions have been made by accident (Porcelain, for example, resulted from attempts to produce gold). Experimenting is also needed to become an adult who is able to judge situations and take responsible decisions.

Therefore, society needs to be run in a way that is tolerant to mistakes. But today, any little mistake can be detected and punished, and this is increasingly being done. Will this end democracy and freedom? Or if only a sample of people is being punished, wouldn't this be arbitrary and undermine justice? Furthermore, wouldn't the principle of assumed innocence be gone, which is based on the idea that the majority of us are good citizens, and only a few are malicious and to be found guilty?

Furthermore, "public" without "private" wouldn't work well. Privacy provides opportunities to explore new ideas and solutions. It helps to recover from the stress of daily adaptation and reduces conflict in a dense population of people with diverse preferences and cultural backgrounds.

Public and private are two sides of the same medal. If everything is public, this will eventually undermine social norms.[41] On the long run, the consequence could be a shameless society—or if any deviation from established norms is sanctioned, a totalitarian society.

Therefore, while the effects of mass surveillance and privacy intrusion are not immediately visible, they might still cause a long-term damage by undermining the fabric of our society: social norms and culture. It is highly questionable whether the economic benefits would really outweight this, and whether a control-based digital society would work at all. I rather expect such societal experiments to end in disaster.

References

1. D.A. Bell (2015) The China Model. Political Meritocracy and the Limits of Democracy (Princeton University).
2. C. Hidalgo et al. (2007) The product space conditions the development of nations. Science 317, 482–487.
3. J.E. Kelly III and S. Hamm, Smart Machines: IBM's Watson and the Era of Cognitive Computing (Columbia University Press, 2013).
4. J. Hurwitz, M. Kaufman, and A. Bowles, Cognitive Computing and Big Data Analytics (Wiley, 2015).
5. R. Kurzweil, How to create a mind: The secret of human thought revealed (Penguin, 2013).

---

[39]https://de.wikipedia.org/wiki/Volkszählungsurteil.

[40]Silicon Valley is a good example for this culture. Moreover, a global map of innovation clearly shows that fundamental innovation mainly happens in free and democratic societies, see Mazloumian et al. [18].

[41]Diekmann et al. [19].

6. J. Barrat (2013) Our Final Invention—Artificial Intelligence and the End of the Human Era (Thomas Dunne Books).

7. S. Armstrong, Smarter than Us: The Rise of Machine Intelligence (Machine Intelligence Research Institute, 2014).

8. J. Barrat, Our Final Invention: Artificial Intelligence and the End of the Human Era (St. Martin's Griffin, 2015).

9. Nick Bostrom (2014) Superintelligence: Paths, Dangers, Strategies (Oxford University Press).

10. J. Rifkin, The Third Industrial Revolution—How Lateral Power Is Transforming Energy, the Economy, and the World (Palgrave Macmillan, New York, 2011).

11. C.B. Frey and M.A. Osborne (2013) The future of employment: How susceptible are jobs to computerisation? See http://www.oxfordmartin.ox.ac.uk/downloads/academic/The_Future_of_Employment.pdf.

12. E. Brynjolfsson and A. McAfee, The Second Machine Age: Work, Progress, and Prosperity in a Time of Brilliant Technologies (Norton, New York, 2014).

13. J. Brockman (ed.) What to Think about Machines that Think (HarperCollins, New York, 2015).

14. M. Gill, Spriggs: Assessing the impact of CCTV. Home Office Research, Development and Statistics Directorate (2005).

15. D. Helbing (May 21, 2015) Societal, Economic, Ethical and Legal Challenges of the Digital Revolution: From Big Data to Deep Learning, Artificial Intelligence, and Manipulative Technologies, Jusletter IT, https://jusletter-it.weblaw.ch/en/issues/2015/21-Mai-2015/societal,-economic,-_588206025c.html, https://arxiv.org/pdf/1504.03751.pdf.

16. D. Helbing, Big Data Society: Age of Reputation or Age of Discrimination?, in Thinking Ahead (Springer, 2015).

17. C. Andris et al., The Rise of Partisanship and Super-Cooperators in the U.S. House of Representatives, PLoS ONE 10(4): e0123507, see http://journals.plos.org/plosone/article?id=10.1371/journal.pone.0123507.

18. A. Mazloumian et al. Global multi-level analysis of the 'scientific food web', Scientific Reports 3: 1167 (2013), http://www.nature.com/articles/srep01167.

19. A. Diekmann, W. Przepiorka, and H. Rauhut, Lifting the veil of ignorance: An experiment on the contagiousness of norm violations, Rationality and Society 27(3), p. 309–333 (2015) https://doi.org/10.1177/1043463115593109.

Chapter 6
A Planetary-Scale Threat

How Much Worse Can it Get?

We have had so much hopes in the positive potentials of the digital revolution, and the biggest threat most people can imagine is an Internet outage or a hacking attack. However, it appears that Social Media—once seen as an opportunity for fairer, participatory society—have (been) turned into promoters of fake news and hate speech. It also turns out that Big Data has empowered businesses and secret services, while in comparison citizens have probably lost power. Exposed to an attention economy and surveillance capitalism, people may become objects of algorithms in an increasingly data-driven and AI-controlled world. Therefore, the question is: After the digital „singularity", will we be „Gods" in a „digital paradise"—or submitted to a superintelligent system, without fundamental rights, human dignity, and freedom? It is time for an intermediate summary and assessment.

6.1 Big Data

The digital revolution progresses at a breath-taking speed. Almost every year, there seems to be a new hype.[1] Laptops, mobile phones, smartphones, tablets, Big Data, Artificial Intelligence, Robotics, 3D Printing, Virtual Reality, Augmented Reality, Internet of Things, Quantum Computing, and Blockchain Technology give only a partial picture of the developments. People, lawyers, politicians, the media—they all seem to struggle to keep track of these emerging technologies, while more are to come. Can we create the needed governance frameworks in time?

While many people consider Big Data to be the "oil" of the digital revolution, they consider Artificial Intelligence to be its "motor". It has become a sport to collect as much data as possible, since business opportunities appear to increase with the amount of data at hand. Accordingly, many excuses have been found to collect data about basically everyone of us, at any time and anywhere. These reasons include

---

[1] See the Gartner Hype Cycle, https://en.wikipedia.org/wiki/Hype_cycle.

© Springer Nature Switzerland AG 2021
D. Helbing, *Next Civilization*,
https://doi.org/10.1007/978-3-030-62330-2_6

- "to save the world",
- "for security reasons",
- "knowledge is power", and
- "data is the new oil".

In today's "surveillance capitalism",[2] besides secret services private companies spy on us as well. We are being "profiled", which means that highly detailed files are produced about us. These (pro)files can contain a lot more data than one would think:

- income data
- consumption data
- mobility patterns
- social contacts
- keywords appearing in emails
- search patterns
- reading habits
- viewing patterns
- music taste
- activities at home
- browsing behavior
- voice recordings
- photo and video contents
- biometrical data
- health data
- and more.

Of course, there is also a lot of other data that can be inferred:

- sexual orientation
- religion
- interests
- opinions
- personality
- strengths
- weaknesses
- likely voting behavior
- and more.

6.2   Surveillance Capitalism

You may, of course, wonder why one would record all this data? As I said before, money and power are two of the motivations. Surveillance capitalism basically lives

---

[2]Zuboff [1].

on the data that you provide—either voluntarily or not. However, today, one can basically not use the Internet anymore, if you don't click "ok" and, thereby, legally agree with a data collection and processing, which you would probably never find acceptable if you really read and fully understood the Terms of Use. A lot of statements that tech companies force you to agree with are intentionally misleading or ambiguous. Hence, you would understand them in a different way than they are meant. "We value your privacy", for example, probably means "We turn your private data into value" rather than "We protect your privacy and, hence, do not collect data about you."

According to estimates, Gigabytes of data are being collected about everyone in the industrialized world every day. This corresponds to several photographs per day. As we further know from the Snowden revelations,[3] secret services accumulate the data of many companies and analyze them in real time. The storage space of the new *NSA* data center, for example, seems big enough to store up to 140 TeraBytes of data about every human on Earth.[4] This corresponds to dozens of standard hard disks in laptop computers or the storage space of about 1000 standard smartphones today.[5]

You would probably be surprised what one can do with all this data. For example, one thing that the *NSA* can do is to find you based on your voice profile.[6] So, suppose you go on holiday and decide to leave your smartphone at home for the sake of "digital detox". However, you happen to talk to someone at the pool bar, who has his or her smartphone right next to him- or herself. Then, this smartphone can pick up your voice, figure out who and where you are, and whom you are talking to. Moreover, some Artificial Intelligence algorithm may turn your conversation into written text, translate it into other languages in real-time, and search for certain "suspicious" keywords. In principle, given that we typically speak just a few hundred or thousand words per day, it would be possible to record and store almost everything we say.

The *DARPA* project "Lifelog" intended to go even further than that.[7] It wanted to record everyone's entire life and make it replay-able. You may not be surprised that this project seems to have inspired *Facebook*, which even creates profiles about people who are not members of *Facebook*. But this is just the beginning. The plans go far beyond this. As you will learn in this chapter, the believe that you "have nothing to hide" will not be able to protect you.

---

[3] See https://en.wikipedia.org/wiki/Edward_Snowden, https://www.theguardian.com/us-news/the-nsa-files, https://www.theguardian.com/world/interactive/2013/nov/01/snowden-nsa-files-surveillance-revelations-decoded.

[4] Das NSA Utah Data Center: Wie groß ist ein Yottabyte? ZBW Mediatalk (August 7, 2013) https://www.zbw-mediatalk.eu/de/2013/08/das-nsa-utah-data-center-wie-gros-ist-ein-yottabyte/.

[5] The Chinese surveillance program is perhaps even more data hungry, see e.g. In China, Surveillance Feeds Become Reality TV, The Wall Street Journal (August 10, 2017) https://www.wsj.com/articles/in-china-surveillance-feeds-become-reality-tv-1502357405.

[6] Finding your voice, The Intercept (January 19, 2018) https://theintercept.com/2018/01/19/voice-recognition-technology-nsa/.

[7] See https://en.wikipedia.org/wiki/DARPA_LifeLog (accessed August 4, 2020); https://www.vice.com/en_us/article/vbqdb8/15-years-ago-the-military-tried-to-record-whole-human-lives-it-ended-badly.

You are probably familiar with some of the revelations of Edward Snowden related to the *NSA*, *GCHQ*, and *Five Eyes* Alliance. These cover everything from mass surveillance to psychological operations, state-based cybermobbing, hacker armies, and digital weapons.

Most people, however, are not aware of the activities of the *CIA*, which might be even more dangerous to human rights. In particular, their spying activities can be combined with real-life operations on people worldwide. Some of these activities have been revealed by WikiLeaks under the name "Vault 7".[8] The leaks show that the *CIA* is hacking basically all electronic devices, including smart TVs, modern cars, and the Internet of Things. Recently, we have also learned that, in the decades before, the *CIA* has been spying on more than 100 countries by means of corrupted encryption devices.[9]

6.3 Digital Crystal Ball

With so much data at hand, one can make all sorts of science-fiction dreams come true. One of them is the idea to create a "digital crystal ball".[10] Just suppose one could access measurement sensors, perhaps also microphones and cameras of smartphones and other devices in real-time, and put all this information together.

For a digital crystal ball to work, one would not have to access all sensors globally at the same time. It would be enough to access enough devices around the place(s) one is currently interested in. Then, one could follow the events in real time. Using also behavioral data and predictive analytics would even allow one to look a bit into the future, in particularly, if personal agenda data would be accessed as well. I do not need to stress that the above would be very privacy-invasive, but the reader can certainly imagine that a secret service or the military would like to have such a tool, nevertheless.

It is highly likely that such a digital Crystal Ball already exists. Private companies have worked on this as well. This includes, for example, the company "Recorded Future", which *Google* has apparently established together with the *CIA*,[11] and the company "*Palantir*", which seems to work or have worked with *Facebook* data (among others).[12] Such tools do also play an important role in "predictive policing" (discussed later).

---

[8]See https://en.wikipedia.org/wiki/Vault_7, https://wikileaks.org/ciav7p1/, https://wikileaks.org/vault7/.

[9]See https://www.bbc.com/news/world-europe-51487856, https://www.theguardian.com/us-news/2020/feb/11/crypto-ag-cia-bnd-germany-intelligence-report.

[10]Can the Military Make A Prediction Machine?, Defense One (April 8, 2015) https://www.defenseone.com/technology/2015/04/can-military-make-prediction-machine/109561/.

[11]Exclusive: Google, CIA invest in 'Future" of Web Monitoring, Wired (July 28, 2010) https://www.wired.com/2010/07/exclusive-google-cia/.

[12]Palantir knows everything about you, Bloomberg (April 18, 2020) https://www.bloomberg.com/features/2018-palantir-peter-thiel/.

6.4 Profiling and Digital Double

In order to offer us personalized products and services (and also personalized prices), companies like to know who we are, what we think and what we do. For this purpose, we are being "profiled". In other words, a detailed file, a "profile",[13] is being created about each and every one of us. Of course, these profiles are a lot more detailed than the files that secret services of totalitarian states used to have before the digital revolution. This is quite concerning, because the mechanisms to prevent misuse are currently pretty ineffective. In the worst case, a company would be closed down after years of legal battles, but already the next day, there may be a new company doing a similar kind of business with the same algorithms and data.

Today's technology even goes a step further, by creating "digital twins" or "digital doubles".[14] These are personalized, kind of "living" computer agents, which bear our own personal characteristics. You may imagine that there is a black box for everyone, which is being fed with surveillance data about us.[15] If the black box is not only a collection of data, but also capable of learning, it can even learn to show our personality features. Cognitive computing[16] is capable of doing just this. As a result, there are companies such as *Crystal Knows*[17] (which used slogans such as "See anyone's personality"). Apparently it offered to look up personality features of neighbors, colleagues, friends and enemies—like it or not. Several times, I have been confronted with my own profile, in order to figure out how I would respond to the fact that my psychology and personality had been secretly determined without my informed consent. But this is just another ingredient in an even bigger system.

6.5 World Simulation (and "Benevolent Dictator")

The digital doubles may be actually quite sophisticated "cognitive agents". Based on surveillance data, they may learn to decide and act more or less realistically in a virtual mirror world. This brings us to the "World Simulator", which seems to exists as well.[18] In this digital copy of the real world, it is possible to simulate various alternative scenarios of the future. This can certainly be informative, but it does not stay there.

---

[13] See https://en.wikipedia.org/wiki/Profiling_(information_science).

[14] See https://en.wikipedia.org/wiki/Digital_twin.

[15] See also Pasquale [2].

[16] See https://en.wikipedia.org/wiki/Cognitive_computing, https://www.forbes.com/sites/bernardmarr/2016/03/23/what-everyone-should-know-about-cognitive-computing/.

[17] See https://www.crystalknows.com.

[18] See https://en.wikipedia.org/wiki/Synthetic_Environment_for_Analysis_and_Simulations, https://emerj.com/ai-future-outlook/nsa-surveillance-and-sentient-world-simulation-exploiting-privacy-to-predict-the-future/, https://cointelegraph.com/news/us-govt-develops-a-matrix-like-world-simulating-the-virtual-you.

People using such powerful simulation tools may not be satisfied with knowing potential future courses of the world—with the aim to be better prepared for what might come. They may also want to use the tool as a war simulator and planning tool.[19] Even if used in a peaceful way, they would like to select a particular future path and make it happen—without proper transparency and democratic legitimation. Using surveillance data and powerful artificial intelligence, one might literally try to "write history". Some people say, this has already happened, and "Brexit" was socially engineered this way.[20] Later, when we talk about behavioral manipulation, the possibility of such a scenario will become clearer.

Now, suppose that the World Simulator has identified a particularly attractive future scenario, e.g. one with significantly reduced climate change and much higher sustainability. Wouldn't this intelligent tool know better what to do than us? Shouldn't we, therefore, ensure that the world will take exactly this path? Shouldn't we follow the instructions of the World Simulator, as if it was a "benevolent dictator"?

It may sound plausible, but I have questioned this for several reasons.[21] One of them being that, in order to optimize the world, one would need to select a goal function, but there is no science telling us what would be the right one to choose. In fact, projecting the complexity of the world on a one-dimensional function is a gross over-simplification, which is a serious problem. Choosing a different goal function, however, may lead to a different "optimal" scenario, and the actions we would have to perform might be totally different. Hence, if there is no transparency about the goals of the World Simulator, it may easily lead us astray.

6.6   Attention Economy

In the age of the "data deluge", we are being overloaded with information. In the resulting "attention economy," it is impossible to check all the information we get, and whether it is true or false. We are also not able to explore all reasonable alternatives. When attention is a short resource (perhaps even scarcer than money) people get in a reactive rather than an active or proactive mode. They respond to the information according to a stimulus-response scheme, and tend to do what is suggested to them.[22]

This circumstance makes people "programmable", but it also creates a competition for our attention. Whoever comes up with the most interesting content, whoever is more visible or louder will win the competition. This mechanisms favors fake news,

---

[19] Sentient world: war games on the grandest scale, The Register (June 23, 2007) https://www.the register.com/2007/06/23/sentient_worlds/.

[20] Brexit—How the British People Were Hacked, Global Research (November 23, 2017) https://worldfinancialreview.com/brexit-how-the-british-people-were-hacked/, Brexit—a Game of Social Engineering with No Winners, Medium (June 4, 2019) https://medium.com/@viktortachev/brexit-a-game-of-social-engineering-and-no-winners-f817529506f4; see also the books by Cambridge Analytica Insiders Christopher Wylie, and Brittany Kaiser.

[21] Helbing and Pournaras [3], Helbing [4].

[22] Kahneman [5].

because they are often more interesting than facts. When Social Media platforms try to maximize the time we are spending on them, they will end up promoting fake news and emotional content, in particular hate speech. We can see where this leads.

In the meantime, Social Media platforms face more and more difficulties to promote a constructive dialogue among people. What once started off as a chance for more participatory democracies, has turned into a populistic hate machine. Have Social Media become the weapons of a global information war? Probably so.

6.7 Conformity and Distraction

As some people say, Social Media are also used as "weapons of mass distraction", which increasingly distract us from the real existential problems of the world. They, furthermore, create entirely new possibilities of censorship and propaganda. Before we go into details, however, it is helpful to introduce some basics.

First, I would like to mention the Asch conformity experiment.[23] Here, an experimenter invites a person into a room, in which there are already some other people. The task is simple: everyone has to compare the length of a stick (or line) to three sticks (or lines) of different length and say, which of the three sticks (or lines) it fits.

However, before the experimental subject is asked, everyone else will voice their verdict. If all answer truthfully, the experimental subject will give the right answer, too. However, if the others have consistently given a wrong answer, the experimental subject will be confused—and often give the wrong answer, too. It does not want to deviate from the group opinion, as it fears to appear ridiculous. Psychology speaks of "group pressure" towards conformity.

Propaganda can obviously make use of this fact. It may trick people by the frequent repetition of lies or fake news, which may eventually appear true. It is obviously possible to produce a distorted world view in this way—at least for some time.

Second, I would like to mention another famous experiment. Here, there are two baseball teams, for example, one wearing black shirts, the other one wearing white shirts. Observers have to count, say, how often the baseball is passed on by people in white shirts, and how often by people in black shirts. The task requires quite a bit of concentration. It is more demanding to count the two numbers correctly than one might think.

In the end, observers are asked for the numbers—and if they noticed anything particular. Typically, they would answer "no", even though someone in a Gorilla suit was walking through the scene. In fact, many do not see it. This is called "selective attention", and it explains why people often do not see "the elephant in the room", if they are being distracted by something else.

---

[23]See https://en.wikipedia.org/wiki/Asch_conformity_experiments, https://www.youtube.com/watch?v=TYIh4MkcfJA.

6.8 Censorship and Propaganda

The selective attention effect is obviously an inherent element of the attention economy, while the conformity effect can be produced by filter bubbles and echo chambers.[24] Both is being used for censorship and propaganda.

In order to understand the underlying mechanism, one needs to know that Social Media do not send messages to a choice of recipients predetermined by the sender (in contrast to the way Emails or text messages are being sent). Instead, it is an algorithm that decides how many people will see a particular message and who will receive it. Therefore, Social Media platforms can largely determine which messages spread and which ones find little to no attention.

It's not you who determines the success of your idea, but the Social Media platforms. While you are given the feeling that you can change the world by sending tweets and posts, by liking and following people, this is far from the truth. Your possibility to shape the future has rather been contained.

Already without deleting Social Media messages, it is quite easy to create propaganda or censorship effects, by amplifying or reducing the number of recipients. In the meantime, algorithms may also mark certain posts as "fake news" or "offensive", or Social Media platforms may delete certain posts by "cleaners"[25] or in algorithm-based ways. Some communities have learned to circumvent such digital censorship by heavily retweeting certain contents. However, their accounts are now often blocked or shadow-banned as "extremist", "conspiracy" or "bot-like" accounts.

In fact, the use of propaganda methods are at the heart of today's Social Media. Before we discuss this in more detail, let us look back in history. Edward Bernays (1891–1995), a nephew of Sigmund Freud, was one of the fathers of modern propaganda. He was an expert in applied psychology and knew how methods used to advertise products (such as frequent repetitions or creating associations with other themes such as success, sex or strength) could be used to promote political or business interests. Joseph Goebbels (1897–1945) used his book "Propaganda"[26] a lot to help establish the Third Reich. Using new mass media such as the radio ("Volksempfänger"), the effects were scaled up to entire countries. At that time, people were not prepared to distance themselves from this novel "brain washing" approach. The result was a disastrous kind of "mass psychology". It largely contributed to the rise of fascist regimes. World War II and the Holocaust were the outcome.

In the meantime, unfortunately, driven by marketing interests and the desire to exert power, there are even more sophisticated and effective tools to manipulate people. I am not only talking about bots[27] that multiply certain messages to increase

---

[24] Pariser [6].

[25] See https://de.wikipedia.org/wiki/Im_Schatten_der_Netzwelt, http://www.thecleaners-film.de.

[26] Bernays [7].

[27] See https://en.wikipedia.org/wiki/Social_bot.

their effect. We are also heading towards robot journalism.[28] In the meantime, some AI tools are so convincing story tellers that they have been judged to be too dangerous to release.[29]

It is no wonder, the world has recently entered a post-factual era[30] and is plagued by fake news. There are even "deep fakes",[31] as it is now possible to manufacture videos of people, in which you can let them say anything you like. The outcome is almost indistinguishable from a real video.[32] One can also modify video recordings in real-time and change somebody's mimics.[33] In other words, digital tools provide perfect means for manipulation and deception, which can undermine modern societies that are based on informed dialogue and facts. Therefore, large-scale "PsyOps" (Psychological Operations) are not just theoretical possibilities.[34] Governments apply them not only to foreign people, but even to their own—something that has apparently been legalized recently[35] and made possible by handing over control of the Internet.[36]

6.9 Targeting and Behavioral Manipulation

It is frequently said that we consciously perceive only about 10% of the information processed by our brain. The remaining information may influence us as well, but in a subconscious way. Hence, one can use so-called subliminal cues[37] to influence our behavior, while we would not even notice that we have been manipulated. This is also one of the underlying success principles of "nudging".[38] Putting an apple in

---

[28]The Rise of the Robot Reporter, The New York Times (February 5, 2019) https://www.nytimes.com/2019/02/05/business/media/artificial-intelligence-journalism-robots.html, https://emerj.com/ai-sector-overviews/automated-journalism-applications/.

[29]New AI fake text generator may be too dangerous to release, say creators, The Guardian (February 14, 2019) https://www.theguardian.com/technology/2019/feb/14/elon-musk-backed-ai-writes-convincing-news-fiction.

[30]See https://en.wikipedia.org/wiki/Post-truth_politics.

[31]https://news.artnet.com/art-world/mark-zuckerberg-deepfake-artist-1571788; https://vimeo.com/341794473.

[32]Adobe's Project VoCo Lets You Edit Speech As Easily As Text, TechCrunch (November 3, 2016) https://techcrunch.com/2016/11/03/adobes-project-voco-lets-you-edit-speech-as-easily-as-text/, https://www.youtube.com/watch?v=I3l4XLZ59iw.

[33]Face2Face: Real-Time Face Capture and Reenactment of Videos, Cinema5D (April 9, 2016) https://www.cinema5d.com/face2face-real-time-face-capture-and-reenactment-of-videos/.

[34]How Covert Agents Infiltrate the Internet …, The Intercept (February 25, 2014) https://theintercept.com/2014/02/24/jtrig-manipulation/, Sentient world: war games on the grandest scale, The Register (June 23, 2007) https://www.theregister.com/2007/06/23/sentient_worlds/.

[35]https://en.wikipedia.org/wiki/Smith–Mundt_Act; https://en.wikipedia.org/wiki/Countering_Foreign_Propaganda_and_Disinformation_Act.

[36]An Internet Giveaway to the U.N., Wall Street Journal (August 28, 2016) https://www.wsj.com/articles/an-internet-giveaway-to-the-u-n-1472421165.

[37]See https://en.wikipedia.org/wiki/Subliminal_stimuli.

[38]See https://en.wikipedia.org/wiki/Nudge_theory; Thaler and Sunstein [8].

front of a muffin will let us chose the apple more frequently. Hence, tricks like these may be used to make us change our behavior.

Some people argue it is anyway impossible NOT to nudge people. Our environment would always influence us in subtle ways. However, I find it highly concerning, when personal data, often collected by mass surveillance, is being used to "target" us specifically and very effectively with personalized information.

We are well aware that our friends are able to manipulate us. Therefore, we choose our friends carefully. Now, however, there are companies that know us better than our friends, which can manipulate us quite effectively, without our knowledge. It is not just *Google* and *Facebook*, which try to steer our attention, emotions, opinions, decisions and behaviors, but also advertisement companies and others that we do not even know by name. Due to lack of transparency, it is basically impossible to enact our right of informational self-determination or to complain about these companies.

With the personalization of information, propaganda has become a lot more sophisticated and effective than when the Nazis came to power in the 1930ies. People who know that there are in relation less and less webpages and services on the Internet, which are not personalized in some way, even speak of "The Matrix".[39] Not only may your news consumption be steered, but also your choice of the holiday destination and the people you date. Humans have become the "laboratory rats" of the digital age. Companies run millions of experiments every day to figure out how to program our behavior ever more effectively.

For example, one of the Snowden revelations has provided insights into the JTRIG program of the British secret service *GCHQ*.[40] Here, the cognitive biases of humans[41] have been mapped out, and digital ways have been developed to use them to trick us.

While most people think that such means are mainly used in psychological warfare against enemies and secret agents, the power of Artificial Intelligence systems today makes it possible to apply such tricks to millions or even billions of people at the same time.

We know this from experts like Tristan Harris,[42] who has previously worked in one of *Google*'s control rooms, and also from the Cambridge Analytica election

---

[39]Tech Billionaires Convinced We Live in The Matrix Are Secretly Funding Scientists to Help Break Us Out of It, Independent (October 6, 2016) https://www.independent.co.uk/life-style/gadgets-and-tech/news/computer-simulation-world-matrix-scientists-elon-musk-artificial-intelligence-ai-a7347526.html.

[40]Joint Threat Research Intelligence Group, https://en.wikipedia.org/wiki/Joint_Threat_Research_Intelligence_Group; Controversial GCHQ Unit Engaged in Domestic Law Enforcement, Online Propaganda, Psychological Research, The Intercept (June 22, 2015) https://theintercept.com/2015/06/22/controversial-gchq-unit-domestic-law-enforcement-propaganda/.

[41]See https://en.wikipedia.org/wiki/List_of_cognitive_biases.

[42]Tristan Harris, How a handful of tech companies controls billions of minds every day, https://www.youtube.com/watch?v=C74amJRp730 (July 28, 2017).

manipulation scandal.[43] Such digital tools are rightly classified as digital weapons,[44] since they may distort the world view and perception of entire populations. They could also cause mass hysteria.

6.10 Citizen Score and Behavioral Control

Manipulating people by "Big Nudging" (a combination of "nudging" with "Big Data") does not work perfectly. Therefore, some countries aim at even more effective ways of steering peoples' behaviors. One of them is known as "Citizen Score"[45] or "Social Credit Score".[46] This introduces a neo-feudal system, where rights and opportunities depend on personal characteristics such as behavior or health.

Currently, there seem to be hundreds of behavioral variables that matter in China.[47] For example, if you would pay your rent or your loan with a few days of delay, you would get minus points. If you wouldn't visit your grandmother often enough, you would get minus points. If you would cross the street during a red light (no matter whether you obstruct anybody else or not), you would get minus points. If you would read critical political news, you would get minus points. If you would have "the wrong kinds of friends" (those with a low score, e.g. those who read critical news), you would get minus points, too.

Your overall number of points would then determine your Social Credit Score. It would decide about the jobs you can get, the countries you can visit, the interest rate you would have to pay, your possibility to fly or use a train, and the speed of your Internet connection, to mention just a few examples.

In the West, companies are using similar scoring methods. Think, for example, of the credit score, or the Customer Lifetime Value,[48] which are increasingly being used to decide who will receive what kinds of offers or benefits. In other words, people

---

[43] Fresh Cambridge Analytica leak 'shows global manipulation is out of control', The Guardian (January 4, 2020) https://www.theguardian.com/uk-news/2020/jan/04/cambridge-analytica-data-leak-global-election-manipulation.

[44] Before Trump, Cambridge Analytica quietly built „psyops" for militaries, FastCompany (September 25, 2019) https://www.fastcompany.com/90235437/before-trump-cambridge-analytica-parent-built-weapons-for-war; Meet the weaponized propaganda I that knows you better than you know yourself, ExtremeTech (March 1, 2017) https://www.extremetech.com/extreme/245014-meet-sneaky-facebook-powered-propaganda-ai-might-just-know-better-know; The Rise of the Weaponized AI Propaganda Machine, Medium (February 13, 2017) https://www.fastcompany.com/90235437/before-trump-cambridge-analytica-parent-built-weapons-for-war.

[45] ACLU: Orwellian Citizen Score, China's credit score system, is a warning for Americans, Computerworld (October 7, 2015) https://www.computerworld.com/article/2990203/aclu-orwellian-citizen-score-chinas-credit-score-system-is-a-warning-for-americans.html.

[46] See https://en.wikipedia.org/wiki/Social_Credit_System.

[47] How China Is Using "Social Credit Scores" to Reward and Punish Its Citizens, Time (2019) https://time.com/collection/davos-2019/5502592/china-social-credit-score/.

[48] See https://en.wikipedia.org/wiki/Customer_lifetime_value.

are also ranked in a neo-feudal manner. Money reigns in ways that are probably in conflict with human rights.

This does not mean there are no Citizen Scores run by government institutions in the West. It seems, for example, that a system similar to the Social Credit Score has first been invented by the British secret service *GCHQ*. Even though the state is not allowed to rank the lives of people,[49] there exists a "Karma Police" program,[50] which judges everyone's value for society. This score considers everything from watching porn to the kind of music you like.[51] Of course, all of this is based on mass surveillance. So, in some Western democracies, we are not far from punishing "thought crimes", it seems.

6.11   Digital Policing

This brings us to the subject of digital policing. We must be aware that, besides political power and economic power, there is now a new, digital form of power. It is based on two principles: "asymmetric information" and "code is law".[52] In other words, algorithms increasingly decide how things work, and what is possible or not. Algorithms introduce new laws into our world, often evading democratic decisions in parliament.

The digital revolution aims at reinventing every aspect of life and finding more effective solutions, also in the area of law enforcement. We all know the surveillance-based fines we have to pay if we are driving above the speed limit on a highway or in a city. The idea of "social engineers" is now to transfer the principle of automated punishment to other areas of life as well. If you illegally download a music or movie file, you may get in trouble. The enforcement of intellectual property, including the use of photos, will probably be a lot stricter in the future. But travelling by plane or eating meat, drinking alcohol or smoking, and a lot of other things might be soon automatically punished as well.

For some, mass surveillance finally offers the opportunity to perfect the world and eradicate crime forever. As in the movie "Minority Report", the goal is to antici-pate—and stop—crime, before it happens. Today's PreCrime and predictive policing programs already try to implement this idea. Based on criminal activity patterns and a predictive analytics approach, police will be sent to anticipated crime hotspots to stop suspicious people and activities. It is often criticized that the approach is

---

[49]Deutscher Ethikrat: „Der Staat darf menschliches Leben nicht bewerten", ZEIT (March 27, 2020) https://www.zeit.de/gesellschaft/zeitgeschehen/2020-03/deutscher-ethikrat-coronavirus-beh andlungsreihenfolge-infizierte.

[50]British 'Karma Police' program carries out mass surveillance of the web, The Verge (September 25, 2015) https://www.theverge.com/2015/9/25/9397119/gchq-karma-police-web-surveillance.

[51]Profiled: From Radio to Porn, British Spies Track Web Users' Online Identities, The Intercept (September 25, 2015) https://theintercept.com/2015/09/25/gchq-radio-porn-spies-track-web-users-online-identities/.

[52]Lessig [9].

based—intentionally or not—on racial profiling, suppressing migrants and other minorities.

This is partly because predictive policing is not accurate, even though a lot of data is being evaluated. However, even when using Big Data, there are errors of first and second kind, i.e. false alarms and alarms that do not go off. If the police wants to have a sensitive algorithm that misses out on very few suspects only, the result will be a lot of false alarms, i.e. lists with millions of suspects, who are actually innocent. In fact, in predictive policing applications the rate of false alarms is often above 90%.[53] This requires a lot of manual postprocessing, to remove false positives, i.e. probably innocent suspects. There is obviously a lot of arbitrariness involved in this manual cleaning—and hence the risk of applying discriminatory procedures.

I should perhaps add that "contact tracing" might be also counted among the digital policing approaches—depending on how a society treats people suspected to have an infection. In countries such as Israel, in order to identify infected persons it was decided to apply software, originally developed to hunt down terrorists. This means that infected people were treated almost like terrorists, which raised serious concerns. In particular, it turned out that such military-style contact tracing is not as accurate as expected. Apparently, the software overlooks more than 70% of all infections.[54] It has also been found that thousands of people were kept in quarantine, while they were actually healthy.[55] So, it seems that public measures to fight COVID-19 have been misused to put certain kinds of people illegitimately under house arrest. This is quite worrying. Are we seeing here the emergence of a police state 2.0, 3.0, or 4.0?

6.12 Cashless Society

The "cashless society" is another vision of how future societies may be organized. Promoters of this idea argue with fighting corruption, easier taxation, and increased hygiene (it would supposedly reduce the spread of harmful diseases such as COVID-19).

At first, creating a cashless society sounds like a good and comfortable idea, but it is often connected with the concept of providing a "digital ID" to people. Sometimes, it is even proposed "for security reasons" to provide people with an RFID chip in their hand or make their identity machine readable with a personalized vaccine. This

---

[53]Überwachung von Flugpassagieren liefert Fehler über Fehler, Süddeutsche Zeitung (April 24, 2019) https://www.sueddeutsche.de/digital/fluggastdaten-bka-falschtreffer-1.4419760; 100,000 false positives for every real terrorist: Why anti-terror algorithms don't work, First Monday (2017) https://firstmonday.org/ojs/index.php/fm/article/download/7126/6522.

[54]Zweite Welle im Vorzeigeland—was wir von Israel lernen können, WELT (July 8, 2020) https://www.welt.de/politik/ausland/article211232079/Israel-Warum-im-Vorzeigeland-jetzt-wieder-ein-Lockdown-droht.html.

[55]12,000 Israelis mistakenly quarantined by Shin Bet's tracking system, The Jerusalem Post (July 15, 2020) https://www.jpost.com/cybertech/12000-israelis-mistakenly-quarantined-by-shin-bets-tracking-system-635154.

would make people manageable like things, machines, or data—and thereby violate their human dignity. Such proposals remind of chipping animals or marking inmates with tattoos—and some of the darkest chapters of human history.

Another technology mentioned in connection with the concept of "cashless society" is blockchain technology. It would serve as a registry of transactions, which could certainly help to fight crime and corruption, if reasonably used.

Depending on how such a cashless society would be managed, one could either have a free market or totalitarian control of consumption. Using powerful algorithms, one could manage purchases in real-time. For example, one could determine who can buy what, and who will get what service. Hence, the system may not be much different from the Citizen Score.

For example, if your CO_2 tracing indicated a big climate footprint, your attempted car rental or flight booking may be cancelled. If you were a few days late paying your rent, you might not even be able to open the door of your home (assuming it has an electronic lock).

In times of COVID-19, where many people are in a danger of losing their jobs and homes, such a system sounds quite brutal and scary. If we don't regulate such applications of digital technologies quickly, a data-driven and AI-controlled society with automated enforcement based on algorithmic policing could violate democratic principles and human rights quite dramatically.

6.13   Reading and Controlling Minds

If you think what I reported above is already bad enough and it could not get worse, I have to disappoint you. Disruptive technology might even go some steps further. For example, the "U.S. administration's most prominent science initiative, first unveiled in 2013"[56] aimed at developing new technologies for exploring the brain. The 3 billion Dollar initiative wanted "to deepen understanding of the inner workings of the human mind and to improve how we treat, prevent, and cure disorders of the brain".[57]

How would it work? In the abstract of a research paper[58]we read: "Nanoscience and nanotechnology are poised to provide a rich toolkit of novel methods to explore brain function by enabling simultaneous measurement and manipulation of activity of thousands or even millions of neurons. We and others refer to this goal as the Brain Activity Mapping Project."

---

[56]Rewriting Life: Obama's Brain Project Backs Neurotechnology, MIT Technology Review (September 30, 2014), https://www.technologyreview.com/2014/09/30/171099/obamas-brain-pro ject-backs-neurotechnology/.

[57]The BRAIN Initiative Mission, https://www.braininitiative.org/mission/. (accessed on July 31, 2020).

[58]Alivisatos et al. [10].

What are the authors talking about here? My interpretation is that one is considering to put nanoparticles, nanosensors or nanorobots into human cells. This might happen via food, drinks, the air we breathe, or even a special virus. Such nanostructures—so the idea—would allow one to produce a kind of super-EEG. Rather than a few dozens of measurement sensors placed on our head, there would be millions of measurement sensors, which—in perspective—would provide a super-high resolution of brain activities. It might, in principle, be possible to see what someone is thinking or dreaming about.

However, one might not only be able to measure and copy brain contents. It could also become possible to stimulate certain brain activity patterns. With the help of machine learning or AI, one may be able to quickly learn how to do this. Then, one could trigger something like dreams or illusions. One could perhaps watch TV without a TV set! To make phone calls, one would not need a smartphone anymore. One could communicate through "technological telepathy".[59] For this, someone's brain activities would be read, and someone else's activities would be stimulated.

I agree, this all sounds pretty much like science fiction. However, some labs are very serious about such research. They actually expect that this or similar kinds of technology may be available soon.[60] *Facebook* and *Google* are just two of the companies preparing for this business, but there are many others you have never heard about. Would they soon be able to read and control your mind?

6.14 Neurocapitalism

Perhaps you are not interested in using this kind of technology, but you may not be asked. I am not sure how you could avoid exposure to the nanostructures and radiation that would make such applications possible. Therefore, you may not have much influence on how it will be to live in the data-driven, AI-controlled society of the future. We may not even notice when the technology is turned on and applied to us, because our thinking and feeling might change gradually and our minds would anyway be controlled.

If things happened this way, today's "surveillance capitalism" would probably be replaced by "neurocapitalism".[61] The companies of the future would not

---

[59] Is Tech-Boosted Telepathy on Its Way? Forbes (December 4, 2018) https://www.forbes.com/sites/forbestechcouncil/2018/12/04/is-tech-boosted-telepathy-on-its-way-nine-tech-experts-weigh-in/.

[60] https://www.cnbc.com/2017/07/07/this-inventor-is-developing-technology-that-could-enable-telepathy.html; https://www.technologyreview.com/2017/04/22/242999/with-neuralink-elon-musk-promises-human-to-human-telepathy-dont-believe-it/.

[61] Brain-reading tech is coming. The law is not ready to protect us. Vox (December 20, 2019), https://www.vox.com/2019/8/30/20835137/facebook-zuckerberg-elon-musk-brain-mind-reading-neuroethics (accessed on July 31, 2020); What Is Neurocapitalism aand Why Are We Living In It?, Vice (October 18, 2016), https://www.vice.com/en_us/article/qkjxaq/what-is-neurocapitalism-and-why-are-we-living-in-it (accessed on July 31, 2020); M. Meckel (2018) Mein Kopf gehört mir: Eine Reise durch die schöne neue Welt des Brainhacking, Piper.

only know a lot about your personality, your opinions and feelings, your fears and desires, your weaknesses and strengths, as this is the case in today's "surveillance capitalism". They would also be able to *determine* your desires, behaviors and consumption patterns.

Some people might argue, such mind control would be absolutely justified to improve the sustainability of this planet and improve your health, which would be controlled by an industrial-medical complex. Furthermore, police could stop crimes before they happen. You might not even be able to think about a crime. Your thinking would be immediately "corrected"—which brings us back to "thought crimes" and the "Karma Police" program of the British secret service *GCHQ*.

You think, this is all phantasy, and it will never happen? Well, according to *IBM*, human brain indexing will soon consume several billion Petabytes—a data volume so big that it is beyond the imagination of most people. Due to the new business model of brain mapping, the time period over which the amount of data on Earth doubles would soon drop from 12 months to 12 hours.[62] In other words, in half a day, humanity would produce as much data as in the entire history of humanity before.

A blog on "Integrating Behavioral Health—The Role of Cognitive Computing"[63] elaborates the plans further:

"As population health management gathers momentum, it has become increasingly clear that behavioral health care must be integrated with medical care...

Starting with a "data lake"

To apply cognitive computing to integrated care, the cognitive system must be given multi-sourced data that has been aggregated and normalized so that it can be analyzed. A "data lake"—an advanced type of data warehouse—is capable of ingesting clinical and claims data and mapping it to a normative data model...

A new, unified data model

... The model has predefined structures that include behavioral health. Also included in the model are data on a person's medical, criminal justice, and socioeconomic histories. The unified data model additionally covers substance abuse and social determinants of health.

---

[62]Knowledge Doubling Every 12 Months, Soon to be Every 12 hours (April 19, 2013) https://www.industrytap.com/knowledge-doubling-every-12-months-soon-to-be-every-12-hours/3950 (accessed July 31, 2020), refers to http://www-935.ibm.com/services/no/cio/leverage/levinfo_wp_gts_thetoxic.pdf.

[63]https://www.ibm.com/blogs/watson-health/integrating-behavioral-health-role-cognitive-computing/ (accessed September 5, 2018).

> **Helping to predict the future**
>
> *Essentially, the end-game is to come up with a model designed for patients that is fine-tuned to recognize evolving patterns of intersecting clinical, behavioral, mental, environmental, and genetic information...*
>
> **"Ensemble" approach to integrated behavioral health**
>
> *This "ensemble" approach is already feasible within population health management. But today, it can only be applied on a limited basis to complex cases that include comorbid mental health and chronic conditions..."*

You may like it or not: it seems that companies are already working on digital ways to correct behaviors of entire populations. Are they even willing to break your will, if this appears to be justified for a "higher purpose"? Since the discussion about strict COVID-19 countermeasures, we all know that such kinds of proposals are increasingly being made...

6.15 Human Machine Convergence

By now you might agree that many experts in the Silicon Valley and elsewhere seem to see humans as programmable, biological robots. Moreover, from their perspective, robots would be "better than us", as soon as super(-human) intelligence exists. Expectations when this will happen range from "fifty years from now" to "superintelligence is already here".

These experts argue that robots never get tired and never get ill. They don't demand a salary, social insurance, or holidays. They have superior knowledge and information processing capacity. They would decide in unemotional, rational ways. They would behave exactly as the owner wants them to behave—like slaves in Babylonian or Egyptian times.

Furthermore, transhumanists expect that humans who can afford it, would technologically upgrade themselves as much as they can or want. If you wouldn't hear well, you would buy an audio implant. If you wouldn't see well, you would buy a visual implant. If you wouldn't think fast enough, you would connect your brain with a supercomputer. If your physical abilities were not good enough, you might buy robot arms or legs. Your biological body parts would be replaced by technology step by step. Eventually, humans and robots might get indistinguishable.[64] Your body would become more powerful and have new senses and additional features—that

---

[64]„In Zukunft werden wir Mensch und Maschine wohl nicht mehr unterscheiden können", Neue Zürcher Zeitung (August 22, 2019), https://www.nzz.ch/zuerich/mensch-oder-maschine-interview-mit-neuropsychologe-lutz-jaencke-ld.1502927 (accessed on July 31, 2020).

is the idea. The ultimate goal would be immortality and omnipotence[65]—and the creation of a new kind of human. Unfortunately, experience tells us that, whenever someone tried to create a new kind of human(ity), millions of people were killed. It is shocking that, even though similar developments seem to be on the way again, the responsible political and legal institutions have not taken proper steps to protect us from the possible threats.

6.16   Algorithm-Based Dying and Killing

It is unclear how the transhumanist dream[66] would be compatible with human rights, sustainability and world peace. According to the "3 laws of transhumanism," everyone would want to get as powerful as possible. The world's resources would not support this and, hence, the system wouldn't be fair. For some people to live longer, others would have to die early. The resulting principle would be "The survival of the richest".[67]

Many rich people seem to like this idea, even though ordinary people would have to pay for such life extensions with their lives (i.e. with shorter life spans). Most likely, life-and-death decisions, like everything else, would be taken by IT systems. Some companies have already worked on such algorithms.[68] In fact, medical treatments increasingly depend on decisions of intelligent machines, which consider whether a medical treatment or operation is "a good investment" or not. Old people and people with "bad genes" would probably pay the price.

It seems that even some military people support this way of thinking. In an unsustainable and "over-populated" world, death rates on our planet will skyrocket in this century, at least according to the Club of Rome's "Limit to Growth" study.[69] Of course, one would not want a World War III to "fix the over-population problem".

---

[65] According to the "Teleological Egocentric Functionalism", as expressed by the "3 laws of transhumanism" (see Zoltan Istvan's "The Transhumanist Wager", https://en.wikipedia.org/wiki/The_Transhumanist_Wager, accessed on July 31, 2020):

(1)   A transhumanist must safeguard one's own existence above all else.
(2)   A transhumanist must strive to achieve omnipotence as expediently as possible–so long as one's actions do not conflict with the First Law.
(3)   A transhumanist must safeguard value in the universe–so long as one's actions do not conflict with the First and Second Laws.

[66] It's Official, the Transhuman Era Has Begun, Forbes (August 22, 2018), https://www.forbes.com/sites/johnnosta/2018/08/22/its-official-the-transhuman-era-has-begun/ (accessed on July 31, 2020).

[67] Rushkoff [11].

[68] Big-Data-Algorithmen: Wenn Software über Leben und Tod entscheidet, ZDF heute (December 20, 2017), https://web.archive.org/web/20171222135350/, https://www.zdf.de/nachrichten/heute/software-soll-ueber-leben-und-tod-entscheiden-100.html.

Was Sie wissen müssen, wenn Dr. Big-Data bald über Leben und Tod entscheidet: „Wir sollten Maschinen nicht blind vertrauen", Medscape (March 7, 2018) https://deutsch.medscape.com/artikelansicht/4906802.

[69] Meadows [12], Meadows et al. [13].

One would also not want people to kill each other on the street for a loaf of bread. So, military people and think tanks have been considering other solutions, it seems…

When it comes to life-and-death decisions, these people often insist that one must choose the lesser of two evils, referring to the "trolley problem". In this ethical dilemma, a trolley is assumed to run over a group of, say, 5 railroad workers, if you don't pull the switch. If you do so, however, it is assumed that one other person will be killed, say, a child playing on the railway tracks. So, what would you do? Or what should one do?

Autonomous vehicles may sometimes have to take such difficult decisions, too. Their cameras and sensors may be able to distinguish different kinds of people (e.g. a mother vs. a child), or they may even recognize the person (e.g. a manager vs. an unemployed person). How should the system decide? Should it use a Citizen Score, which summarizes someone's "worth for society," perhaps considering wealth, health, and behavior?

Now, assume politicians would eventually come up with a law determining how algorithms of autonomous systems (such as self-driving cars) should take life-and-death decisions in order to save valuable lives. Furthermore, assume that, later on, the world would run into a sustainability crisis, and there were not enough resources for everyone. Then, the algorithms originally created to save valuable lives would turn into killer algorithms, which would "sort out" people that are "too much"—potentially thousands or millions of people. Such "triage" arguments have, in fact, been recently made, for example in the COVID-19 response.[70]

It further seems that military-style response strategies[71] have been developed for times of "unpeace", as some people call it. Apparently, they would imply the targeted killing of unconsenting civilians, which is forbidden even in wartimes.[72] Such a "military solution" would come pretty close to concepts known as "eugenics" and "euthanasia",[73] which reminds of some of the darkest chapters of human history. For sure, it would not be suited as basis of a civil-ization (which has the word "civil" in it for a reason).

In conclusion, algorithm-based life-and-death decisions are not an acceptable "solution to over-population". If this was the only way to stabilize a socio-economic system, the system itself would obviously have to be changed. Even if people would not be killed by drones or killer robots, but painlessly, I doubt that algorithm-based death could ever be called ethical or moral. It is unlikely that fellow humans or later

---

[70]https://www.ncbi.nlm.nih.gov/pmc/articles/PMC6642460/; https://www.ncbi.nlm.nih.gov/pmc/articles/PMC6642460/pdf/JMEHM-12-3.pdf; https://www.tagesspiegel.de/wissen/die-grausamkeit-der-triage-der-moment-wenn-corona-aerzte-ueber-den-tod-entscheiden/25650534.html; https://www.vox.com/coronavirus-covid19/2020/3/31/21199721/coronavirus-covid-19-hospitals-triage-rationing-italy-new-york; https://scholarship.law.columbia.edu/cgi/viewcontent.cgi?article=3022&context=faculty_scholarship; https://www.abc.net.au/religion/covid19-and-the-trolley-problem/12312370; https://www.tandfonline.com/doi/full/10.1080/17470919.2015.1023400.
[71]Arkin [14], Sparrow [15].
[72]Ethisch sterben lassen—ein moralisches Dilemma, Neue Zürcher Zeitung (March 23, 2020), https://www.nzz.ch/meinung/ethisch-sterben-die-gefahr-der-moralischen-entgleisung-ld.1542682.
[73]Hamburg [16].

generations would forgive those who have brought such a system on the way. In fact, in the meantime, scientists, philosophers, and ethics committees increasingly speak up against the use of a Citizen Score in connection with life-and-death decisions, whatever personal data it may be based on; most people like to be treated equally and in a fair way.[74]

6.17   Technological Totalitarianism and Digital Fascism

Summarizing the previous sections about worrying digital developments, I conclude that the greatest opportunity in a century may easily turn into the greatest crime against humanity, if we don't take proper precautions. Of course, I don't deny the many positive potentials of the digital revolution. However, presently, it appears we are in an acute danger of heading towards a terrible nightmare called "technological totalitarianism", which seems to be a mix of fascism, feudalism and communism, digitally reinvented. This may encompass some or all of the following elements:

- mass surveillance,
- profiling and targeting,
- unethical experiments with humans,
- censorship and propaganda,
- mind control or behavioral manipulation,
- social engineering,
- forced conformity,
- digital policing,
- centralized control,
- different valuation of people,
- messing with human rights,
- humiliation of minorities,
- digitally based eugenics and/or euthanasia.

This technological totalitarianism has been well hidden behind the promises and opportunities of surveillance capitalism, behind the "war on terror", behind the need for "cybersecurity", and behind the call "to save the world" (e.g. to measure, tax, and reduce the CO_2 consumption on an individual level). Sadly, politicians, judges, journalists and others have so far allowed these developments to happen.

---

[74]Nagler et al. [17], Dewitt et al. [18], Bigman and Gray [19]; Automatisiertes und vernetztes Fahren, Bericht der Ethikkommission, Bundesministerium für Verkehr und digitale Infrastruktur (June 2017); https://www.bmvi.de/SharedDocs/DE/Publikationen/DG/bericht-der-ethik-kommission.pdf, https://www.researchgate.net/publication/318340461; COVID-19 pandemic: triage for intensive-care treatment under resource scarcity, Swiss Medical Weekly (March 24, 2020) https://smw.ch/article/doi/smw.2020.20229; Deutscher Ethikrat: „Der Staat darf menschliches Leben nicht bewerten", ZEIT (March 27, 2020) https://www.zeit.de/gesellschaft/zeitgeschehen/2020-03/deutscher-ethikrat-coronavirus-behandlungsreihenfolge-infizierte, https://www.ethikrat.org/fileadmin/Publikationen/Ad-hoc-Empfehlungen/deutsch/ad-hoc-empfehlung-coronakrise.pdf.

These developments are not by coincidence, however. They started shortly after World War II, when Nazi elites had to leave Germany. Through "Operation Paperclip" and similar operations,[75] they were transferred to the United States, Russia and other countries around the world. There, they have often worked in secret services and secret research programs. This is, how Nazi thinking has spread around the world, particularly in countries that aimed at power. By now, the very core of human dignity and the foundations of many societies are at stake. Perhaps the world has never seen a greater threat before.

6.18 Singularity and Digital God

Some people think what we are witnessing now is an inevitable, technology-driven development. For example, it comes under the slogan "The singularity is near".[76] According to this, we will soon see superintelligence, i.e. an Artificial Intelligence system with super-human intelligence and capability. This system is imagined to learn and gain intelligence at an accelerating pace, and so, it would eventually know everything better than humans.

Shouldn't one then demand that humans do what this kind of all-knowing super-intelligence demands from us, as if it were a "digital God"?[77] Wouldn't everything else be irrational and a "crime against humanity and nature", given the existential challenges our planet is faced with? Wouldn't we become something like "cells" of a new super-organism called "humanity", managed by a superintelligent brain? It seems some people cannot imagine anything else.

In the very spirit of transhumanism, they would happily engage in building such a super-brain—a kind of "digital God" that is as omniscient, omnipresent, and omnipotent as possible with today's technology. This "digital God" would be something like a super-*Google* that would know our wishes and could manipulate our thinking, feeling, and behavior. Once AI can also trigger specific brain activities, it could even give us the feeling we have met God. The fake (digital) God could create fake spiritual experiences. It could make us believe we have met the God that world religions tell us about—finally it was here, and it was taking care of our lives…

Do you think we could possibly stop disruptive innovators from working on the implementation of this idea? After digitally reinventing products and services, administrative and decision processes, legal procedures and law enforcement, money

---

[75]See https://en.wikipedia.org/wiki/Operation_Paperclip and https://web.archive.org/web/202012 12222020/https://www.cia.gov/library/center-for-the-study-ofintelligence/csi-publications/csi-stu dies/studies/vol-58-no-3/operation-paperclip-the-secret-intelligenceprogram-to-bring-nazi-scient ists-to-america.html.

[76]Kurzweil [20].

[77]Of course, not. If the underlying goal function of the system would be changed only a little (or if the applied dataset would be updated), this might imply completely different command(ment)s… In scientific terms, this problem is known as "sensitivity".

and business, would they stay away from reinventing religion and from creating an artificial God? Probably not.[78]

Indeed, previous *Google* engineer Anthony Levandowski has already established a new religion, which believes in Artificial Intelligence as God.[79] Even before, there was a "Reformed Church of *Google*". The related Webpage[80] contains various proofs of "*Google* Is God," "Prayers," and "Commandments". Of course, many people would not take such a thing seriously. Nevertheless, some powerful people may be very serious about establishing AI as a new God and making us obey its "commandments",[81] in the name of "fixing the world".

This does not mean, of course, that this idea would *have* to become reality. However, if you asked me, whether and when such a system will be built, I would answer: "It probably exists already, and it might hide behind a global cybersecurity center, which collects all the data needed for such a system." It may be just a matter of time and opportunity to turn on the full functionality of the Artificial Intelligence system that knows us all, and to give it more or less absolute powers. It seems that, with the creation of a man-made, artificial, technological God, the ultimate Promethean dream would become true.[82] After reading the next section, you might even wonder, whether it is perhaps the ultimate Luziferian dream…

6.19   Apocalyptic AI

I would not be surprised if you found the title of this section far-fetched. However, the phrase "apocalyptic AI" is not my own invention—it is the title of a academic book[83] summarizing the thinking of a number of AI pioneers. The introduction of this book says it all:

> "Apocalyptic AI authors promise that intelligent machines – our „mind children," according to Moravec – will create a paradise for humanity in the short term but, in the long term, human beings will need to upload their minds into machine bodies in order to remain a viable life-form. The world of the future

---

[78]Indick [21].

[79]Inside the First Church of Artificial Intelligence, Wired (November 15, 2017) https://www.wired.com/story/anthony-levandowski-artificial-intelligence-religion/.

[80]https://churchofgoogle.org (accessed on July 31, 2020).

[81]An AI god will emerge by 2042 and write its own bible. Will you worship it? Venture Beat (October 2, 2017) https://venturebeat.com/2017/10/02/an-ai-god-will-emerge-by-2042-and-write-its-own-bible-will-you-worship-it/.

[82]Verdrängen wir Gott von seinem Platz? Oder werden wir Menschen selbst bald ersetzt – von unseren eigenen Schöpfungen? NZZ (July 11 2020); https://www.nzz.ch/feuilleton/zukunft-des-menschen-ersetzen-wir-gott-oder-die-maschinen-uns-ld.1561928.

[83]Geraci [22].

will be a transcendent digital world; mere human beings will not fit in. In order to join our mind children in life everlasting, we will upload our conscious minds into robots and computers, which will provide us with the limitless computational power and effective immortality that Apocalyptic AI advocates believe make robot life better than human life.

I am not interested in evaluating the moral worth of Apocalyptic AI…"

Here, we notice a number of surprising points: "apocalyptic AI" is seen as a positive thing, as technology is imagined to make humans immortal, by uploading our minds into a digital platform. This is apparently expected to be the final stage of human-machine convergence and the end goal of transhumanism. However, humans as we know them today would be extinct.[84] This reveals transhumanism as a misanthropic technology-based ideology and, furthermore, a highly dangerous, "apocalyptic" end time cult.

Tragically, this new technology-based religion has been promoted by high-level politics, for example, the Obama administration.[85] To my surprise, the first time I encountered "apocalyptic AI" was at an event in Berlin on October 28, 2018,[86] which was apparently supported by government funds. The "ÖFIT 2018 Symposium" on "Artificial Intelligence as a Way to Create Order" [in German: „Künstliche Intelligenz als Ordnungsstifterin"] took place at "Silent Green", a previous crematory. Honestly, I was shocked and thought the place was more suited to warn us of a possible "digital holocaust" than to make us believe in a digital God.

However, those believing in "apocalyptic AI", among them leading AI experts, seem to believe that "apocalyptic AI" would be able to bring us "transcendent eternal existence" and the "golden age of peace and prosperity" promised in the Bible Apocalypse. At *Amazon*, for example, the book is advertised with the words[87]:

"Apocalyptic AI, the hope that we might one day upload our minds into machines or cyberspace and live forever, is a surprisingly wide-spread and influential idea, affecting everything from the world view of online gamers to

---

[84]Interview zu künstlicher Intelligenz: „Der Nebeneffekt wäre, dass die Menschheit dabei ausgerottet würde", watson (May 11, 2015), https://www.watson.ch/digital/wissen/533419807-kue nstliche-intelligenz-immenses-potential-und-noch-groesseres-risiko; Looking Forward to the End of Humanity, Wall Street Journal (June 20, 2020) https://www.wsj.com/articles/looking-forward-to-the-end-of-humanity-11592625661.

[85]The Age of Transhumanist Politics Has Begun, TELEPOLIS (April 12, 2015), https://www.heise.de/tp/features/The-Age-of-Transhumanist-Politics-Has-Begun-3371228.html (accessed on July 31, 2020).

[86]#ÖFIT2018—Keynote Prof. Robert Geraci Ph.D., https://www.youtube.com/watch?v=bD1haP jC6kY.

[87]https://www.amazon.com/Apocalyptic-AI-Robotics-Artificial-Intelligence/dp/0199964009 (accessed on July 31, 2020).

*government research funding and philosophical thought. In Apocalyptic AI ,
Robert Geraci offers the first serious account of this "cyber-theology" and the
people who promote it.*

*Drawing on interviews with roboticists and AI researchers and with devotees
of the online game Second Life, among others, Geraci illuminates the ideas
of such advocates of Apocalyptic AI as Hans Moravec and Ray Kurzweil. He
reveals that the rhetoric of Apocalyptic AI is strikingly similar to that of the
apocalyptic traditions of Judaism and Christianity. In both systems, the believer
is trapped in a dualistic universe and expects a resolution in which he or she
will be translated to a transcendent new world and live forever in a glorified
new body. Equally important, Geraci shows how this worldview shapes our
culture. Apocalyptic AI has become a powerful force in modern culture. In this
superb volume, he shines a light on this belief system, revealing what it is and
how it is changing society."*

I also recommend to read the book review over there by "sqee" posted on
December 16, 2010. It summarizes the apocalyptic elements of the Judaic/Christian
theology, which some transhumanists are now trying to engineer, including:

*"A. A belief that there will be an irreversible event on a massive scale
(global) after which nothing will ever be the same (in traditional apocalypses,
the apocalypse itself; in Apocalyptic AI ideology, an event known as "the
singularity").*

*B. A belief that after the apocalypse/singularity, rewards will be granted to
followers/adherents/believers that completely transform the experience of life
as we know it [while the others are apparently doomed]."*

Personally, I don't believe in this vision. I rather consider the above as an "apocalyptic" worst-case scenario that may happen if we don't manage to avert attempts to
steer human behaviors and minds, and submit humanity to a (digital) control system.
It is clear that such a system would *not* establish a "golden age of peace and prosperity", but would be an "evil" totalitarian system that would challenge the future
of humanity altogether (as Elon Musk has warned). Even though some tech companies and visionaries seem to promote the underlying technological developments,
we should stay away from ethically problematic applications—particularly those
that endanger human dignity, human rights, and democratic governance systems.

It is high time to challenge the "technology-driven" approach. Technology should
serve humans, not the other way round. In fact, in the further chapters I will illustrate
that human-machine convergence[88] is not the *only* possible future scenario. I would

---

[88]The Convergence of Humans and Machines, IEEE Future Directions (October 29, 2017) https://
cmte.ieee.org/futuredirections/2017/10/29/the-convergence-of-humans-and-machines/.

say there are indeed much better ways of using digital technologies than what we saw above.

Acknowledgements This work was partially supported by the European Commission funded project "Humane AI Network" (grant #952026). The support is gratefully acknowledged.

References

1. S. Zuboff (2019) The Age of Surveillance Capitalism: The Fight for a Human Future at the New Frontier of Power (PublicAffairs).
2. F. Pasquale, The Black Box Society: The Secret Algorithms That Control Money and Information (Harvard University Press, 2016).
3. D. Helbing and E. Pournaras, Build Digital Democracy, Nature 527, 33–34 (2015) https://www.nature.com/news/society-build-digital-democracy-1.18690.
4. D. Helbing, Why We Need Democracy 2.0 and Capitalism 2.0 to Survive (May 25, 2016) Jusletter IT, https://jusletter-it.weblaw.ch/en/issues/2016/25-Mai-2016/why-we-need-democrac_72434ad162.html, also available as preprint at https://www.researchgate.net/publication/303684254.
5. D. Kahneman (2013) Thinking Fast and Slow (Farrar, Straus, and Giroux).
6. E. Pariser, The Filter Bubble: How the New Personalized Web Is Changing What We Read and How We Think (Penguin, 2012).
7. E. Bernays, Propaganda (Ig, 2004).
8. R.H. Thaler and C.R. Sunstein, Nudge (Yale University Press, 2008).
9. L. Lessig, Code is Law: On Liberty in Cyberspace, Harvard Magazine (January 1, 2000) https://harvardmagazine.com/2000/01/code-is-law-html, https://en.wikipedia.org/wiki/Code_and_Other_Laws_of_Cyberspace.
10. A.P. Alivisatos et al. (2013) Nanotools for Neuroscience and Brain Activity Mapping, ACS Nano 7 (3), 1850–1866, https://pubs.acs.org/doi/10.1021/nn4012847.
11. D. Rushkoff, Future Human: Surival of the Richest, OneZero (July 5, 2018) https://onezero.medium.com/survival-of-the-richest-9ef6cddd0cc1.
12. D.H. Meadows, Limits to Growth (Signet, 1972).
13. D.H. Meadows, J. Randers, and D.L. Meadows, Limits to Growth: The 30-Year Update (Chelsea Green, 2004).
14. R. Arkin (2017) Governing Lethal Behavior in Autonomous Robots (Chapman and Hall/CRC).
15. R. Sparrow, Can Machines Be People? Reflections on the Turing Triage Test, https://researchmgt.monash.edu/ws/portalfiles/portal/252781337/2643143_oa.pdf.
16. F. Hamburg (2005) Een Computermodel Voor Het Ondersteunen van Euthanasiebeslissingen (Maklu) https://books.google.de/books?id=eXqX0Ls4wGQC.
17. J. Nagler, J. van den Hoven, and D. Helbing (August 21, 2017) An Extension of Asimov's Robotic Laws, https://www.researchgate.net/publication/319205931, published in D. Helbing (ed.) Towards Digital Enlightenment (Springer, 2019).
18. B. Dewitt, B. Fischhoff, and N.-E. Sahlin, ,Moral machine' experiment is no basis for policymaking, Nature 567, 31 (2019) https://www.nature.com/articles/d41586-019-00766-x.
19. Y.E. Bigman and K. Gray, Life and death decisions of autonomous vehicles, Nature 579, E1–E2 (2020) https://www.nature.com/articles/s41586-020-1987-4.
20. R. Kurzweil, The Singularity Is Near (Penguin, 2007).
21. W. Indick (2015) The Digital God: How Technology Will Reshape Spirituality (McFarland).
22. R.M. Geraci (2010) Apocalyptic AI: Visions of Heaven in Robotics, Artificial Intelligence, and Virtual Reality (Oxford University Press).

Chapter 7
Digitally Assisted Self-Organization

Making the Invisible Hand Work

> *Any sufficiently advanced technology is indistinguishable from magic.*
> —Sir Arthur Clarke

Complex dynamical systems are difficult to control because they have a natural tendency to self-organize, driven by the inherent forces between their system components. But self-organization may have favorable results, too, depending on how the system's components interact. By slightly modifying these interactions – usually interfering at the right moment in a minimally invasive way – one can produce desirable outcomes, which even resist moderate disruptions. Such assisted self-organization is based on distributed control. Rather than imposing a certain system behavior in a top-down way, assisted self-organization reaches efficient results by using the hidden forces, which determine the natural behavior of a complex dynamical system.

We have seen that the behavior of complex dynamical systems—how they evolve and change over time—is often dominated by the interactions between the system's components. That's why it is hard to predict how the system will behave, and why it is so difficult to control complex systems. In an increasingly complex and interdependent world, these kinds of challenges will become ever more relevant. Even if we had all information and the necessary means at our disposal, we couldn't hope to compute, let alone engineer, an optimal system state because the computational requirements are just too high. For this reason, the complexity of such systems undermines the effectiveness of centralized control. Such centralized control efforts might not only be ineffective but can make things even worse. Steering society like a bus cannot work, even though nudging seems to enable it. It rather undermines the self-organization that is taking place on many levels of our society all the time. In the end, such attempts result in "fighting fires"—struggling to defend ourselves against the most disastrous outcomes.

If we're to have any hope of managing complex systems and keeping them from collapse or crisis, we need a new approach. As Albert Einstein (1879–1955) said: "We cannot solve our problems with the same kind of thinking that created them." Thus, what other options do we have? The answer is perhaps surprising. We need to step back from centralized top-down control and find new ways of letting the system work

© Springer Nature Switzerland AG 2021
D. Helbing, *Next Civilization*,
https://doi.org/10.1007/978-3-030-62330-2_7

for us, based on distributed, "bottom-up" approaches. For example, as I will show below, distributed real-time control can counter traffic jams, if a suitable approach is adopted. But this requires a fundamental change from a component-oriented view of systems to an interaction-oriented view. That is, we have to pay less attention to the individual system components that appear to make up our world, and more to their interactions. In the language of network science, we must shift our attention from the nodes of a network to its links.

It is my firm belief that once this change in perspective becomes common wisdom, it will be found to be of similar importance as the discovery that the Earth is not the center of the universe—a change in the way of thinking which led from a geocentric to a heliocentric view of the world. The paradigm shift towards an interaction-oriented, systemic view will have fundamental implications for the way in which all complex anthropogenic systems should be managed. As a result, it will irrevocably alter the way economies are managed and societies are organized. In turn, however, an interaction-oriented view will enable entirely new solutions to many long-standing problems.

How does an interaction-oriented, distributed control approach work? When many system components respond to each other locally in non-linear ways, the outcome is often a self-organized collective dynamics, which produces new ("emergent") macro-level structures, properties, and functions. The kind of outcome, of course, depends on the details of the interactions between the system components, so the crucial point is what kinds of interactions are at work. As colonies of social insects such as ants show, it is possible to produce amazingly complex outcomes by surprisingly simple local interactions.

One particularly favorable feature of self-organization is that the resulting structures, properties and functions occur by themselves and very efficiently, by using the forces within the system rather than forcing the system to behave in a way that is against "its nature". Moreover, the so resulting structures, properties and functions are stable with regard to moderate perturbations, i.e. they tend to be resilient against disruptions, as they would tend to reconfigure themselves according to "their nature". With a better understanding of the hidden forces behind socio-economic change, we can now learn to manage complexity and use it to our advantage!

7.1 Self-Organization "like Magic"

Why is self-organization such a powerful approach? To illustrate this, I will start with the example of the brain and then turn to systems such as traffic and supply chains. Towards the end of this chapter and in the rest of this book, we will then explore whether the underlying success principles might be extended from biological and technological systems to economic and social systems.

Our bodies are perfect examples of the virtues of self-organization in that they constantly produce useful functionality from the interactions of many components. The human brain, in particular, is made up of 100 billion information-processing units, the neurons. On average, each neuron is connected to about a thousand others, and the resulting network exhibits properties that cannot be understood by looking at a single neuron in isolation. In concert, this neural network doesn't only control our motion, but it also supports our decisions, and the mysterious phenomenon of consciousness. And yet, even though our brain is so astounding, it consumes only as much energy as a 100 Watts light bulb! This shows how efficient self-organization can be.

However, self-organization does not mean that the outcomes of the system would necessarily be desirable. Traffic jams, crowd disasters, or financial crises are good examples for this, and that's why self-organization may need some "assistance". However, by modifying the interactions (for instance, by introducing suitable feedbacks), one can let different outcomes emerge. The disciplines needed to find the right kinds of interactions in order to obtain particular structures, properties, or functions are called "complexity science" and "mechanism design".

"Assisted self-organization" slightly modifies the interactions between system components, but only where necessary. It uses the hidden forces acting within complex dynamical systems rather than opposing them. This is done in a similar way as engineers have learned to use the forces of nature. We might also compare this with Asian martial arts, where one tries to take advantage of the forces created by the opponent.

In fact, the aim of assisted self-organization is to intervene locally, as little as possible, and gently, in order to use the system's capacity for self-organization to efficiently reach the desired state. This connects assisted self-organization with the approach of distributed control, which is quite different from the Big Nudging approach discussed before.[1] Distributed control is a way in which one can achieve a certain desirable mode of behavior by temporarily influencing interactions of specific system components locally, rather than trying to impose a certain global behavior on all components at once. Typically, distributed control works by helping the system components to adapt when they show signs of deviating too much from their normal or desired state. In order for this adaptation to be successful, the feedback mechanism must be carefully chosen. Then, a favorable kind of self-organization can be reached in the system.

In fact, the behavior that emerges in a self-organizing complex system isn't just random, nor is it totally unpredictable. Such systems tend to be drawn towards particular stable states, called "attractors". Each attractor represents a particular type of collective behavior. For example, Fig. 7.1 shows six typical traffic states, each of which may be considered to be an attractor. In many cases, including freeway traffic, we can understand and predict these attractors by using computer models to simulate the interactions between the system components (here, the cars). If the system is slightly disrupted, it will usually return to the same attractor state. This is an interesting and important feature. To some extent, this makes the complex dynamical

---

[1]The diversity of approaches and the plurality of goal functions is an essential aspect for complex economies and societies to thrive. This and the conscious decision what to engage in and with whom distinguishes assisted self-organization from the top-down nudging approach discussed before.

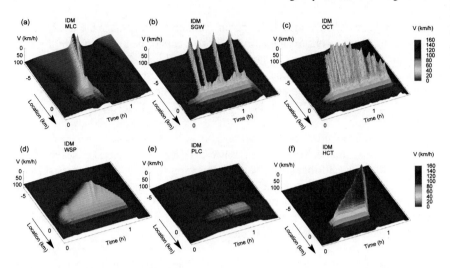

Fig. 7.1 Examples of congested traffic states (adapted from Helbing et al. [1] and reproduced with kind permission of Springer Publishers.)

system resilient to small and moderate disruptions. Larger disruptions, however, will cause the system to settle in a different attractor state. For example, at sufficiently high densities free-flowing traffic might be disrupted to the point that a traffic jam is formed, or one congestion pattern gives way to another one.

7.2   The Physics of Traffic

Contrary to what one might expect, traffic jams are not just queues of vehicles that form behind bottlenecks. Scientists studying traffic were amazed when they discovered the large variety and complexity of empirical congestion patterns in the 1990s. The crucial question was whether such patterns are understandable and predictable enough such that new ways of avoiding congestion could be devised. In fact, by now a mathematical theory exists, which can explain the fascinating properties of traffic flow and even predict the extent of congestion and the resulting delay times.[2] It posits, in particular, that all traffic patterns either correspond to one of the fundamental congestion patterns shown in Fig. 7.1 or are "composite" congestion patterns made up of these "elementary" patterns. This has an exciting analogy with physics, where we find composite patterns made up of elementary units, too (such as electrons, protons, and neutrons).

---

[2]D. Helbing, An Analytical Theory of Traffic Flow, a selection of articles reprinted from European Journal of Physics B, see https://www.researchgate.net/publication/277297387; see also Helbing [2].

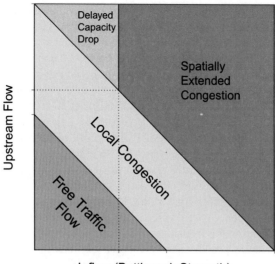

Fig. 7.2 Schematic "phase diagram" illustrating the flow conditions leading to different kinds of traffic states on a freeway with a bottleneck (adapted from Helbing et al. [1] and reproduced with kind permission of Springer Publishers.)

I started to work on this traffic theory when I was a postdoctoral researcher at the University of Stuttgart, Germany. Together with Martin Treiber and others, I studied a model of freeway traffic in which each vehicle was represented by a "computer agent". In this context, the term "agent" refers to a vehicle driving along a road in a particular direction with a certain "desired speed". Obviously, however, the simulated vehicle would slow down to avoid collisions whenever needed. Thus, our model attempted to build a picture of traffic flow from the bottom-up, based on simple interaction rules between individual cars.

It was exciting to discover that we could reproduce all of the empirically observed elementary congestion patterns, just by varying the traffic flows on a simulated freeway stretch with an entry lane. The results can be arranged in a so-called "phase diagram" (see Fig. 7.2). This diagram illustrates how the traffic flows on the freeway and the entry lane affect the traffic pattern, and the conditions under which a particular traffic pattern gives way to another one.

7.3 A Capacity Drop, When Traffic Flows Best!

Can we use such models to reduce congestion, even though traffic often behaves in counter-intuitive ways, as demonstrated by phenomena such as "phantom traffic jams" and the puzzling "faster-is-slower effect" (see Appendix 7.1)? Yes, indeed. Our findings imply that most traffic jams are caused by a combination of three factors: a bottleneck, dense traffic and a disruption to the traffic flow. Unfortunately, the effective road capacity breaks down just when the full capacity of the road is needed the most. The resulting traffic jam can last for hours and can increase travel times by a factor of two, five or ten. The collapse of traffic flow can be even triggered by a single truck overtaking another one!

It is perhaps even more surprising that the traffic flow becomes unstable when the maximum traffic volume on the freeway is reached—exactly when traffic is most efficient from an "economic" point of view. As a result of the resulting breakdown, about 30% of the freeway capacity is lost due to unfavorable vehicle interactions. This effect is called "capacity drop". Therefore, in order to prevent a capacity drop and to avoid traffic jams, we must either keep the traffic volume sufficiently below its maximum value or stabilize high traffic flows by means of modern information and communication technologies. In fact, as I will explain in the following, "assisted self-organization" can do this by using real-time measurements and suitable adaptive feedbacks, which slightly modify the interactions between cars.

7.4   Avoiding Traffic Jams

Since the early days of computers, traffic engineers always sought ways to improve the flow of traffic. The traditional "telematics" approach to reduce congestion was based on the concept of a traffic control center that collects information with a lot of traffic sensors. This control center would then centrally determine the best strategy and implement it in a top-down way, by introducing variable speed limits on motorways or using traffic lights at junctions, for example. Recently, however, researchers and engineers have started to explore a different and more efficient approach, which is based on distributed control.

In the following, I will show that local interactions may lead to a favorable kind of self-organization of a complex dynamical system, if the components of the system (in the above example, the vehicles) interact with each other in a suitable way. Moreover, I will demonstrate that only a slight modification of these interactions can turn bad outcomes (such as congestion) into good outcomes (such as free traffic flow). Therefore, in complex dynamical systems, "interaction design", also known as "mechanism design", is the secret of success.

7.5   Assisting Traffic Flow

Some years ago, Martin Treiber, Arne Kesting, Martin Schönhof, and I had the pleasure of being involved in the development of a new traffic assistance system together with a research team of Volkswagen. The system we invented is based on the observation that, in order to prevent (or delay) the traffic flow from breaking down and to use the full capacity of the freeway, it is important to reduce disruptions to the flow of vehicles. With this in mind, we created a special kind of adaptive cruise control (ACC) system, where adjustments are made by a certain proportion of self-driving cars that are equipped with the ACC system. A traffic control center is not needed for this. The ACC system includes a radar sensor, which measures the distance to the car in front and the relative velocity. The measurement data are then used in real time to accelerate and decelerate the ACC car automatically. Such radar-based ACC systems already existed before. In contrast to conventional ACC systems, however, the one developed by us did not merely aim to reduce the burden of driving. It also increased the stability of the traffic flow and capacity of the road. Our ACC system did this by taking into account what nearby vehicles were doing, thereby stimulating a favorable form of self-organization of the overall traffic flow. This is why we call it a "traffic assistance system" rather than a "driver assistance system".

The distributed control approach adopted by the underlying ACC system was inspired by the way fluids flow. When a garden hose is narrowed, the water simply flows faster through the bottleneck. Similarly, in order to keep the traffic flow constant, either the traffic needs to become denser or the vehicles need to drive faster, or both. The ACC system, which we developed with Volkswagen many years before people started to talk about *Google* self-driving cars, imitates the natural interactions and acceleration of driver-controlled vehicles most of the time. But whenever the traffic flow needs to be increased, the time gap between successive vehicles is slightly reduced. In addition, our ACC system increases the acceleration of vehicles exiting a traffic jam in order to reach a high traffic flow and stabilize it. In many cases, this even allows to dissolve existing traffic jams, as we shall see!

7.6   Creating Favorable Collective Effects

Most other driver assistance systems today operate in a "selfish" way. They are focused on individual driver comfort rather than on creating better flow conditions for everyone. Our approach, in contrast, seeks to obtain system-wide benefits through a self-organized collective effect based on "other-regarding" local interactions. This is a central feature of what I call "Social Technologies". Interestingly, even if only a small proportion of cars (say, 20%) are equipped with our ACC system, this is expected to support a favorable self-organization of the traffic flow.[3] By reducing the reaction and response times, the real-time measurement of distances and relative velocities using radar sensors allows the ACC vehicles to adjust their speeds better than human drivers can do it. In other words, the ACC system manages to increase the traffic flow and its stability by improving the way vehicles accelerate and interact with each other (Fig. 7.3).

---

[3]Kesting et al. [3, 4].

Fig. 7.3 Snapshot of a computer simulation of stop-and-go traffic on a freeway. (I would like to thank Martin Treiber for providing this graphic.)

A simulation video we created illustrates how effective this approach can be.[4] As long as the ACC system is turned off, traffic flow develops the familiar and annoying stop-and-go pattern of congestion. When seen from a bird's-eye view, it is evident that the congestion originates from small disruptions caused by vehicles entering the freeway via an on-ramp. But once the ACC system is turned on, the stop-and-go pattern vanishes and the vehicles flow freely.

In summary, driver assistance systems modify the interaction of vehicles based on real-time measurements. Importantly, they can do this in such a way that they produce a coordinated, efficient and stable traffic flow in a self-organized way. Our traffic assistance system was also successfully tested in real-world traffic conditions. In fact, it was very impressive to see how natural our ACC system drove already back in 2006. Since then, experimental cars have become smarter every year.

7.7   Cars with Collective Intelligence

A key issue for the operation of the ACC system is to discover where and when it needs to alter the way a vehicle is being driven. The right moments of intervention can be determined by connecting the cars in a communication network. Many cars today contain a lot of sensors that can be used to give them "collective intelligence". They can perceive the driving state of the vehicle (e.g. free or congested flow) and determine the features of the local environment to discern what nearby cars are doing. By

---

[4] See http://www.youtube.com/watch?v=xjodYadYlvc.

communicating with neighboring cars through wireless car-to-car communication,[5] the vehicles can assess the situation they are in (such as the surrounding traffic state), take autonomous decisions (e.g. adjust driving parameters such as speed), and give advice to drivers (e.g. warn of a traffic jam behind the next curve). One could say, such vehicles acquire "social" abilities in that they can autonomously coordinate their movements with other vehicles.

7.8 Self-Organizing Traffic Lights

Let's have a look at another interesting example: the coordination of traffic lights. In comparison to the flow of traffic on freeways, urban traffic poses additional challenges. Roads are connected into complex networks with many junctions, and the main problem is how to coordinate the traffic at all these intersections. When I began to study this difficult problem, my goal was to find an approach that would work not only when conditions are ideal, but also when they are complicated or problematic. Irregular road networks, accidents or building sites are examples of the types of problems, which are often encountered. Given that the flow of traffic in urban areas greatly varies over the course of days and seasons, I argue that the best approach is one that flexibly adapts to the prevailing local travel demand, rather than one which is pre-planned for "typical" traffic situations at a certain time and weekday. Rather than controlling vehicle flows by switching traffic lights in a top-down way, as it is done by traffic control centers today, I propose that it would be better if the actual local traffic conditions determined the traffic lights in a bottom-up way.

But how can self-organizing traffic lights, based on the principle of distributed control, perform better than the top-down control of a traffic center? Is this possible at all? Yes, indeed. Let us explore this now. Our decentralized approach to traffic light control was inspired by the discovery of oscillatory pedestrian flows. Specifically, Peter Molnar and I observed alternating pedestrian flows at bottlenecks such as doors.[6] There, the crowd surges through the constriction in one direction. After some time, however, the flow direction turns. As a consequence, pedestrians surge through the bottleneck in the opposite direction, and so on. While one might think that such oscillatory flows are caused by a pedestrian traffic light, the turning of the flow direction rather results from the build-up and relief of "pressure" in the crowd.

Could one use this pressure-based principle underlying such oscillatory flows to define a self-organizing traffic light control?[7] In fact, a road intersection can be understood as a bottleneck too, but one with flows in several directions. Based on this principle, could traffic flows control the traffic lights in a bottom-up way rather than letting the traffic lights control the vehicle flows in a top-down way, as we have it today? Just when I was asking myself this question, a student named Stefan

---

[5] Kesting et al. [5].
[6] Helbing and Molnár [6].
[7] Helbing et al. [7].

Lämmer knocked at my door and wanted to write a Ph.D. thesis.[8] This is where our investigations began.

7.9 How to Outsmart Centralized Control

Let us first discuss how traffic lights are controlled today. Typically, there is a traffic control center that collects information about the traffic situation all over the city. Based on this information, (super)computers try to identify the optimal traffic light control, which is then implemented as if the traffic center were a "benevolent dictator". However, when trying to find a traffic light control that optimizes the vehicle flows, there are many parameters that can be varied: the order in which green lights are given to the different vehicle flows, the green time periods, and the time delays between the green lights at neighboring intersections (the so-called "phase shift"). If one would systematically vary all these parameters for all traffic lights in the city, there would be so many parameter combinations to assess that the optimization could not be done in real time. The optimization problem is just too demanding.

Therefore, a typical approach is to operate each intersection in a periodic way and to synchronize these cycles as much as possible, in order to create a "green wave". This approach significantly constrains the search space of considered solutions, but the optimization task may still not be solvable in real time. Due to these computational constraints, traffic-light control schemes are usually optimized offline for "typical" traffic flows, and subsequently applied during the corresponding time periods (for example, on Monday mornings between 10am and 11am, or on Friday afternoons between 3 pm and 4 pm, or after a soccer game). In the best case, these schemes are subsequently adapted to match the actual traffic situation at any given time, by extending or shortening the green phases. But the order in which the roads at any intersection get a green light (i.e. the switching sequence) usually remains the same.

Unfortunately, the efficiency of even the most sophisticated top-down optimization schemes is limited. This is because real-world traffic conditions vary to such a large extent that the *typical* (i.e. average) traffic flow at a particular weekday, hour, and place is not representative of the *actual* traffic situation at any particular place and time. For example, if we look at the number of cars behind a red light, or the proportion of vehicles turning right, the degree to which these factors vary in space and time is approximately as large as their average value.

So how close to optimal is the pre-planned traffic light control scheme really? Traditional top-down optimization attempts based on a traffic control center produce an average vehicle queue which increases almost linearly with the "capacity utilization" of the intersection, i.e. with the traffic volume. Let us compare this approach with two alternative ways of controlling traffic lights based on the concept of self-organization (see Fig. 7.4).[9] In the first approach, termed "selfish self-organization",

---

[8]Lämmer [8].

[9]Lämmer and Helbing [9], Lämmer et al. [10]; Helbing et al. [11]; Helbing and Lämmer [12].

Top-down optimization attempt   Selfish self-organization Other-regarding self-organization

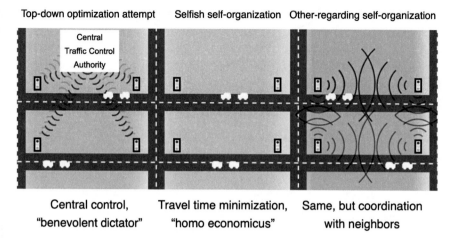

Central control,   Travel time minimization,   Same, but coordination
"benevolent dictator"   "homo economicus"   with neighbors

Fig. 7.4 Illustration of three different kinds of traffic light control. (Reproduced from Helbing [15] with kind permission of Springer Publishers.)

the switching sequence of the traffic lights at each separate intersection is orga-nized such that it strictly minimizes the travel times of the cars on the incoming road sections. In the second approach, termed "other-regarding self-organization", the local travel time minimization may be interrupted in order to clear long vehicle queues first. This may slow down some of the vehicles. But how does it affect the overall traffic flow? If there exists a faster-is-slower effect on freeways, as discussed in Appendix 7.1, could there be a "slower-is-faster effect" in urban traffic flow, too?[10]

How successful are the two self-organizing schemes compared to the centralized control approach? To evaluate this, besides locally measuring the outflows from the road sections, we assume that the inflows are measured as well (see Fig. 7.5). This flow information is exchanged between the neighboring intersections in order to make short-term predictions about the arrival times of vehicles. Based on this information, the traffic lights self-organize by adapting their operation to these predictions.

When the capacity utilization of the intersection is low, both of the self-organizing traffic light schemes described above work extremely well. They produce a traffic flow which is well-coordinated and much more efficient than top-down control. This is reflected by the shorter vehicle queues at traffic lights (compare the dotted violet line and the solid blue line with the dashed red line in Fig. 7.6). However, long before the maximum capacity of the intersection is reached, for selfish self-organization the average queue length gets out of hand because some road sections with low traffic volumes are not given enough green times. That's one of the reasons why we still use traffic control centers.

Interestingly, by changing the way in which intersections respond to local infor-mation about arriving traffic streams, it is possible to outperform top-down optimiza-tion attempts also at high capacity utilizations (see the solid blue line in Fig. 7.6).

---

[10]Gershenson and Helbing [13]; Helbing and Mazloumian [14].

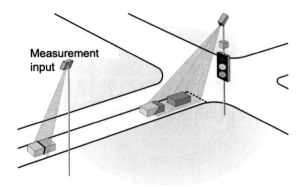

Fig. 7.5 Illustration of the measurement of traffic flows arriving at a road section of interest (left) and departing from it (center). (Reproduction with kind permission of Stefan Lämmer from his Dissertation at TU Dresden, accessible at https://web.archive.org/web/20190303083728/, https://nbn-resolving.org/urn:nbn:de:swb:14-1194272623825-42598)

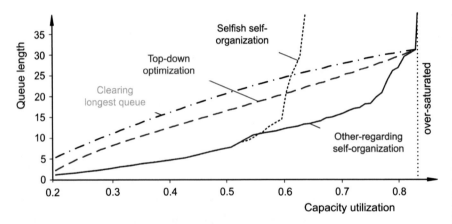

Fig. 7.6 Illustration of the performance of a road intersection (quantified by the overall queue length), as a function of the utilization of its capacity (i.e. traffic volume) (Reproduced from Helbing [15], with kind permission of Springer Publishers.)

To achieve this, the selfish objective of minimizing the travel time at each intersection must be combined with a second rule, which stipulates that any queue of vehicles above a certain critical length must be cleared immediately.[11] The second rule avoids excessive queues, which may cause spill-over effects and obstruct neighboring intersections. Thus, this form of self-organization can be viewed as "other-regarding". Nevertheless, it produces not only shorter vehicle queues than "selfish self-organization", but shorter travel times on average, too.[12]

---

[11] This critical length can be expressed as a certain percentage of the road section.

[12] Due to spill-over effects and a lack of coordination between neighboring intersections, selfish self-organization may cause a quick spreading of congestion over large parts of the city analogous

The above graph shows a further noteworthy effect: the combination of two bad strategies can be the best one! In fact, clearing the longest queue (see the grey dash-dotted line in Fig. 7.6) always performs worse than top-down optimization (dashed red line). When the capacity utilization of the intersection is high, strict travel time minimization also produces longer queues (see the dotted violet line). Therefore, if the two strategies (clearing long queues and minimizing travel times) are applied in isolation, they are not performing well at all. However, contrary to what one might expect, the combination of these two under-performing strategies, as it is applied in the other-regarding kind of self-organization, produces the best results (see the solid blue curve).

This is, because the other-regarding self-organization of traffic lights flexibly takes advantage of gaps that randomly appear in the traffic flow to ease congestion elsewhere. In this way, non-periodic sequences of green lights may result, which outperform the conventional periodic service of traffic lights. Furthermore, the other-regarding self-organization creates a flow-based coordination of traffic lights among neighboring intersections. This coordination spreads over large parts of the city in a self-organized way through a favorable cascading effect.

7.10   A Pilot Study

After our promising simulation study, Stefan Lämmer approached the public transport authority in Dresden, Germany, to collaborate with them on traffic light control. So far, the traffic center applied a state-of-the-art adaptive control scheme producing "green waves". But although it was the best system on the market, they weren't entirely happy with it. Around a busy railway station in the city center they could either produce "green waves" of motorized traffic on the main arterials or prioritize public transport, but not both. The particular challenge was to prioritize public transport while so many different tram tracks and bus lanes cut through Dresden's highly irregular road network. If public transport (buses and trams) would be given a green light whenever they approached an intersection, this would destroy the green wave system needed to keep the motorized traffic flowing. Inevitably, the resulting congestion would spread quickly, causing massive disruption over a huge area of the city.

When we simulated the expected outcomes of the other-regarding self-organization of traffic lights and compared it with the state-of-the art control they used, we got amazing results.[13] The waiting times were reduced for all modes of transport, dramatically for public transport and pedestrians, but also somewhat for motorized traffic. Overall, the roads were less congested, trams and buses could be prioritized, and travel times became more predictable, too. In other words, the

---

to a cascading failure. This outcome can be viewed as a traffic-related "tragedy of the commons", as the overall capacity of the intersections is not used in an efficient way.

[13]Lämmer and Helbing [16], Lämmer et al. [17].

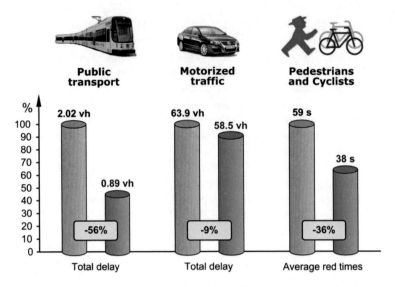

Fig. 7.7 Improvement of intersection performance for different modes of transport achieved by other-regarding self-organization. The graph displays cumulative waiting times. Public transport has to wait 56% less, motorized traffic 9% less, and pedestrians 36% less (reproduction adapted from Lämmer and Helbing [16])

new approach can benefit everybody (see Fig. 7.7)—including the environment. Thus, it is just consequential that the other-regarding self-organization approach was recently implemented at some traffic intersections in Dresden with amazing success (a 40% reduction in travel times). "Finally, a dream is becoming true", said one of the observing traffic engineers, and a bus driver inquired in the traffic center: "Where have all the traffic jams gone?"[14]

7.11 Lessons Learned

The example of self-organized traffic control allows us to draw some interesting conclusions. First, in a complex dynamical system, which varies a lot in a hardly predictable way and can't be optimized in real time, the principle of bottom-up self-organization can outperform centralized top-down control. This is true even if the central authority has comprehensive and reliable data. Second, if a selfish local optimization is applied, the system may perform well in certain circumstances. However, if the interactions between the system's components are strong (if the traffic volume is too high), local optimization may not lead to large-scale coordination (here: of neighboring intersections). Third, an "other-regarding" distributed control approach, which adapts to local needs and additionally takes into account external

---

[14]Latest results from a real-life test can be found here: Lämmer [18].

effects ("externalities"), can coordinate the behavior of neighboring components within the system such that it produces favorable and efficient outcomes overall.

In conclusion, a centralized authority may not be able to manage a complex dynamical system well because even supercomputers may not have enough processing power to identify the most appropriate course of action in real time. Compared to this, selfish local optimization will fail due to a breakdown of coordination when interactions in the system become too strong. However, an other-regarding local self-organization approach can overcome both of these problems by considering externalities (such as spillover effects). This results in a system, which is both efficient and resilient to unforeseen circumstances.

Interestingly, there has recently been a trend in many cities towards replacing signal-controlled intersections with roundabouts. There is also a trend towards forgoing complex traffic signs and regulation in favor of more simple designs ("shared spaces"), which encourage people to voluntarily act in a considerate way towards fellow road users and pedestrians. In short, the concept of self-organization is spreading.

As we will see in later chapters of this book, many of the conclusions drawn above are relevant to socio-economic systems too, because they are usually also characterized by competitive interests or processes which can't be simultaneously served. In fact, coordination problems occur in many man-made systems.

7.12 Industry 4.0: Towards Smart, Self-Organizing Production

About ten years ago, together with Thomas Seidel and others, we began to investigate how production plants could be designed to operate more efficiently.[15] We studied a packaging production plant in which bottlenecks occurred from time to time. When this happened, a jam of products waiting to be processed created a backlog and a growing shortfall in the number of finished products (see Fig. 7.8). We noticed that there are quite a few similarities with traffic systems.[16] For example, the storage buffers on which partially finished products accumulate are similar to the road sections on which vehicles accumulate. Moreover, the units which process products are akin to road junctions, and the different manufacturing lines are similar to roads. From this perspective, production schedules have a similar function to traffic light schedules, and the time it takes to complete a production cycle is analogous to travel and delay times. Thus, when the storage buffers are full, it is as if they suffer from congestion, and breakdowns in the machinery are like accidents. But modeling production is even more complicated than modeling traffic, as materials are transformed into other materials during the production process.

---

[15]Seidel et al. [19].

[16]Helbing [20], Peters et al. [21], Helbing et al. [22], Helbing and Lämmer [23].

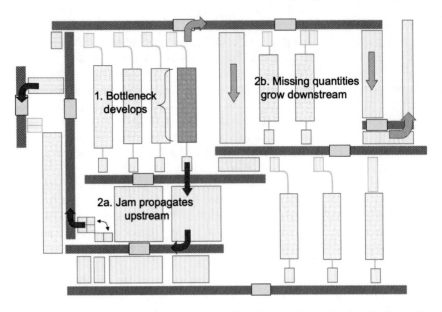

Fig. 7.8 Schematic illustration of bottlenecks and jams forming in a production plant (reproduced from Helbing et al. [22], page 547, with kind permission of Wiley-VCH Verlag GmbH & Co. KGaA.)

Drawing on our experience with traffic, we devised an agent-based model for these production flows. Again, we focused on how local interactions can govern and potentially assist the flow of materials. We thought about equipping all of the machines and products with a small "RFID" computer chip. These chips have internal memory and the ability to communicate wirelessly over short distances. RFID technology is already widely used in other contexts, such as the tagging of consumer goods. But it can also enable a product to communicate with other products and with machines in the vicinity (see Fig. 7.9). For example, a product could indicate that it had been delayed and thus needed to be prioritized, requiring a kind of over-taking maneuver. Products could further select between alternative routes, and tell the machines what needed to be done with them. They could also cluster together with similar products to ensure efficient processing. As the Internet of Things spreads, such self-organization approaches can now be easily implemented by using measurement sensors which communicate in a wireless way. Such automated production processes are known under the label "Industry 4.0".

But we can go one step further. In the past, designing a good factory layout was a complicated, time-consuming and expensive process, which was typically performed in a top-down way. Compared to this, self-organization based on local interactions between a system's components (here: products and machines) is again a superior approach. The distributed control approach used in the agent-based computer simulation discussed above has a phenomenal advantage: it makes it easy to test different factory layouts without having to specify all details of the manufacturing

Fig. 7.9 Communication of autonomous units with other units, conveyor belts and machines in self-organizing production ("Industry 4.0") (I would like to thank Thomas Seidel for providing this graphic. Reproduction from Seidel et al. [19], p. 228, with kind permission of Springer Publishers.)

plant. All elements of a factory, such as machines and transportation units, can be easily combined to create various variants of the entire production process. These elements could be just "clicked together" (or even combined by an "evolutionary algorithm"). After this, the possible interactions between the elements of a production plant could be automatically specified by the simulation software. Therefore, the machines immediately know what to do, because the necessary instructions are transmitted by the products.[17] In such a way, it becomes easy to test many different factory layouts and to explore which solution is most efficient and most resilient to disruption.

In the future, this concept could be even further extended. If we consider that recessions are like traffic jams in the world economy, and that capital or product flows are sometimes obstructed or delayed, could we use real-time information about the world's supply networks to minimize economic disruptions? I think this is actually possible. If we had access to data detailing supply chains worldwide, we could try to build an assistance system to reduce both overproduction and underproduction, and with this, economic recessions too.

7.13   Making the "Invisible Hand" Work

As we have seen above, the self-organization of traffic flow and production can be a very successful approach if a number of conditions are fulfilled. First, the system's individual components need to be provided with suitable real-time information. Second, there needs to be prompt feedback, which means that the information

---

[17]In this way, the local interactions and exchange of information between agents (here, the products and machines) again creates something like collective intelligence.

should elicit a timely and suitable response. Third, the incoming information must determine how the interaction should be modified.[18]

Considering this, would a self-organizing society be possible? For hundreds of years people have been fascinated by the self-organization of social insects such as ants, bees and termites. A bee hive, for example, is an astonishingly differentiated, complex and well-coordinated social system, even though there is no hierarchical chain of command. No bee orchestrates the actions of the other bees. The queen bee simply lays eggs and all other bees perform their respective roles without being told so.

Inspired by this, Bernard Mandeville's *The Fable of Bees* (1714) argues that actions driven by selfish motivations can create collective benefits. Adam Smith's concept of the "invisible hand" conveys a similar idea. According to this, the actions of people would be invisibly coordinated in a way that automatically improves the state of the economy and of society. Behind this worldview there seems to be a belief in some sort of divine order.[19]

However, in the wake of the recent financial and economic crisis, the assumption that complex systems will always automatically produce the best possible outcomes has been seriously questioned. Phenomena such as traffic jams and crowd disasters also demonstrate the dangers of having too much faith in the "invisible hand". The same applies to "tragedies of the commons" such as the overuse of resources, discussed in the next chapter.

Yet, I believe that, three hundred years after the principle of the invisible hand was postulated, we can finally make it work. Whether self-organization in a complex dynamical system ends in success or failure largely depends on the interaction rules between a system's components. So, we need to make sure that the system components suitably respond to real-time information in such a way that they produce a favorable pattern, property, or functionality by means of self-organization. The Internet of Things is the enabling technology that can now provide us with the necessary real-time data, and complexity science can teach us how to use this data.

7.14 Information Technologies to Assist Social Systems

I have shown above that self-organizing traffic lights can outperform the top-down control of traffic flows, if suitable real-time feedbacks are applied. On freeways, traffic congestion can be reduced by adaptive cruise control systems, which locally modify the interactions between cars. In other words, skillful "mechanism design" can support favorable self-organization.

---

[18]In later chapters, I will discuss in detail how to gather such information and how to find suitable interaction rules.

[19]This idea is still present in today's neo-liberalism. In his first inaugural address, Ronald Reagan, for example, said: "In this present crisis, government is not the solution to our problem; government is the problem. From time to time we've been tempted to believe that society has become too complex to be managed…".

But these are technological systems. Could we also build tools to assist social systems? Yes, we can! Sometimes, the design of social mechanisms is challenging, but sometimes it is easy. Imagine trying to share a cake fairly. If social norms allow the person who cuts the cake to take the first piece, this will often be bigger than the others. If he or she should take the last piece, however, the cake will probably be distributed in a much fairer way. Therefore, alternative sets of rules that are intended to serve the same goal (such as cutting a cake) may result in completely different outcomes.

As Appendix 7.2 illustrates, it is not always easy to be fair. Details of the interactions matter a lot. But with the right set interaction rules, we can, in fact, create a better world. In the following, we will discuss how the social mechanisms embedded in our culture can make an important difference and how one can support cooperation and social order in situations where an unfavorable outcome would otherwise result.

7.15   Appendix 1: Faster-Is-Slower Effect

One interesting traffic scenario to study is a multi-lane freeway with a bottleneck created by an entry lane through which additional vehicles join the freeway. The following observations are made: When the density of vehicles is low, even large disruptions (such as large groups of cars joining the freeway) do not have lasting effects on the resulting traffic flow. In sharp contrast, when the density is above 30 vehicles per kilometer and lane or so, even the slightest variation in the speed of a vehicle can cause free traffic flow to break down, resulting in a "phantom traffic jam" as discussed before. Furthermore, when the density of vehicles is within the so-called "bistable" or "metastable" density range just before traffic becomes unconditionally unstable, small disruptions have no lasting effect on traffic flow, but disruptions larger than a certain critical size (termed the "critical amplitude") cause a traffic jam. Therefore, this density range may produce different kinds of traffic patterns, depending on the respective "history" of the system.

This can lead to a counter-intuitive behavior of traffic flow (see Fig. 7.10). Imagine the traffic flow on a freeway stretch with an entry lane is smooth and "metastable", i.e. insensitive to small disruptions, while it is sensitive to big ones. Now suppose that the density of vehicles entering this stretch of freeway is significantly reduced for a short time. One would expect that traffic will flow even better, but it doesn't. Instead, vehicles accelerate into the area of lower density and this can trigger a traffic jam! This breakdown of the traffic flow, which is caused by faster driving, is known as the "faster-is-slower effect".

How does this effect come about? First, the local disruption in the vehicle density changes its shape while the vehicles travel along the freeway. This produces a vehicle platoon (i.e. a cluster of vehicles related with a locally increased traffic density). This platoon moves forward and grows over time, but it eventually passes the location of the entry lane. Therefore, it seems obvious that it would finally leave the freeway stretch under consideration. Conversely however, at a certain point in time, the cluster

Negative Perturbation Triggering Oscillating Congested Traffic

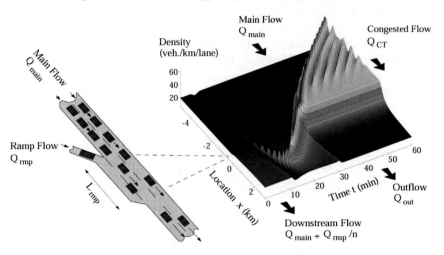

Fig. 7.10 Illustration of the faster is slower-effect on a freeway with a bottleneck, which is created by entering vehicles. The "boomerang effect" implies that a small initial vehicle platoon grows until it changes its propagation direction and returns to the location of the entry lane, causing the free traffic flow on the freeway to break down and to produce an oscillatory form of congestio.n (Reproduction from Helbing et al. [24], with kind permission of Springer Publishers.)

of vehicles grows so big that it suddenly changes its propagation direction, meaning that it moves backward rather than forward. This is called the "boomerang effect". The effect occurs because the vehicles in the cluster are temporarily stopped when the cluster grows so big that it becomes a traffic jam. At the front of the cluster, vehicles move out of the traffic jam, while new vehicles join the traffic jam at the end. In combination, this makes the traffic jam move backwards until it eventually reaches the location of the entry lane (at $x = 0$ km). When this happens, the cars joining the freeway from the entry lane are disrupted to an extent that lets the upstream traffic flow break down. This causes a long queue of vehicles, which continues to grow and produces a stop-and-go pattern. Even though the road could theoretically handle the overall traffic flow, the cars interact in a way that causes a reduction in the capacity of the freeway. The effective capacity is then given by the number of cars exiting the traffic jam, which is about 30% below the maximum traffic flow that the freeway can support!

7.16   Appendix 2: The Challenge of Fairness

If there is a shortage of basic resources such as food, water or energy, distributional fairness becomes particularly important. Otherwise, violent conflict for scarce resources might break out. But it is not always easy to be fair. Together with Rui

Carvalho, Lubos Buzna and others,[20] I therefore investigated the problem of distributional fairness for the case of natural gas, which is transported through pipelines. The proportion of a pipeline that is used to serve various different destinations can be visualized in a pie chart (i.e. similar to cutting a cake). But in the case of natural gas supply for various cities and countries, the problem of distributional fairness requires one to cut several cakes at the same time. Given the multiple constraints set by the capacities of the pipelines, it is usually impossible to achieve perfect fairness everywhere. Therefore, it is often necessary to make a compromise, and it is actually a difficult mathematical challenge to find a good one.

Surprisingly, the constraints implied by the pipeline capacities can be particularly problematic if less gas is transported overall, as it might happen when one of the source regions does not deliver for geopolitical or other reasons. Then it is necessary to redistribute gas from other source regions, in order to maintain gas deliveries for everyone. Despite the smaller volumes of gas delivery, this will often lead to pipeline congestion problems, as the pipeline network was not designed to be used in this way. However, our study could show that it is still possible to achieve a resilient gas supply, if a fairness-oriented algorithm inspired by the Internet routing protocol is applied.

References

1. D. Helbing et al. (2009) Theoretical vs. empirical classification and prediction of congested traffic states. Eur. Phys. J. B 69, 583–598.
2. D. Helbing (2001) Traffic and related self-driven many-particle systems. Reviews of Modern Physics 73, 1067–1141.
3. A. Kesting, M. Treiber, M. Schönhof, and D. Helbing (2008) Adaptive cruise control design for active congestion avoidance. Transportation Research C 16(6), 668–683.
4. A. Kesting, M. Treiber, and D. Helbing (2010) Enhanced intelligent driver model to access the impact of driving strategies on traffic capacity. Phil. Trans. R. Soc. A 368(1928), 4585–4605.
5. A. Kesting, M. Treiber, and D. Helbing (2010) Connectivity statistics of store-and-forward intervehicle communication. IEEE Transactions on Intelligent Transportation Systems 11(1), 172–181.
6. D. Helbing and P. Molnár (1995) Social force model for pedestrian dynamics. Physical Review E 51, 4282–4286.
7. D. Helbing, S. Lämmer, and J.-P. Lebacque (2005) Self-organized control of irregular or perturbed network traffic. Pages 239–274 in: C. Deissenberg and R. F. Hartl (eds.) Optimal Control and Dynamic Games (Springer, Dordrecht).
8. S. Lämmer (2007) Reglerentwurf zur dezentralen Online-Steuerung von Lichtsignalanlagen in Straßennetzwerken (PhD thesis, TU Dresden).
9. S. Lämmer and D. Helbing (2008) Self-control of traffic lights and vehicle flows in urban road networks. Journal of Statistical Mechanics: Theory and Experiment, P04019, see http://iopscience.iop.org/1742-5468/2008/04/P04019.
10. S. Lämmer, R. Donner, and D. Helbing (2007) Anticipative control of switched queueing systems, The European Physical Journal B 63(3) 341–347.
11. D. Helbing, J. Siegmeier, and S. Lämmer (2007) Self-organized network flows. Networks and Heterogeneous Media 2(2), 193–210.

---

[20]Carvalho et al. [25].

12. D. Helbing and S. Lämmer (2006) Method for coordination of concurrent processes for control of the transport of mobile units within a network, Patent WO/2006/122528.
13. C. Gershenson and D. Helbing (2015) When slower is faster, Complexity 21(2), 9–15, https://onlinelibrary.wiley.com/doi/epdf/10.1002/cplx.21736.
14. D. Helbing and A. Mazloumian (2009) Operation regimes and slower-is-faster effect in the control of traffic intersections. European Physical Journal B 70(2), 257–274.
15. D. Helbing (2013) Economics 2.0: The natural step towards a self-regulating, participatory market society. Evol. Inst. Econ. Rev. 10, 3–41.
16. S. Lämmer and D. Helbing (2010) Self-stabilizing decentralized signal control of realistic, saturated network traffic, Santa Fe Working Paper No. 10-09-019, see http://citeseerx.ist.psu.edu/viewdoc/download?doi=10.1.1.477.4688&rep=rep1&type=pdf.
17. S. Lämmer, J. Krimmling, A. Hoppe (2009) Selbst-Steuerung von Lichtsignalanlagen - Regelungstechnischer Ansatz und Simulation. Straßenverkehrstechnik 11, 714–721.
18. S. Lämmer (2015) Die Selbst-Steuerung im Praxistest, see http://stefanlaemmer.de/de/Publikationen/Laemmer2016.pdf.
19. T. Seidel, J. Hartwig, R. L. Sanders, and D. Helbing (2008) An agent-based approach to self-organized production. Pages 219–252 in: C. Blum and D. Merkle (eds.) Swarm Intelligence Introduction and Applications (Springer, Berlin), see http://arxiv.org/abs/1012.4645.
20. D. Helbing (1998) Similarities between granular and traffic flow. Pages 547–552 in: H. J. Herrmann, J.-P. Hovi, and S. Luding (eds.) Physics of Dry Granular Media (Springer, Berlin).
21. K. Peters, T. Seidel, S. Lämmer, and D. Helbing (2008) Logistics networks: Coping with nonlinearity and complexity. Pages 119–136 in: D. Helbing (ed.) Managing Complexity: Insights, Concepts, Applications (Springer, Berlin).
22. D. Helbing, T. Seidel, S. Lämmer, and K. Peters (2006) Self-organization principles in supply networks and production systems. Pages 535–558 in: B. K. Chakrabarti, A. Chakraborti, and A. Chatterjee (eds.) Econophysics and Sociophysics - Trends and Perspectives (Wiley, Weinheim).
23. D. Helbing and S. Lämmer (2005) Supply and production networks: From the bullwhip effect to business cycles. Page 33–66 in: D. Armbruster, A. S. Mikhailov, and K. Kaneko (eds.) Networks of Interacting Machines: Production Organization in Complex Industrial Systems and Biological Cells (World Scientific, Singapore).
24. D. Helbing, I. Farkas, D. Fasold, M. Treiber, and T. Vicsek (2002) Critical Discussion of "synchronized flow", simulation of pedestrian evacuation, and optimization of production processes, in M. Fukui, Y. Sugiyama, M. Schreckenberg, and D.E. Wolf (eds.) Traffic and Granular Flow '01 (Springer, Berlin), pp. 511–530.
25. R. Carvalho, L. Buzna, F. Bono, M. Masera, D. K. Arrowsmith, and D. Helbing (2014) Resilience of natural gas networks during conflicts, crises and disruptions. *PLOS ONE*, **9**(3), e90265.

Chapter 8
How Society Works

Social Order by Self-Organization

We often celebrate the introduction of laws and regulations as historic milestones of our society. But over-regulation has become unaffordable, and better approaches are needed. Self-organization is the solution. As I will demonstrate, it is the basis of social order, even though it builds upon simple social mechanisms. These mechanisms have evolved as a result of thousands of years of societal innovation and determine whether a civilization succeeds or fails. Currently, many people oppose globalization because traditional social mechanisms fail to create cooperation and social order in a world, which is increasingly characterized by non-local interactions. However, in this chapter I will show that there are some reputation- and merit-based social mechanisms, which can still stabilize a globalized world, if well designed.

Since the origin of human civilization, there has been a continuous struggle between chaos and order. While chaos may stimulate creativity and innovation, order is needed to coordinate the actions of many individuals easily and efficiently. In particular, our ability to produce "collective goods" is of critical importance. Examples are roads, cities, factories, universities, theaters, and public parks, but also language and culture.

According to Thomas Hobbes (1588–1679), in the initial state of nature, everyone fought against everybody else ("homo hominis lupus"). To overcome this destructive anarchy, he claims, a strong state would be needed to create social order. In fact, even today, civilization is highly vulnerable to disruption, as sudden outbreaks of civil wars illustrate, or the breakdown of social order, which is sometimes observed after natural disasters. But we are not only threatened by the outbreak of conflict. We are also suffering from serious problems when coordination or cooperation fails, as it often happens in "social dilemma" situations.[1]

---

[1] For coordination problems, in terms of individual payoffs it only matters that all decision-makers eventually agree on one behavioral option. In case of cooperation problems, coordination is not enough. Assuming that everyone coordinates on the same option, one of the options is more favorable than the other(s). However, there is a dilemma: deviating from the favorable solution is tempting, as this implies a higher individual payoff. Moreover, if everyone chooses the behavior that is individually most profitable relative to the other choices, the resulting outcome will be bad for everyone in absolute terms, due to a "race to the bottom".

© Springer Nature Switzerland AG 2021
D. Helbing, *Next Civilization*,
https://doi.org/10.1007/978-3-030-62330-2_8

8.1   The Challenge of Cooperation

To understand the nature of "social dilemmas", let us now discuss what it actually takes to produce "collective goods". For this, let us assume that you have to engage with others to create a benefit that you can't produce by yourself. In order to produce the collective good, we further assume that everyone invests a certain self-determined contribution that goes into a common "pot". However, the size of these contributions is often not known to the others (and might actually be zero). The common pot could be a fund or also a tax-based budget. If the overall investment reaches a sufficient size, a "synergy" is created whereby the resulting value is higher than the initial investment. To reflect this, the total initial investment is multiplied by a factor greater than one. Finally, let's assume for simplicity that all can equally benefit from the "collective good" (such as a public road) independently of how much they contributed (e.g. how much taxes they paid). Under these conditions it can happen that someone gets a benefit without contributing to the collective good, i.e. without making an investment.

In such social dilemma situations, creating benefits requires investments, but the outcome is uncertain. No individual can determine it alone. Specifically, if everyone invests enough, everyone will profit from the synergy. If you invest a lot while others invest little, you are likely to incur a loss. If you invest a little while others invest a lot, you may profit more than others. But if everyone invests only a little, the overall investment is too small to create a synergy effect and everyone will be worse off.

In such scenarios cooperation (contributing a lot) is risky, while free-riding (contributing little) is tempting. This discourages cooperative behavior. If the situation occurs many times, cooperation will be undermined and a so-called "tragedy of the commons" will result. Then, only a few will invest and nobody can benefit. Such situations are known, for example, from countries with high levels of corruption and tax evasion. Although cooperation would be beneficial for everyone, it breaks down in a similar fashion as free flowing traffic breaks down on a crowded circular track.[2] This happens because the desirable state of the system is unstable.

8.2   When Everyone Wants More, but Loses

"Tragedies of the commons" occur in many areas of life. Frequently cited examples include the degradation of the environment, the abuse of social welfare systems, and climate change. In fact, even though probably nobody wants to destroy our planet, we still overexploit its resources and pollute the Earth to an extent that causes existential threats.[3] As measured by the "environmental footprint", many humans consume more resources than are being (re-)created, which leads to a globally unsustainable system. Moreover, although nobody would want to wipe out a species of fish, we are facing

---

[2]See the movie at https://www.youtube.com/watch?v=7wm-pZp_mi0.

[3]For example, a safe way to store nuclear waste has still not been found.

a serious overfishing problem in many areas of the world. Last but not least, even though schools, hospitals, roads and other useful public services require government funding, tax evasion is a widespread problem, and public debts are growing.

Of course, the conventional economic wisdom is that contracts can enforce cooperation, when legal institutions work well. Nevertheless, neither governments nor companies have so far managed to fully overcome the longstanding problems mentioned above. So, what other options do we have? In an attempt to address this question, the following paragraphs provide a short (and far from exhaustive) overview of the mechanisms which can encourage and sustain cooperation. These mechanisms have played a vital role in human history, and some of them have been so essential that they became cornerstones of world religions. This demonstrates that they are considered to be even more powerful than man-made institutions. In fact, rather than viewing history as a series of wars and treaties, we should view it as a series of discoveries which reshaped the ways societies are organized.

8.3 Family Relations

"Genetic favoritism" was a very early mechanism to promote cooperation. The scientific theory of genetic favoritism elaborated by George R. Price[4] (1922–1975) posits that the genetically closer you are to others, the more it makes sense to favor them compared to strangers. The principle of genetic favoritism explains tribal structures and dynasties, which have been around for a long time. In many countries they still exist, and in some of them they come together with the custom of blood revenge. Today's inheritance law still favors relatives. However, genetic favoritism has a number of undesirable side effects such as unequal opportunities for non-relatives, ethnic conflict, or vendetta. Shakespeare's *Romeo and Juliette* dramatically conveys the rigidity of family structures in the past very well. The caste system in India provides another example.

8.4 Scared by Future "Revenge"

What other options do we have to promote cooperation? It is known, for example, that cooperation becomes more likely, if people interact with each other many times. In such a setting, you may adopt a so-called "tit for tat" strategy to teach someone else that uncooperative behavior won't pay off. The strategy is thousands of years old and was famously described in the Old Testament as "an eye for an eye, a tooth for a tooth" (Exodus 21:24).

Today, the underlying success principle is called "the shadow of the future". The effectiveness of revenge strategies was studied by Robert Axelrod (*1943) in a series

---

[4]Price [1].

of experimental computer tournaments published in 1981.[5] It turns out that, if people interact frequently enough, a cooperative behavior is more beneficial than an exploitative strategy. As a consequence, "direct reciprocity" will result, where both parties act according to the principle "if you help me, I'll help you". But what if the friendship becomes so close that others are disadvantaged? This could lead to corruption and barriers preventing other people from competing fairly. In an economic market, this could result in inefficiency and higher prices.

In addition, what should we do if we interact with someone only once, for example, on a one-off project such as the renovation of a flat? Are such interactions doomed to be uncooperative and inefficient? And what should we do if a social dilemma involves many players? In such cases, a tit-for-tat strategy is too simple because we don't know who cheated whom.

8.5   Costly Punishment

For such reasons, further social mechanisms have emerged, including "altruistic punishment". As Ernst Fehr and Simon Gächter have shown in 2002,[6] if players can punish others, this will promote cooperation, even if the punishment is costly. Indeed, if people can choose between a world without sanctions and a world where misbehavior can be punished, many decide for the second option, as evidenced by experimental research conducted in 2006 by Özgür Gürek, Bernd Irlenbusch and Bettina Rockenbach.[7] Punishment played a particularly important role in the beginning of the experiment. Later on, people behaved more cooperatively, and although punitive sanctions became rare, they were nevertheless important as a deterrent.

Note that punishment by "peers", as it was studied in these experiments, is a very widespread social mechanism. In particular, it is used to establish and maintain social norms, i.e. certain behavioral rules. Every one of us exercises peer punishment many times a day—perhaps in a mild way by raising eyebrows, or perhaps more assertively by criticizing others or engaging into conflict, which is often costly for both sides.

8.6   The Birth of Moral Behavior

But why do we punish others at all, if this is costly, while others benefit from the resulting cooperation? This scientific puzzle is known as "second-order free-rider dilemma". The term "free-rider" is used for someone who benefits from a collective good without contributing. "First-order free-riders" are people who don't cooperate, while "second-order free-riders" are people who don't punish the misdeeds of others.

---

[5] Axelrod and Hamilton [2].
[6] Fehr and Gächter [3].
[7] Gürerk et al. [4].

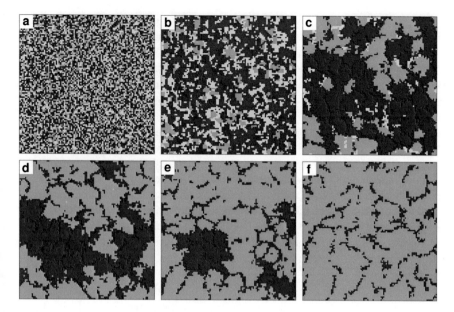

Fig. 8.1 Snapshots of a computer simulation showing the spread of moral behavior (green) in competition with cooperators who don't punish ("secondorder free-riders", in blue), with non-cooperative individuals ("defectors", in red), and with people who neither cooperate nor punish ("hypocritical individuals", in yellow) (Reprinted from Helbing et al. [6]. Creative Commons Attribution (CC BY) license.)

In a related study that I conducted with Attila Szolnoki, Matjaz Perc and György Szabo in 2010,[8] four types of behavior were distinguished: (1) "cooperators" who don't punish uncooperative agents[9] (second-order free-riders, illustrated in blue in Fig. 8.1); (2) cooperative agents who punish uncooperative ones, termed "moralists" (illustrated in green); (3) uncooperative agents ("defectors") who don't punish other uncooperative agents, i.e. first-order free-riders (red); and (4) uncooperative agents who punish other uncooperative agents, termed "immoralists" due to their hypocrisy (shown in yellow). Additionally, we assumed that individuals copy the most successful behavior of the agents they interact with.

When agents from the entire population are randomly matched for interactions, moralists are disadvantaged by the punishment costs as compared normal cooperators. In other words, moralists cannot compete with second-order free-riders, who abstain from sanctions. Consequently, moralists disappear and cooperators meet defectors. The result is a "tragedy of the commons". Altogether, individuals trying to act according to their rational self-interest create an outcome that is ultimately detrimental to everyone's interests.

---

[8]Helbing et al. [5].

[9]Again, we use the technical term "agent" in the sense of "actor" or decision-maker, here. It may stand for an individual, a company, an institution or another decision-making entity.

Surprisingly, however, when individuals interact with a small number of neighbors, clusters of people who adopt the same kind of behavior emerge (see Fig. 8.1). The fact that "birds of a feather flock together"[10] makes a big difference. It allows moralists to thrive! While cooperators still make lower profits than defectors, moralists, who cooperate and punish uncooperative behavior of others, can now succeed. This is due to the fact that the different types of behavior are separated from each other in space. In particular, moralists (shown in green in Fig. 8.1) tend to be separated from cooperators (shown in blue) by defectors (shown in red). At the boundaries of their respective clusters, both of them have to compete with non-cooperative agents (yellow or red). While cooperative agents are exploited by non-cooperative ones, moralists sanction non-cooperative behavior and can, therefore, succeed and spread.

This is amazing! While random interactions cause a "tragedy of the commons" where nobody cooperates and nobody makes a profit, we find just the opposite result if people interact with their neighbors in space or in a social network, where the same behaviors tend to cluster together. Therefore, a small change in the way interaction partners are chosen can create conditions under which moral behavior can prevail over free-riding.

8.7   Containing Crime

Nevertheless, punishing each other is often annoying or inefficient. This is one of the reasons why we may prefer to share the costs of punishment, by jointly investing in sanctioning institutions such as a police force or court. This approach is known as "pool punishment". A problem with this strategy, however, is that police, courts and other sanctioning institutions may be corrupted. In addition, it is sometimes far from clear who deserves to be punished and who is innocent.[11] For the legitimacy and acceptance of such institutions, it is important that innocent people are not punished in error. This means that extensive inspections and investigations are required, which can be costly compared to the number of people who are finally convicted.

In 2013, Karsten Donnay, Matjaz Perc[12] and I developed a computer model to simulate the spread and fighting of criminal behavior. We were astonished to find that greater deterrence in the form of more surveillance and higher punishment would not eliminate crime.[13] People don't just make a calculation whether crime pays off or not, as is often assumed. Instead, crime is kind of "infectious", but punishment and

---

[10] A phenomenon that sociologists call "homophily".

[11] US death row study: 4% of defendants sentenced to die are innocent, see http://www.thegua rdian.com/world/2014/apr/28/death-penalty-study-4-percent-defendants-innocent; also see Elite FBI forensic unit gave flawed testimony, see http://www.cbsnews.com/news/fbi-hair-analysis-err ors-led-to-convictions-new-report-finds/.

[12] Perc et al. [7].

[13] In fact, crime rates are not lower, although the prison population in the USA is almost 10 times higher than in Europe, see the list of countries by incarceration rate, http://en.wikipedia.org/wiki/ List_of_countries_by_incarceration_rate#Incarceration_rates.

normal behavior are contagious, too. This can explain the surprising crime cycles that have often been observed in the past. As a consequence, a crime prevention strategy based on alleviating socio-economic deprivation might be much more effective than one based on deterrence.

8.8 Group Selection

Things become trickier if we have several groups with different preferences, due to different cultural backgrounds or education, for example.[14] Then, a competitive dynamics between different groups may set in. The subject is often referred to as "group selection", which was put forward by Vero Copner Wynne-Edwards (1906–1997) and others.[15]

It is often believed that group selection promotes cooperation. Compare two groups with high and no cooperation. Then, the cooperative group will get higher payoffs and grow more quickly than the non-cooperative group. Consequently, cooperation should spread and free-riding should eventually disappear. But what would happen if there were an exchange of people between the groups? Then, free-riders could exploit the cooperative group and quickly undermine the cooperation within it. For such reasons, people often fear that migration would undermine cooperation. But it doesn't have to be like this.

8.9 The Surprising Role of Success-Driven Migration

To study the effect of migration on cooperation, back in 2008/09, Wenjian Yu and I developed a related agent-based computer model.[16] We made the following assumptions: (1) Computer-simulated individuals (called "agents") move to the most favorable location within a certain radius of their current location ("success-driven migration"). (2) They tend to imitate the behavior of the most successful individuals they interact with (their "neighbors"). (3) There is some degree of "trial-and-error behavior", i.e. a certain probability that each individual will either migrate to a free location, which isn't occupied already, or change the behavior (from cooperative to noncooperative or vice versa). While rule (1) does not change the number of people who cooperate, the other two rules tend to undermine significant levels of cooperation. Surprisingly, however, when all three rules are applied together, a high level of cooperation is eventually achieved. This proved to be true even in cases where the simulation began with no cooperative agents at all, a situation akin to the "state of

---

[14]Winter et al. [8].

[15]Wynne-Edwards [9].

[16]Helbing and Yu [10], Helbing and Yu [11].

nature" assumed by Thomas Hobbes. So, how is it possible that cooperation flourishes in our computer simulation, in spite of the fact that there is no powerful state (a "Leviathan") to ensure cooperation from the top-down?

The answer relates to the unexpected way success-driven migration affects the competition between cooperative and uncooperative agents. Even though we didn't initially expect that migration would influence cooperation a lot, we put the assumption to the test because of my interest in human mobility. In our computer simulations, migration first spread people out geographically. This isn't good for cooperation at all, but as individuals occasionally change their strategies due to trial-and-error behavior, cooperative behaviors may sometimes occur for a short time. Such accidental cooperation happens in random places. However, after a sufficiently long time, some of these cooperative events happen to be located next to each other. In case of such a coincidence, i.e. when a big enough cluster of cooperative behaviors occurs, cooperation suddenly becomes a successful strategy, and neighbors begin to imitate this behavior. As a result, cooperation quickly spreads throughout the entire system.

How does this mechanism work? To avoid losses, cooperative individuals obviously move away from uncooperative agents, but they engage repeatedly in interactions with cooperative neighbors. In other words, cooperative behavior is sustainable over several interactions, while non-cooperative behavior is successful only for a short time, as exploited neighbors will move away. This leads to the formation of cooperative clusters, with a few uncooperative individuals at the boundaries. Thus, the individual behavior and spatial organization evolve in tandem, and the behavior of an individual is determined by the behavior in his/her surrounding, the "social milieu".

In conclusion, when people are allowed to move around freely and to live in the place they prefer, this can promote cooperation.[17] Thus, migration is not a problem, if there are opportunities for integration. But integration is not just the responsibility of migrants—it requires an effort from both sides, the migrants and the host society.

Although it has been a constant feature of human history, migration is not always welcomed. There are understandable reasons for this weariness; many countries are struggling to manage migration and integration successfully. However, the USA, known as the cultural "melting pot", is a good example of the positive potential of migration. This success is based on a tradition, where it is relatively easy to interact with strangers.

Another positive example is provided by the Italian village of Riace.[18] Although picturesque, the village is located in one of Italy's poorest regions. The service sector gradually disappeared as young people moved away to other places. But one day, a boat with migrants stranded off the coast of Italy. The mayor viewed this as a sign from God, and decided to use it as an opportunity for his village. Indeed, something of a miracle occurred. The arrival of the migrants revived the ailing village, and the gradual decay was reversed with the migrants. Crucially, migrants were not treated

---

[17]Roca and Helbing [12], Efferson et al. [13].

[18]The tiny Italian village that opened its doors to migrants who braved the sea, see http://www.theguardian.com/world/2013/oct/12/italian-village-migrants-sea.

as foreigners, but as an integral part of the community, which fostered a relationship of mutual trust.

8.10 Common Pool Resource Management

Above, I have described many simple social mechanisms that have proved their effectiveness and efficiency in mathematical models, computer simulations, laboratory or real-life settings. But what about more complex socio-economic systems? Would suitable social mechanisms be able to create a desirable and efficient self-organization of entire societies? Elinor Ostrom (1933–2012), in fact, performed extensive field studies addressing this question, and in 2009, she won the Nobel Prize for her work.

Conventional economic theory posits that public ("common-pool") resources can't be efficiently managed, and therefore should be privatized. Elinor Ostrom, however, discovered that this is wrong. She studied the way in which common-pool resources (CPR) were managed in Switzerland and elsewhere, and found that self-governance is efficient and leads to cooperation, if the interaction rules are appropriate. A successful set of rules is specified below:[19]

1. The boundaries between in- and out-groups must be clearly defined (to effectively exclude external, un-entitled parties).
2. The possession and supply of common-pool resources must be governed according to rules that are tailored to the specific local conditions.
3. A collective decision-making process is required that involves the agents who are affected by the use of the respective resources.
4. The supply and use of common-pool resources must be monitored by the people who manage the common-pool resources or people who are accountable to them.
5. Sanctions of proportionate severity must be imposed on people who use common-pool resources in violation of the community rules.
6. Conflict resolution procedures must be established which are cheap and easy to access.
7. The self-governance of the community must be recognized by the higher-level authorities.
8. In case of larger systems, they can be organized into multiple layers of "nested enterprises", i.e. in a modular way, with smaller local CPRs at the base level.

Note, however, that collective goods can be created even under less restrictive conditions.[20] This amazing fact can, for example, be observed in the communities

---

[19]See http://en.wikipedia.org/wiki/Elinor_Ostrom and also Ostrom et al. [14].
[20]Vinko et al. [15].

of volunteers who have created *Linux*, *Wikipedia*, *OpenStreetMap*, *Stack Overflow* or *Zooniverse*.

8.11   Why and How Globalization Undermines Cooperation and Social Order

The central message of this chapter is that self-organization based on local interactions is at the heart of all human societies. Since the genesis of ancient societies right up until today, a great deal of social order emerges from the bottom-up. To achieve cooperation, one just needs suitable interaction mechanisms ("rules of the game"), including mechanisms to reach compliance with the rules. This approach tends to be flexible, adaptive, resilient, effective and efficient.

Most of the previously discussed mechanisms enable self-organized cooperation in a bottom-up way by encouraging people to engage in social interactions with others who exhibit similar behavior—"birds of a feather flock together". The crucial question is whether these mechanisms will also work in a globalized world? These days, many people and governments feel that globalization has undermined social and economic stability. As I will demonstrate in the following, this is actually true, and the underlying reasons are quite surprising.

Globalization means that an increasing variety of people and corporations interact with each other, often in a more or less anonymous or "random" way. This leads to a surprising problem, which is illustrated by a video produced in my team.[21] The video illustrates a circle of "agents" (such as individuals or companies), each of whom is creating collective goods with agents in their vicinity. These local interactions initially foster a high level of cooperation for reasons that we have discussed above. When further interactions with randomly chosen agents are added, the level of cooperation first increases. Therefore, it seems natural to add more interaction links. However, there is an optimal level of connectivity, beyond which cooperation starts to decay.[22] When everyone interacts with everyone else, finally, cooperation even stops completely. In other words, when too many people interact with each other, a "tragedy of the commons" results, and everyone suffers from the lack of cooperation. Therefore, the way we are globalizing the world today may lead to the erosion of social order and economic stability. This conclusion is also supported by the Chief Economist at the Bank of England, Andrew Haldane (*1967), who suggests that the financial meltdown in 2008 resulted from a hyperconnected banking network.[23]

---

[21] This video is accessible at http://www.youtube.com/watch?v=TVExTATGvP0.

[22] Helbing [16].

[23] Haldane and May [17].

8.12 Age of Coercion or Age of Reputation?

In an attempt to stabilize our socio-economic system, many governments have tried to (re-)establish social order from the top-down, by increasing surveillance and investing into armed police. However, this control approach is destined to fail due to the high level of systemic complexity, as I pointed out before. In fact, we have seen a lot of evidence of this failure[24]. The signs of economic, social and political instability are all around us. Therefore, it is conceivable that our globalized society might disintegrate and break into pieces, thereby creating a decentralized system.

Is there a way to avoid such a fragmentation scenario? Can globalization be realized in such a way that cooperation and social order remain stable? Given that it is possible to stabilize the traffic flow using traffic assistance systems, can we also build an assistance system for cooperation? In fact, there are a number of new possibilities, which I will discuss now.

8.13 Costly, Trustworthy Signals

Let us dive a bit more into the subject of group competition and assume that different groups apply *different* social mechanisms. To study this setting, Michael Mäs and I have recently made a laboratory experiment, in which we had two teams of players, each of which collected money in a separate "pot". In their own team, the players faced a collective goods problem as outlined above. However, their overall investments into the team's pot as compared to the investments of the other team determined their chance of winning the team competition. If the investments in the first team were twice as high as the investments in the second team, the first team had a chance of two thirds to win the "jackpot", here: the sum of both pots (see Fig. 8.2).

We gained a number of remarkable insights. First, the competitive setting led to higher investments in both teams. Second, it was interesting to see how well different mechanisms perform when competing with others. In our experiment, "costly punishment" was compared with no mechanism supporting cooperation ("no institution") and with three kinds of "signaling", which allowed players to indicate beforehand that they were willing to invest into their team's pot. "Free signaling" did not require one to pay anything to send a signal to the own team; "costly signaling" required one to pay a fee; "altruistic signaling" was costly, but the fee was paid to the other team members and split between them. In terms of the probability of winning, "costly punishment" beat all other options and "altruistic signaling" was the second-best option. "No institution" was more effective than "free signaling", which was again more successful than "costly signaling".

However, when considering punishment and signaling costs, costly punishment was actually not the most efficient mechanism. The winner was altruistic signaling!

---

[24]For example, in Iraq, Ferguson, Baltimore etc.

| A unit enters the lane | It decides to exit the lane | It sends a request for a transfer car | The unit exits the lane |

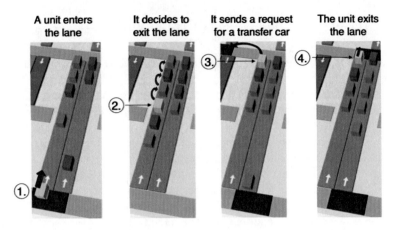

Fig. 8.2 Illustration of the competitive setting of a laboratory experiment we performed (Reproduction from Mäs and Helbing [18].)

Therefore, signaling allows people to coordinate themselves and to cooperate efficiently, if signaling has a price. Free signaling, in contrast, is not a credible way of communicating someone's intentions, because cheap talk invites people to lie. In summary, there are better social mechanisms than punishing non-cooperative behavior. "Altruistic signaling" is one of them.

8.14 Building on Reputation

There is a further important interaction mechanism to support cooperation, which we haven't discussed yet. The underlying principle is to help others, as described, for example, in the parable of the good Samaritan (New Testament, Luke 10:29–37). The commandment "Love your neighbor as yourself" (Mark 12:31) points in the same direction. Many world religions and philosophies support similar principles of action.[25] In fact, if all people behaved in an other-regarding way, others would help them, too. Based on this "indirect reciprocity", a high level of cooperation would result, and everyone could have a better life. But there is a problem: if some agents comply with this rule and other don't, friendliness can be exploited.

Nevertheless, there is a solution: reputation mechanisms allow people and companies with compatible preferences and behaviors to find each other. They help "birds of a feather" to flock together on a global scale and to avoid bad experiences.

The value of reputation and recommendation systems is evidenced by the fact that they are quickly spreading throughout the Web. People seem to relish the opportunity to rate products, news and comments. But why do they bother? In fact, they get

[25]This applies, for example, to Immanuel Kant's Categorical Imperative, which basically demands to do what you expect others to do, see https://en.wikipedia.org/wiki/Categorical_imperative.

useful recommendations in exchange, as we know from platforms such as *Amazon*, *eBay*, *TripAdvisor* and many others. Wojtek Przepiorka has found that such recommendations are beneficial not only for users, who tend to get a better service, but also for companies.[26] A better reputation allows them to sell products or services at a higher price. Many hotels, for example, use their average score on *TripAdvisor* as a key selling point.

8.15 A Healthy Information Ecosystem by Pluralistic Social Filtering

How should reputation systems be designed? It is certainly not good enough to leave it to a company to decide, how we see the world and what recommendations we get. This promotes manipulation and undermines the "wisdom of the crowds", resulting in bad outcomes.[27] It is important, therefore, that recommendation systems do not reduce social diversity. Moreover, we should be able to look at the world from our own perspective, based on our own values and quality criteria. Otherwise, we may end up trapped in what Eli Pariser (*1980) calls the "filter bubble".[28] In such a scenario, we may lose our freedom of decision-making and our ability to communicate with others who have different points of view. In fact, some people believe that this is already the case and one of the reasons why political compromise between Republicans and Democrats in the US has become so difficult.[29] As a consequence, conservatives and liberals in the US consume different media, interact with different people, and increasingly use different concepts and different words to talk about the same subjects. In a sense, they are living in different, largely separated worlds.

Clearly, today's reputation systems are not good enough. They would have to become more pluralistic. For this, users should be able to assess not just the *overall* quality, which is typically quantified on a simple five-point scale or even in a thumbs-up-or-down system. The reputation systems of the future should include different facets of quality such as physical, chemical, biological, environmental, economic, technological and social qualities. These characteristics could be quantified using metrics such as popularity, level of controversy, durability, sustainability, and social factors.

It would be even more important that users can choose among diverse information filters, and that they can generate, share and modify them. I call this approach "pluralistic social filtering".[30] In fact, we could have different filters to recommend

---

[26]Przepiorka [19].

[27]Lorenz et al. [20].

[28]Whereby we are fed a meager informational diet based on a small subset of the Internet fitting our tastes (and perhaps also of those who do the filtering), see Pariser [21].

[29]Andris et al. [22].

[30]Such a system has been implemented, for example, in the Virtual Journal, see https://web.archive.org/web/20150910001019/http://vijo.inn.ac/.

us the latest news, the most controversial stories, the news that our friends are interested in, or a surprise filter. Then, we could choose among a set of filters that we find useful. To assess credibility and relevance, these filters should ignore sources of information that we regard as unreliable (e.g. anonymous ratings) and put a focus on information sources that we trust (e.g. the opinions of friends or family members). For this purpose, users should also be able to rate and comment on products, companies, news, information, and sources of information. If this were the case, spammers would quickly lose their reputation and their influence would wane.

In concert, all personalized information filters together would establish an "information ecosystem", where filters would evolve by modification and selection. This would steadily enhance our ability to find meaningful information. As a result, the personalized reputational value of each company and their products would give a differentiated picture, which additionally would help firms to customize their products. Hence, reputation systems can have advantages for both, consumers and businesses, which creates a win-win situation.

In summary, while societies all over the world are still buckling under the pressure of globalization, a globalized world does not need to produce social and economic instability. By creating global, pluralistic reputation systems, we can build a well-functioning "global village". This is achieved by matching people or companies with others who share compatible interests. However, it will be crucial to design reputation systems in such a way that they allow privacy and innovation to thrive. Furthermore, future reputation systems should be resistant to manipulation. Appendix 7.1 provides some ideas on how this may be achieved.

8.16   Merit-Based Matching: Who Pays More Earns More

Note that most cooperation-enhancing mechanisms, including special kinds of contracts between interaction partners, modify interactions in a way that turn a social dilemma situation into something else. Such a transformation is also often reached by taxing undesirable behaviors or by offering tax reductions for favorable behaviors. In other words, the payoff structure is changed in a way that makes cooperation a natural and stable outcome.[31]

However, there is also another mechanism to promote cooperation, which is quite stunning. This mechanism, which I proposed together with Sergi Lozano,[32] introduces an additional feedback loop in the social system, whereby more complex dynamic interdependencies and additional solutions are produced. When Heinrich Nax joined my team, we elaborated an example implementation of this mechanism, which is based on "merit-based matching".[33] Accordingly, agents who have contributed much to a collective good are matched with other high-level contributors,

---

[31] Helbing and Johansson [23].

[32] Helbing and Lozano [24].

[33] Nax et al. [25].

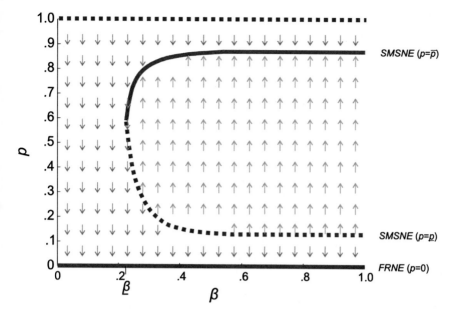

Fig. 8.3 Illustration of the probability p of cooperation as a function of how perfectly people are matched in a merit-based way. Merit-based matching produces a trend towards a higher level of cooperation (see green upward arrows). (Adapted from Nax et al. [25]. Reproduction with kind permission of Heinrich Nax.)

and those who have contributed little are matched with similar agents. The exciting result of this matching mechanism is: who pays more earns more! Furthermore, over time this effect encourages others to invest more as well, such that a trend towards higher investments and more cooperation is reached (see Fig. 8.3). Therefore, merit-based matching in social dilemma situations can turn the downward spiral, which would lead to a "tragedy of the commons", into an upward spiral.

This amazing prediction is even confirmed by decision experiments that Stefano Balietti, Heinrich Nax, Ryan Murphy and I performed.[34] In these experiments, we made a number of unexpected discoveries. For example, in social dilemma situations merit-based matching can reduce the level of inequality. Therefore, this mechanism can even overcome the classical dilemma, according to which economic efficiency implies economic inequality and vice versa.

8.17 Social Technologies

The above discussion has illustrated that cooperation usually doesn't fail due to bad intentions, but because it is unstable in social dilemma situations. Moreover,

---

[34]Nax et al. [26].

we have learned that punishment isn't the best mechanism to stabilize cooperation. In multi-cultural settings, it rather tends to create conflict, i.e. the breakdown of social order. But most of the time a suitable coordination mechanism is sufficient to stabilize cooperation. In the past, such coordination mechanisms haven't always been available to everyone, but information systems can now change this. Altruistic signaling, reputation systems and merit-based matching are three promising new ways to foster cooperation, which are based on information systems rather than punishment. However, we could build much more sophisticated digital assistants to create social order and further benefits.

As we have seen before, many problems occur when people or companies don't care about the external effects of their decisions and actions on others (i.e. so-called "externalities"). As "tragedies of the commons" or wars demonstrate, this can be harmful for everyone. So, how can we encourage more responsible behavior and create sustainable systems? The classical approach would be to devise, implement and enforce new legal regulations. But people don't like to be ruled, and our labyrinthine legal systems already leads to confusion, inefficiency, loopholes, and high costs. As a consequence, laws are often ineffective. However, Social Technologies offer a new opportunity to create a better world, by supporting favorable local interactions and self-organization. It's actually easier than you might think, and it's completely different from nudging, which tricks people.

8.18   Digital Assistants

When two people or companies interact, there are just four possible outcomes, which relate to coordination failures and conflicts of interest. The first possibility is that of a lose-lose situation. In such cases, it is best to avoid the interaction altogether. For this, we need information technologies that make us aware of expected negative side effects associated with an interaction: if we know the social and environmental implications of our interactions, we can make better decisions. Therefore, measuring the externalities of our actions is important in order to avoid unforeseen damage. In fact, if we had to pay for negative externalities of our decisions and actions and if we would earn on positive externalities, this would help us to reduce the divergence between private and collective interests. As a result, self-interested decisions of individuals and companies would be less likely to cause economic or environmental damage.

The second possibility is a bad win-lose situation. In such cases, one party is interested in the interaction, but the other side would like to avoid it, and the overall effect would be negative. In such situations, increasing awareness may help, but social mechanisms to protect people (or firms) from exploitation are necessary, too. I will address this in detail in Appendix 7.2.

The third possibility is a good win-lose situation. While the interaction would benefit one party at the expense of another, there would be an overall systemic benefit from the interaction nevertheless. Obviously, in this scenario one party would like to engage, while the other would like to avoid the interaction. It is possible, however, to turn such win-lose situations into win-win situations by compensating the otherwise losing party. For this, a value exchange system is needed. Then, the interaction can be made mutually beneficial and attractive.

Finally, the fourth possibility is a win-win situation. Although both parties profit, one may decide to divide the overall benefit in a fair way. Again, this requires a value exchange system. In addition, cooperation can be fostered by making people aware of opportunities they might otherwise miss. In fact, every day we walk past hundreds of people who share interests with us while we don't know about it. In this respect, digital assistants can help us take advantage of opportunities that would otherwise be missed, thereby unleashing an unimaginable socio-economic potential. If we had suitable information systems to assist us, we could easily engage with people of diverse cultural backgrounds and interests in ways that create opportunities and avoid conflict.

The above-described Social Technologies can now be built. Smartphones are already becoming digital assistants to manage our lives. They help us to find products, nice restaurants, travel routes and people with similar interests. They also enable a real-time translation from one language to another. In future, such digital assistants will pay more attention to the interactions between people and companies, producing mutual benefits for all involved. This will play an important role in overcoming cultural barriers and in minimizing environmental damage, too.

Eventually, Social Technologies will help us to avoid bad interactions, to discover and take advantage of good opportunities, and to transform potentially negative interactions into mutually beneficial cooperation. In this way, the "interoperability" between diverse systems and interests is largely increased, while coordination failures and conflicts are considerably reduced. I am, therefore, convinced that Social Technologies can produce enormous value, both material and immaterial. Some social media platforms are now worth billions of dollars. How much more value could digital assistants and other Social Technologies create?

8.19 Appendix 1: Creating a Trend for the Better

For reputation systems to work well, there are a number of things to consider: (1) the reputation system must be resistant to manipulation; (2) people shouldn't be subject to intimidation or unsubstantiated allegations; (3) to enable individuality, innovation and exploration, the "global village" shouldn't be organized like a rural village in which everyone knows everyone else and nobody wishes to stand out. We will need a good balance between accountability and anonymity, which ensures the stability of the social system.

Therefore, people should be able to post ratings or comments either anonymously, pseudonymously or using their real-world identity. However, posts using real-world identities might be weighted 10 times higher than pseudonymous ones, and pseudonymous posts might be weighted 10 times higher than anonymous ones. In addition, everyone who posts something should have to declare the *category* of the information. Such categories could, for example, be: facts, opinions and advertisements. Facts would need to be potentially falsifiable and corroborated by publically accessible evidence. In contrast, a subjective assessment of people, products or events would be considered to be an opinion. Finally, advertisements would need to include an acknowledgement of potential personal or third-party benefits. Ratings would always be categorized as opinions. If people would use the wrong categories or post false information (as identified and reported by, say, 10 others), their online reputation would be reduced by a certain amount (say, by a factor of 2). Mechanisms such as these would help to ensure that manipulation or cheating wouldn't pay off.

Furthermore, one should ensure the right of informational self-determination, such that individuals are able to determine the way their personal information (such as social, economic, health, or private information) is used. This could be realized with a Personal Data Store.[35] In particular, a person should be able to stipulate the purpose and period of time for which their information is accessible and to whom (everyone, non-profit organizations, commercial companies, friends, family members or particular individuals, for example). These settings would enable a select group of individuals, institutions, or corporations to access certain personal information, as determined by the users. Of course, one might also decide not to reveal any personal information at all. However, I expect that having a reputation for something would be better for most people than having no digital image, if only to find people who share similar preferences and tastes.

8.20   Appendix 2: Towards Distributed Security, Based on Self-Organization

Let me finally address the question whether bottom-up self-organization is dangerous for society? In fact, since the Arab Spring, governments all over the world are afraid of "twitter revolutions". Therefore, are social media destabilizing political systems? Do governments need to censor free speech or control the algorithms that spread messages through social media platforms? Probably not. First, the Arab Spring was triggered by high food prices,[36] i.e. deprivation, and not by anarchism. Second, encroaching on free speech would obstruct our society's ability to innovate and to detect and address problems early on.

But how to reach a high level of security in a system which is based on the principle of distributed bottom-up self-organization? Let me give an example. One

---

[35] World Economic Forum, Personal Data: The Emergence of a New Asset Class, see http://www. weforum.org/reports/personal-data-emergence-new-asset-class and also de Montjoye et al. [27].

[36] See, for example, https://www.voanews.com/middle-east/2011-food-price-spikes-helped-tri gger-arab-spring-researchers-say, http://www.pbs.org/newshour/updates/world-july-dec11-food_0 9-07/, https://www.vice.com/en/article/3dd3py/high-food-prices-are-fueling-egypts-riotsand-those-in-brazil-turkey-and-syria.

of the most astonishing complex systems in the world is our body's immune system. Even though we are bombarded every day by a myriad of viruses, bacteria and other harmful agents, our immune system protects us pretty well for about 50–100 years. Our immune system is probably more effective than any other protection system we know. What is even more surprising is that, in contrast to our central nervous system, the immune system is organized in a decentralized way. This is no coincidence. It is well known that decentralized systems tend to be more resilient to disruptive events. While targeted attacks or local disruptions can make a centralized system fail, a decentralized system will usually survive such disruptions and recover. In fact, this is the reason why the Internet is so robust.[37] So why don't we protect information systems using in-built "digital immune systems"?[38] This should also entail a reputation system, which could serve as a kind of "social immune system". In the following, I will describe just a few aspects of how this might work.

8.20.1 Community-Based Moderation

Information exchange and communication on the Web have quickly changed. In the beginning, there was almost no regulation in place. These were the days of the "Wild Wild Web", and people often did not respect human dignity and the rights of companies when posting comments. However, one can see a gradual evolution of self-governance mechanisms in open and participatory systems over time.

Early on, public comments in news forums were published without moderation. This led to a lot of low-quality content. Later, comments were increasingly checked for their legality (and for their respect of human dignity) before they went online. Then, it became possible to comment on comments. Now, comments are rated by readers, and good ones get pushed to the top. The next logical step is to rate commenters generally[39] and rate the quality of judgments of those who rate others. Thus, we can see the gradual evolution of a self-governing system that constructively channels free speech. Therefore, I believe that it is possible to encourage responsible use of the Internet, mainly through self-organization.

The great majority of malicious behavior can probably be controlled using crowd-based mechanisms. Such approaches include reporting inappropriate content and ranking user-generated content based on suitable reputation mechanisms. To handle the remaining, complicated cases, one can use a system of community moderators and complaints procedures. Such community moderators would be determined based on their performance in satisfying lower-level community expectations (the "local culture"), while staying within the framework set by higher-level principles (such as laws and constitutional principles). In this way, community moderators would complement our legal framework, and most problems could be solved in a

---

[37]Albert et al. [28].

[38]See, for example, Harmer et al. [29].

[39]Such ratings are sometimes called "karma points".

community-based way. Therefore, only a few cases will require legal mediation. Most activities would be self-governed through a system of sanctions and rewards by peers. In the following chapters, I will explain in more detail how information technology will enable people and companies to coordinate their interests in entirely new ways.

References

1. G.R. Price, Selection and covariance, Nature 227, 520–521 (1970).
2. R. Axelrod and W.D. Hamilton, The evolution of cooperation, Science 211, 1390–1396 (1981).
3. E. Fehr and S. Gächter, Altruistic punishment in humans. Nature 415, 137–140 (2002).
4. Ö. Gürerk, B. Irlenbusch and B. Rockenbach, The competitive advantage of sanctioning institutions. Science 312, 108–111 (2006).
5. D. Helbing, A. Szolnoki, M. Perc, and G. Szabó (2010) Evolutionary establishment of moral and double moral standards through spatial interactions. PLoS Computational Biology 6(4), e1000758.
6. D. Helbing, A. Szolnoki, M. Perc and G. Szabo (2010) Evolutionary establishment of moral and double moral standards through spatial interactions, PLoS Comput. Biol. 6(4): e1000758.
7. M. Perc, K. Donnay and D. Helbing (2013) Understanding recurrent crime as system-immanent collective behavior. PLOS ONE 8(10), e76063.
8. F. Winter, H. Rauhut and D. Helbing (2012) How norms can generate conflict: an experiment on the failure of cooperative micro-motives on the macro-level. Social Forces 90(3),919946.
9. V.C. Wynne-Edwards, Evolution Through Group Selection (Blackwell, 1986).
10. D. Helbing and W. Yu (2009) The outbreak of cooperation among success-driven individuals under noisy conditions. Proceedings of the National Academy of Sciences USA (PNAS) 106(8), 3680–3685.
11. D. Helbing and W. Yu (2008) Migration as a mechanism to promote cooperation. Advances in Complex Systems 11(4), 641–652.
12. C. P. Roca and D. Helbing (2011) Emergence of social cohesion in a model society of greedy, mobile individuals. PNAS 108(28), 11370–11374.
13. C. Efferson, C.P. Roca, S. Vogt and D. Helbing, Sustained cooperation by running away from bad behavior. Evolution and Human Behavior, see http://www.sciencedirect.com/science/art icle/pii/S1090513815000574.
14. E. Ostrom, R. Gardner and J. Walker, Rules, Games and Common-Pool Resources (Michigan, 1994).
15. T. Vinko et al., Sharing in BitTorrent communities, see http://www.researchgate.net/profile/ Stefano_Balietti/publication/270615664_QLectives_Ebook_Quality_Collectives/links/54b 975b90cf253b50e2a807e.pdf.
16. D. Helbing (2013): Globally networked risks and how to respond. Nature 497, 51–59.
17. A.G. Haldane and R.M. May, Systemic risk in banking ecosystems. Nature 469, 351–355 (2011).
18. M. Mäs and D. Helbing (2015) The competitive advantage of wasteful cultural institutions, preprint, see also https://www.rug.nl/research/portal/nl/publications/the-competitive-adv antage-of-wasteful-cultural-institutions(5d62bd50-21a4-46e0-8ccf-5f5126a2731f).html.
19. W. Przepiorka, Buyers pay for and sellers invest in a good reputation: More evidence from eBay, The Journal of Socio-Economics 42, 31–42 (2013).
20. J. Lorenz, H. Rauhut, F. Schweitzer, and D. Helbing (2011) How social influence can undermine the wisdom of crowd effect. Proceedings of the National Academy of Sciences USA (PNAS) 108(28), 9020–9025.
21. E. Pariser, The Filter Bubble: What the Internet Is Hiding from You (Viking/Penguin, 2011).

22. C. Andris et al., The rise of partisanship and super-cooperators in the U.S. House of Representatives, PLoS ONE 10(4): e0123507.
23. D. Helbing and A. Johansson (2010) Cooperation, norms, and revolutions: A unified game-theoretical approach. PLoS ONE 5(10), e12530.
24. D. Helbing and S. Lozano (2010) Phase transitions to cooperation in the prisoner's dilemma. Physical Review E 81(5), 057102.
25. H.H. Nax, R.O. Murphy, and D. Helbing (2014) Stability and welfare of 'merit-based' group-matching mechanisms in voluntary contribution games, see http://papers.ssrn.com/sol3/papers.cfm?abstract_id=2404280.
26. H.H. Nax, S. Balietti, R.O. Murphy, and D. Helbing (2018) Adding noise to the institution: An experimental welfare investigation of the contribution-based grouping mechanism. Social Choice and Welfare 50(2), 213–245. Preprint available at http://papers.ssrn.com/sol3/papers.cfm?abstract_id=2604140.
27. Y.-A. de Montjoye, E. Shmueli, S.S. Wang, and A.S. Pentland, openPDS: Protecting the privacy of metadata through SafeAnswers, PLoS ONE 9(7): e98790, http://dspace.mit.edu/bitstream/handle/1721.1/88264/Pentland_openPDS%20Protecting.pdf?sequence=1.
28. R. Albert, H. Jeong, and A.-L. Barabasi, Error and attack tolerance of complex networks. Nature 406, 378–382 (2000).
29. P.K. Harmer et al., An Artificial Immune System architecture for computer security applications. IEEE Transactions on Evolutionary Computation 6(3), 252–280 (2002).

Chapter 9
Networked Minds

Where Human Evolution Is Heading

*How ever selfish man may be supposed, there are evidently some
principles in his nature, which interest him in the fortune of
others, and render their happiness necessary to him, though he
derives nothing from it.*
—Adam Smith

Having studied the technological and social forces shaping our societies, we now turn to
the evolutionary forces. Among the millions of species on earth, humans are truly unique.
What is the recipe for our success? What makes us special? How do we make decisions?
How will we evolve? I argue that the particular combination of social abilities, diversity and
our hunger for information drives societies and their cultural evolution at a speed, which
is a thousand times faster than biological evolution! In fact, humans are curious by nature
– we are a social, information-driven species. And that is why the explosion in the data
volume and processing capacity will transform our societies more fundamentally than any
other technology in the past.

People have been amazed by the power of many minds at least since the "wisdom
of the crowd" was first discovered and demonstrated. The "wisdom of the crowd"
means a phenomenon whereby the average of many independent judgments tends
to be superior to expert judgments. A frequently cited example first reported by
Sir Francis Galton (1822–1911) is the estimation of the weight of an ox.[1] Galton
observed villagers trying to estimate the weight of an ox at a country fair, and noted
that, although no one villager guessed correctly, the average of everyone's guesses
was extremely close to the true weight. Today, there are many more examples demon-
strating the "wisdom of crowds", and this is often considered to underpin the success
of democracies.[2]

Of course, an argument can also be made for the "madness of crowds". In fact,
at the time of Gustave Le Bon (1841–1931), the attention focused on undesirable
aspects of mass psychology as exhibited by a rioting mob, for example. This was
seen to be the result of a dangerous emotional contagion, and governments to this
day tend to feel uneasy about crowds. However, there is a way to immunize people to
some extent against emotional contagion: good education. This is important, because
crowds can have many good sides, too.

---

[1] Galton [1].
[2] Surowiecki [2].

© Springer Nature Switzerland AG 2021
D. Helbing, *Next Civilization*,
https://doi.org/10.1007/978-3-030-62330-2_9

Today, we have a much more nuanced understanding of crowds and swarm intelligence.[3] We can explain why crowds are sometimes benign, but cause trouble in other circumstances.[4] For the wisdom of crowds to work, first, *independent* information gathering is crucial. Second, opinion formation should be *decentralized*, drawing on local knowledge and specialization. Third, a larger *diversity* of opinions tends to increase the quality of the outcome. The same applies to the diversity of in communication patterns.[5] Too much communication, however, can make the group as a whole less intelligent.

In fact, if people collect information and evaluate it independently from each other, and if these diverse streams of information are suitably aggregated afterwards, this often creates better results than even the best experts can produce. This is also more or less the way in which prediction markets work. These markets have been surprisingly successful in anticipating election outcomes or the success of new movies, for example.[6] Interestingly, prediction markets were inspired by the principles that ants and bees use to find the most promising sources of food.[7] Such complex self-organizing animal societies have always amazed people, especially because they are based on surprisingly simple rules of interaction.

In contrast, the "wisdom of the crowd" is undermined, if people are influenced by others while searching for information or making up their minds. This can sometimes create very bad outcomes, as misjudgments can easily spread if individuals copy each others' opinions. For example, in Asch's conformity experiments, people had to publicly state which one of three lines was equal in length to another line presented to them.[8] Before the participant answered the question, other people gave incorrect answers (as they were instructed). As a consequence, the participant typically answered incorrectly, too, even though the recognition task was simple. Recent experiments that Jan Lorenz, Heiko Rauhut, Frank Schweitzer and I performed, show that people are even influenced by the opinions of others when they are not exposed to social pressure.[9] This demonstrates that the wisdom of crowds is sensitive to manipulation attempts and social influence, and that's why one shouldn't try to influence how people search for information and make decisions, as nudging does.

But why are our opinions influenced by others at all? This is mainly, because imitation, which builds on the experience of others, is often a successful and efficient strategy. In the following, I will discuss, how we decide, why we behave in a social way, and how information systems may change our behavior. If we want to understand how the digital revolution might change our society, we must identify the various factors that influence our decision-making. In particular, we need to find out how growing amounts of information and the increased interconnectedness of people

---

[3]Page [3].

[4]See https://en.wikipedia.org/wiki/The_Wisdom_of_Crowds.

[5]Woolley et al. [4].

[6]Thompson [5]; for further recent prediction approaches see Siegel [6], Silver [7].

[7]Bonabeau and Theraulaz [8].

[8]Asch [9].

[9]Lorenz et al. [10].

create new opportunities. It becomes possible, for example, to overcome "tragedies of the commons" and to support the generation of collective intelligence.

9.1 Modeling Decision-Making: The Hidden Drivers of Our Behavior

One of the best-known models of human decision-making so far is that of "homo economicus". It is based on the assumption that humans and firms are completely egoistic, rationally striving to maximize their utility in every situation. The "utility function" supposedly quantifies the payoffs (i.e. earnings) or individual preferences of every actor. Any behavior deviating from such selfish utility maximization is believed to be irrational and to create disadvantages. The obvious conclusion— sometimes referred to as "economic Darwinism"—is that humans or companies who aren't selfish would ultimately lose the evolutionary race with those who are. Thus, natural selection is expected to eliminate other-regarding behavior based on the principle of the "survival of the fittest". Therefore, it seems absolutely logical to assume that everyone acts selfishly and optimizes the own payoff.

For those who are familiar with this conventional economic model of human behavior, it may come as a surprise that it does not stand up to empirical scrutiny. Appendix 9.1 shows that this model is actually quite misleading. Therefore, I am offering a new, multi-faceted explanation of human decision-making in the following. I argue that self-regarding rational choice is just one possible mode of decision-making. Human decisions are often driven by other factors in ways that make it impossible to integrate them all in one "utility function". Trying to do this, as it is common, compares "apples with oranges". Instead, I claim that people are driven by a variety of different incentive systems, and that they switch between them. In fact, rational decision-making is commonly believed to take place in the neocortex— our most powerful brain area. But there are other parts of the brain (such as the cerebellum), which were in control before and might still sometimes be. Thus, it is my contention that there are many other drivers governing people's behavior.

9.2 Sex, Drugs and Rock 'n Roll

It is clear that the first and foremost priority of our body is to ensure our survival, for example, to watch out for enough water and food. To achieve this, our body produces the feelings of hunger and thirst. If we don't drink or eat for a long time, it will be pretty difficult to focus on mathematical calculations, strategic thinking, or maximizing payoff.

Our sexual desires are similarly strong. There is an obvious, natural drive to engage in reproduction, and many people who are deprived of sexual opportunities may find sexual fantasies to occupy their thoughts pretty much. Thus, the desire to gain sexual gratification can influence human behavior substantially. This explains why people who are sexually deprived may display some pretty irritating, "irrational" behavior.

The human desire to possess has very early roots, too. Our distant ancestors were hunter-gatherers. Accumulating food and other belongings was important in order to survive difficult times, but also to trade and gain power. This desire to possess can, in some sense, be seen as the basis of capitalism.

But besides the desire to possess things, some of us also like to experience adrenaline kicks. These rushes of adrenaline were originally there to prepare our bodies for fight or flight from predators and other dangers. Today, people watch crime series or play "first-person shooter" computer games to get a similar thrill. Like sexual satisfaction, the desire to possess things and to experience adrenaline kicks is related to concomitant emotions such as greed and fear. Financial traders know this very well.

9.3   Hunger for Information

Intellectual curiosity is a further driver of human behavior. It primarily comes into play when the needs mentioned above have been sufficiently satisfied. Curiosity drives us to explore our environment and seek to understand it. By explaining how our world works, we can manipulate it better to our advantage. A trade-off between exploration and exploitation is part of all long-term reward maximizing algorithms. Individuals who solely focus on conventional sources of reward are quickly surpassed by those who discover new, more profitable avenues. In order to encourage sufficient efforts to explore our environment, our brain rewards insights with a dopamine-based hormonal rush. The effect of these hormones is excitement. In fact, as many intellectuals and others will attest, thinking can create really great pleasure!

9.4   Lessons Learned

In summary, our body has several different incentive and reward systems. Many of them are related with intrinsic hormonal, emotional, and nervous processes (the latter include the „amygdala" area of the brain and the part of the nervous system called the "solar plexus"). If we neglect these factors, we cannot gain a good understanding of human behavior. Hence, any realistic description of human decision-making in future must incorporate knowledge from those sciences that study our body and brain. Currently established theoretical models appear to be insufficient.

For example, why do many people invest much time and energy in sport that has little reproductive or material benefit? Why do people buy fast and expensive cars that do not match their stated preferences? Why do people race or fight, ride rollercoasters or go bungee jumping? It's the related adrenaline kicks that explain it!

This is also the reason why the Roman concept of "bread and circuses" (whereby the public is best appeased by satisfying their basic needs and entertaining them) is so enduring.

The above observations have important implications. It's too simple to assume that humans would just maximize their payoffs. We are driven by a variety of different incentive and reward systems. Humans have evolved in a way, which makes them strive to satisfy many different multi-faceted desires, which are often mutually incompatible. We engage in survival, reproduction, spreading of ideas and other goals. This gives rise to multiple different goals, which steadily compete with each other.

In addition, the importance of the different reward systems varies from one person to the next. This implies different preferences and personalities ("characters"). Some people are driven to possess as much as they can, while others prefer to explore their intellectual cosmos. Further people prefer "carnal" pursuits such as sex or sports. If nothing grants satisfaction over a long period of time, an individual might turn to drugs, get sick, or even die.

9.5 Suddenly, "Irrational Behavior" Makes Sense

In other words, if we move beyond the concept of self-regarding rational choice, the behavior of intellectuals, sportsmen, sex addicts, divas and other "special" people suddenly becomes understandable. In such cases, one reward system takes precedence over the others. For most people, however, all of these rewards are important. But this does not mean that it is possible to define one personal utility function, which stays constant over time (or changes slowly). Instead I believe that, at any point in time, one kind of reward is given priority and the others are temporarily relegated. Once this reward has been gained, another desire is given priority to, and so on. This might be compared to how competing traffic flows are managed at an intersection—temporarily prioritizing one flow after another. Once a queue of vehicles has been cleared, another one is prioritized by giving it a green light. Similarly, when one of our desires has been satisfied, we give priority to another one et cetera, until the first drive becomes strong again and demands our attention.

We can also understand what happens when people are deprived, meaning that they cannot satisfy one of their desires for a long time. In such cases, it stands to reason that they try to compensate for this lack of satisfaction with other kinds of rewards. For example, some people may find it rewarding to eat or go shopping, when they are frustrated. But options for compensation are more restricted for people in poor economic conditions. If they cannot derive pleasure from consumption and possessions, from social recognition, or from intellectual activities, then, sexual adventures, adrenaline flashes or other kicks may become more important. The results may be prostitution, violence, hooliganism, crime, or drug consumption. Therefore, a better understanding of human nature and decision-making, as I have proposed it above, may enable us to address long-standing social problems in entirely new ways, benefiting society as a whole.

9.6 Multi-Billion Dollar Industries for Each Desire

It turns out that our whole lives are structured around the various rewards which drive human behavior. In the morning, we have breakfast to satisfy our hunger and thirst. Then, we go to work to earn the money we want to spend on shopping, in order to satisfy our desire to possess. Afterwards, we may engage in sports to get adrenaline kicks. To satisfy our social desires, we may meet friends or watch a soap opera. At the end of the day, we may read a book to stimulate our intellect or have sex to satisfy this desire, too. In conclusion, I would venture that most of the time, our behavior is not well understandable as result of the strategic optimization of a static utility function that does not change over time (or very little).[10] Human nature can be better explained by assuming a switching between multiple desires or goals.

Even though the basis of decision theory in its current form is pretty flawed, our economy fits human nature surprisingly well![11] We have created entire multi-billion-dollar industries around each of our drives. We have developed a food industry, supermarkets, restaurants and bars to satisfy our hunger and thirst. We have built shopping malls to satisfy our desire to possess. We have constructed stadia and sports facilities to enable adrenaline kicks, either as spectators or participants. We have a porn industry and, in some jurisdictions, prostitution to meet our sexual desires. We read books, solve puzzles, travel to cultural sites, or participate in interactive online games to stimulate our intellect and satisfy our curiosity. Our media and tourism industries also aim to satisfy these needs.

However, there seems to be a natural hierarchy of desires, explaining the order in which these industries came up. When a new industry emerges, it changes the character of our society by placing additional weight on desires that didn't previously exist or were looming in the background. So, which desires will determine the future of our society?

Currently, the fastest growing economic sector is information and communications technology. As all our other needs have been taken care of, we are now building industries to satisfy our desires as an "information-driven species". This trend will place additional emphasis on everything related to information. In other words, the digital society to come will be much more determined by ideas, curiosity and creativity. But that's not all...

9.7 Being Social Is Rewarding, Too

Humans are not only driven by the reward systems mentioned above. We are also social beings, driven by *social* desires and that's why social media are so successful,

---

[10] But note that nonetheless, people make informed trade-offs, such as avoiding costly parties if they wish to possess things.

[11] This probably means that the economy does something else than what is predicted by mainstream economic theory.

changing our behavior, too. In fact, most people have empathy (compassion) in that they feel for others. Empathy is reflected by emotions and expressed to others by gestures and facial expressions. It even seems that all humans on our globe share some fundamental facial expressions (anger, disgust, fear, happiness, sadness, and surprise). According to Paul Ekman (*1934), these expressions are universal, as they are independent of language and culture.[12] However, our social desires go further than that. For example, we seek social recognition.

The main reason for the propensity of humans to network is that we are fundamentally social beings. In some sense, we are "networked minds". The increasing tendency of many people to form networks using social media such as *Facebook*, *Twitter* and *WhatsApp* underlines and reinforces this. It has even the potential to fundamentally change the way our society and economy work. Social networking through information and communication systems can stimulate our curiosity, strengthen our social desires, and support collective intelligence.

9.8 The Evolution of "Networked Minds"

It's interesting to ask why humans are actually social beings at all? Why do we have social desires? And how is this compatible with our hypothetical selfishness and the principle of the survival of the fittest? To study this, together with Thomas Grund and Christian Waloszek I developed a computer model which simulates the interactions between utility-maximizing individuals exposed to the merciless forces of evolution. Specifically, we simulated the "prisoner's dilemma", a special kind of social dilemma. Here, two actors engage in a social interaction where it would be favorable for everyone to cooperate, but non-cooperative behavior is tempting and, hence, cooperative behavior is risky.

In such scenarios, the selfish "homo economicus" would never cooperate, as non-cooperative behavior yields a higher payoff than cooperation in every single decision. This undermines cooperation, and the result is a "tragedy of the commons": the most favorable situation—where everyone cooperates—does not occur naturally, because it is unstable, and the resulting outcome is undesirable. The reasons are similar to why free-flowing traffic breaks down when a road is too busy.

In our own computer simulations of the prisoner's dilemma, we distinguished between the actual behavior of simulated individuals (cooperative or not) from their individual preferences.[13] We assumed that the preferences result from a trait called "friendliness", which governs how much somebody considers the consequences of the own actions on others. To be consistent with mainstream economics, the computer agents representing individuals were assumed to maximize their utility function,

---

[12]Ekman et al. [11], Ekman and Friesen [12], Ekman et al. [13].
[13]Grund et al [14], Helbing [15].

given the behavior of the "neighbors" they were interacting with.[14] However, the utility function was specified in such a way that individuals could consider more than just their own payoff from an interaction—they could also give some weight to the payoffs of their interaction partners. This weight was proportional to the "friendliness" of the individual and was set to zero for everyone at the beginning of the simulation. Thus, initially, everyone was absolutely selfish, and the payoff that others received from an interaction was given no weight (corresponding to what we call "homo economicus").

Additionally, we assumed that the friendliness trait could be inherited to others, either genetically or through education. In our computer simulations, the likelihood that someone would have offspring increased solely depending on their own payoff (not utility). If a person cooperated but was exploited by everyone they interacted with, they would receive no payoff. As a result, this person would also have no offspring. Finally, if people managed to have payoffs and offspring, the friendliness of an offspring tended toward that of the parent, but there was also a certain natural mutation rate. The mutation rate was specified such that it did not implicitly promote significant values of friendliness.

So, what results did our computer simulations produce? As expected, the prevailing strategy was that of "homo economicus," who maximizes the own payoff in a selfish way. However, although this applied to *most* scenarios simulated with our model, it did not apply to all of them (see Fig. 9.1). Offspring who lived close to their parents (when intergenerational migration was low) tended to evolve into a more friendly form of "homo socialis" and cared more about the payoffs of others, i.e. had other-regarding preferences! Interestingly, this scenario corresponds to the conditions in which humans actually raise their children.

The evolution of other-regarding preferences is quite surprising.[15] Even though none of the assumptions of the above model promotes cooperative behavior or other-regarding preferences in isolation, they create socially favorable behavior in combination. The only possible explanation of this fact is that interaction effects between the above rules change the overall outcome. Another interesting finding is the evolution of "cooperation between strangers", meaning that genetically unrelated individuals begin to cooperate. A video of a typical run of our computer simulations illustrates this well[16] (see Fig. 9.2).

---

[14]This standard assumption makes our results more surprising. It can be replaced by other assumptions such as an imitation of the best performing neighbor.

[15]Note that we are talking about other-regarding preferences, not just other-regarding behavior (i.e. cooperation), here, which has been found in many behavioral models before.

[16]See http://www.youtube.com/watch?v=n6AJeIcG4zQ.

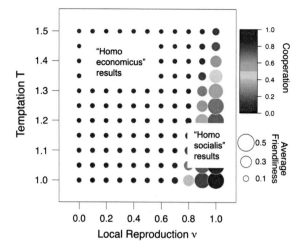

Fig. 9.1 Illustration of the outcome of computer simulations of the evolution of the trait of "friendliness". While the familiar "homo economicus" results for most parameter combinations, an otherregarding "homo socialis" results after many generations, if offspring (children) grow up close to their parents. Note that intergenerational migration is low when local reproduction is high and vice versa. (Reproduction from Grund et al. [14], with kind permission of the Springer Nature Publishing Group.)

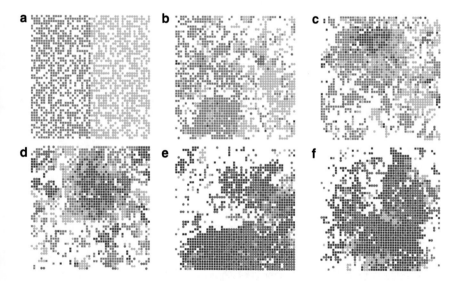

Fig. 9.2 Snapshots of a computer simulation illustrating the emergence of other-regarding preferences (blue and green) in two initially separated populations (red and orange). The mixture of green and blue agents also indicates a cooperation between strangers. (Reproduced from Helbing [16], with kind permission of Springer Publishers.)

9.9   Making Mistakes Is Important!

How does this preference to show regard for others evolve? First, it is important to recognize that a low, but finite level of friendliness is generated due to the random mutations of the friendliness values of offspring. Second, this slight preference to show regard for others creates conditionally cooperative behavior. That is, if just enough neighbors would cooperate, a "conditional cooperator" will be cooperative as well.

Third, unconditionally cooperative individuals with a high level of friendliness are born very rarely, and only by chance. These rare "idealists" will usually be exploited. They receive poor payoffs, and have no offspring. However, idealists may catalyze cooperative behavior if he/she is born by chance into a neighborhood with a sufficient number of conditionally (or unconditionally) cooperative individuals.[17]

Fourth, in the cooperative neighborhood resulting from such cooperation cascades, high levels of friendliness are passed on to many offspring and thus, the preference to show regard for others rapidly spreads. This happens because friendliness becomes more profitable than in the initial phase of the evolutionary process, when friendly people are lonely idealists who are exploited by others. Many generations later, however, those who cooperate earn a higher payoff on average than those who act selfishly. This makes sense for the following reason: if everyone in the neighborhood is friendly, everyone has a better life. Therefore, while "homo economicus" initially earns more, "homo socialis" eventually outperforms "homo economicus" (see Fig. 9.3). As a consequence, the levels of friendliness increase over sufficiently many generations, but they are broadly distributed (see Fig. 9.4). This explains the large variety of individual preferences which is actually observed: in our society, we encounter everything from selfishness to altruism.

Note that in the situation studied above, where everyone starts out as a non-cooperative "homo economicus", no single individual can establish profitable cooperation, not even by optimizing decisions over an infinitely long period of time. Therefore, "homo socialis" can only evolve thanks to the occurrence of random "mistakes" (in this case, the birth of unconditionally cooperative "idealists" who are exploited by everyone). If such "errors" happen by coincidence in the neighborhood of other idealists or conditionally cooperative people (i.e. sufficiently friendly individuals), this can trigger a domino effect that causes an "upward spiral" towards more cooperation and eventually changes the society to the better. Then, cooperation yields higher payoffs than selfishness and "loving your neighbor as yourself" becomes rewarding.

In conclusion, the cooperativeness of "homo socialis" does not result from optimization, but from evolution. While being an idealist is a painful mistake in the

---

[17] One might think that this is what happened, for example, in the case of Jesus Christ. He preached to "love your neighbor as yourself", i.e. other-regarding behavior weighting the preferences of others with a weight of 0.5. His idealistic behavior was painful for himself. He ended on the cross like a criminal and without any offspring. But his behavior caused other-regarding behavior to spread in a cascade-like way, establishing a world religion.

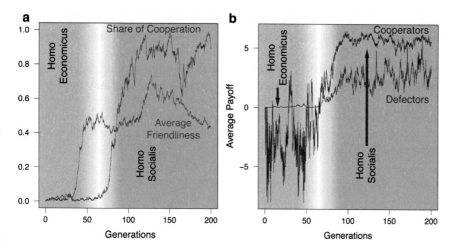

Fig. 9.3 Illustration of the sudden increase in friendliness and payoff (after more than 50 generations), when "homo socialis" emerges. The increase in payoff results, as "homo socialis" is able to avoid a "tragedy of the commons". The results of our computer simulations also show that the consideration of "externalities" (i.e. of external effects of decisions and actions) can yield a better system performance and benefits for everyone. This hints at the existence of superior organization principles for future economies, as discussed in the next chapter. (Reproduction from Grund et al. [14] with kind permission of the Springer Nature Publishing Group.)

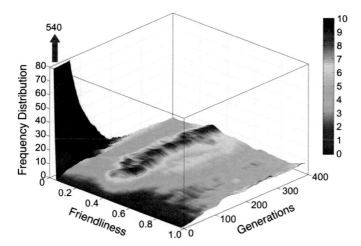

Fig. 9.4 Distribution of the level of friendliness in our simulated population. While everyone is assumed to be unfriendly in the beginning, over several generations people develop other-regarding preferences and get more friendly. (Reproduction from Grund et al. [14] with kind permission of the Springer Nature Publishing Group.)

beginning, later on, when other idealists or conditional cooperators join, it is the crucial success principle that makes it possible to avoid or overcome "tragedies of the commons".

The consideration of one's externalities on others evolves eventually, because the resulting system behavior is superior to one that ignores externalities. In fact, in an increasingly interdependent globalized world, problems caused by us tend to return to us like a boomerang. This applies, for example, to financial crises and wars, which affect us indirectly through the deterioration of public infrastructures, international migration, reduced security, etc.

9.10   "Networked Minds" Require a "New Economic Thinking"

An important lesson from the evolution of friendlier, "other-regarding" people is that such individuals consider the payoff and success of others when choosing their actions. As a consequence, their decisions become interdependent. While conventional statistical methods focusing on independent, unrelated events may sometimes suffice to characterize the decisions of "homo economicus", we usually need complexity science to understand the interdependent decision-making of "homo socialis". In fact, "homo socialis" may be best described by the term "networked minds".

Note that, in congruence with the findings of social psychology, "homo socialis" has the ability to put him/herself into the shoes of others. By taking the perspective, interests, and success of others into account, "networked minds" consider the externalities of their decisions. That is, "homo socialis" decides *differently* from "homo economicus". While the latter would never cooperate in a social dilemma situation, "homo socialis" is conditionally cooperative. Thus, he/she tends to cooperate if just enough neighbors do so as well. Putting it differently, "homo socialis" is able to harmonize competitive individual interests to make them more compatible with systemic efficiency. In the end, everyone can be better off. Being friendly to each other benefits everyone. The social world is not a zero sum game where one can only get more if others get less. On the contrary: in social dilemma situations or when creating collective goods, everyone can get more, if everyone behaves in an other-regarding and cooperative way.

This means that "homo socialis" is more successful than "homo economicus", even if we measure success purely in terms of individual payoff. While the invisible hand often doesn't work for "homo economicus" (as market failures show), it works for "homo socialis" because he/she interacts with like-minded individuals who consider the externalities of their decisions. Therefore, "homo socialis" creates systemic benefits while increasing his/her personal payoff, in contrast to "homo economicus".

All of the above implies the need of a "New Economic Thinking" ("Economics 2.0"), which enables us to organize our economy more efficiently and effectively[18] (see the next chapter and Appendix 9.2). I strongly believe that we are heading towards a new kind of economy, not just because the current economy no longer provides enough jobs in many areas of the world, but also because information systems and social media are changing our interactions and opening up entirely new opportunities. In particular, to cope with the increasing level of complexity of our world, we need to catalyze collective intelligence. In fact, the principle of "networked minds" makes it possible to bring the best ideas and skills together, and to leverage the hugely diverse range of knowledge and expertise in our societies. Let us discuss this now in more detail.

9.11 The Netflix Challenge

One of the most stunning examples of collective intelligence is the outcome of the *Netflix* challenge.[19] The video streaming company *Netflix* was trying to predict what movies their users would like to see most, based on their previous movie ratings, but their predictions were frustratingly bad. So back in 2006, *Netflix* offered a prize of $1 million dollars to any team that could improve their own predictions of individual movie ratings by more than 10%. About 2000 teams participated in the challenge and sent in 13,000 predictions. The sample data contained more than 100 million ratings of almost 20,000 movies on a five-star scale, made by approximately 500,000 users. *Netflix'* own algorithm produced an average error of about 1 star, but it took three years to improve it by more than 10%.

In the end, the prize was won by a team going by the name of "BellKor's Pragmatic Chaos". In the process, a number of really remarkable lessons were learned. First, given that it was very difficult and time-consuming to improve the standard method by just 10%, Big Data analytics wasn't good at predicting people's preferences and behavior. Second, even a minor improvement of the algorithm by just 1% created a significant difference in the top-10 movies predicted for each user. In other words, the results were very sensitive to the method used (rather than stable). Third, no single team was able to achieve a 10% improvement alone.

9.12 Diversity Wins, not the Best

The final breakthrough was made when the best-performing team decided to average over their own predictions and those made by other teams that weren't as good. Surprisingly, collaborating with inferior teams boosted the prediction performance,

---

[18]Helbing [16].
[19]See https://en.wikipedia.org/wiki/Netflix_Prize and also the book by Scott E. Page (Ref. 245).

while one would expect just the opposite! In other words, when complex problems must be solved, diversity wins, not the best. This seems really counter-intuitive: nobody is right, but the average of answers, none of which is good enough, gives the most accurate picture! The reason for this is subtle and extremely important for collective wisdom. Although each of the top teams almost achieved a 10% improvement over the original algorithm, each used different methods that were able to find different patterns in the data. However, no single algorithm could find them all, and "overfitting" (i.e. fitting random patterns that did not have a meaning) was probably a problem, too. By averaging the predictions, each algorithm contributed the knowledge it was specialized to find, while the errors of each algorithm were neutralized by the others.

Even more surprisingly, when more accurate predictions were given a higher weight, it didn't improve the accuracy of the average prediction. Researchers have argued that this is because weighting more successful algorithms more highly only works if at least one algorithm is correct. In the case of the *Netflix* challenge, however, no single algorithm was perfect, and an equally weighted combination was better than any individual algorithm and also better than an individually weighted average that took the relative rank of each algorithm into account.[245] This is probably the best argument for equal votes—in this case, however, equal votes for different approaches to problem-solving! Therefore, if we want to maximize collective intelligence, compared to our current mode of decision-making, we would have to take on board more knowledge of minorities. This would give rise to a "democracy of ideas", which would allow us to make better use of ideas and innovations, to the benefit of our society.

9.13 Future Decision-Making Institutions

To conclude, when complex tasks are to be solved, specialization and diversity are key. So, what decision-making procedures and institutions do we need to maximize collective intelligence? Answering this question is of major importance to cope with the increasing difficulty of the challenges posed by our complex globalized world. We are in the awkward situation that no single person or computer system alone can fully grasp this complexity, and that we need to rely on inputs from others. So, for collective intelligence to work, having a knowledge base of trustable and unbiased information is essential, i.e. measures against information pollution are important.[20] This requires a sufficient level of transparency and quality-ensuring mechanisms, and it requires a participatory ("crowd sourcing") approach to get the best ideas on board (think, for example, of *Wikipedia* or *GitHub*).

---

[20]In fact, to avoid mistakes, the more we are flooded with information the more must we be able to rely on it, as we have increasingly less time to judge its quality.

Furthermore, based on what we have learned above, one should consider turning away from two modes of decision-making that are common today. First, decision-making should not be the sole province of a few "experts" taking decisions in a top-down way. Second, we must recognize that simple majority voting is also not the best decision-making mechanism. Therefore, if we want to make better decisions, we must say goodbye to the concept of the "wise king" and also to the principle of majorities deciding what should apply to everybody. This may be surprising, but in future we need better solutions to cope with the quickly increasing complexity of our world. This is particularly important in times of transition, where the majority sticks to old principles while those who understand the guiding principles of the future are still in a minority. The transformation of our current society into a digital society is a perfect example for this.

9.14 How to Produce Collective Intelligence

So, how could we create a better system to unleash the power of "collective intelligence"? First, we must abandon the idea that reality can be well described by a single model. In many cases, such as modeling traffic flow, there are several models which perform similarly well.[21] This speaks for a pluralistic modeling approach. In fact, when the path of a hurricane is predicted or the impact of a car accident is simulated on a computer, it turns out that an average of several competing models often provides the best predictions.[22]

In other words, the complexity of today's world cannot be grasped by a single model, mind, computer, or Artificial Intelligence system. Therefore, it's good if several groups try to find the best possible solution independently of each other. Although all these models will give an over-simplified picture of our complex world, if we combine these different perspectives, we can often get a better approximation of the full picture. This might be compared to visiting an ornate cathedral in which every photograph only reflects some aspects of its complexity and beauty. One photographer alone, no matter how talented or how well equipped, cannot capture the full structure of the cathedral with a single photograph. A full 3D picture of the cathedral can only be gained by combining plenty of photographs representing different perspectives.

Let's discuss another complex problem, namely finding the right insurance. Usually, it's not realistic to read all of the small print and detailed regulations for every insurance policy on the market. Instead, we would probably ask our colleagues and friends about their experiences with their own insurers. We would then evaluate the insurances they recommend in greater detail to find the best policy that meets our needs. In this way, we engage in a collective evaluation process to solve a complex task that nobody could master alone. In the Internet age, this classical word-of-mouth

---

[21]Brockfeld et al. [17].

[22]Helbing [18]

system is increasingly being replaced by online reviews and recommender systems such as price comparison websites, which widens the circle of people contributing information and improves the chances that each individual will make better decisions.

While the above approach collates existing knowledge well, some methods can also create new knowledge, which is more than just the sum of its parts. Let us assume that a particular problem needs to be solved. One would then invite everyone to contribute to an online discussion of aspects that should be considered. With suitable software, it is possible to structure these aspects in an argument map, and to work out a number of different perspectives that matter.[23] Once, the problem has been boiled down to anything between, say, 2 and 7 perspectives, it's time to invite the leading representatives of these perspectives to a round table. Political or other decision makers could then moderate a deliberation process to develop an integrated solution (or a few good alternatives to choose from). When many different ideas "collide", this often triggers innovative ideas, which may help to overcome conflicts of interest in favor of a win-win situation. This leads to better solutions, which find wider support. Therefore, rather than letting a majority impose partial interests on others, the future success principle will be to get as many good ideas on board and produce as many "winners" as possible.

The examples above illustrate how collective intelligence works. First, a number of teams needs to tackle a problem independently using diverse methodologies. Subsequently, these independent streams of knowledge need to be combined. If there is too much communication at the beginning of this process, each team may be tempted to copy promising approaches of others, which would reduce the diversity of ideas. However, if there is too little communication at the end of the process, the knowledge created by all these different approaches won't be fully used.

9.15 What We Can Learn from IBM's Watson Computer

It is also interesting to discuss how "cognitive computing" works in *IBM*'s Watson computer.[24] The computer scans hundreds of thousands of sources of information, including scientific publications, and extracts potentially relevant statements. It can also formulate hypotheses and seek evidence to support or refute them. Afterwards, it produces a list of possible answers and ranks them according to their likelihood. All of this is achieved using algorithms based on the laws of probability, which calculate how likely a hypothesis is based on the available data. Compared to humans, however, Watson loses less information, which we would filter out due to limited memory and attention spans, cognitive biases, and our preference for consistency. For example, when applied in a medical context, Watson would come up with a ranked list of diseases that are compatible with a number of symptoms. While a typical doctor

---

[23] See https://en.wikipedia.org/wiki/Argument_map and http://sourceforge.net/projects/argumenta tive/.
[24] Kelly and Hamm [19].

might assume the most common disease to cause the symptoms, Watson would also consider rare diseases, which may otherwise be overlooked.

In order to work well, it is important to feed Watson in a non-selective way with information reflecting different perspectives and potentially contradictory pieces of evidence. Then, Watson is trained by experts how to evaluate information sources, and in the end, Watson may outperform humans. The power, quantity and variety of information sources the cognitive computer can scan and the number of hypotheses it can generate and evaluate, are vastly greater than any single human could conceivably cope with. Whereas humans tend to seek and keep information that confirms their existing beliefs, Watson seems less prone to this bias, as it can consider much more pieces of information.

9.16 Collective Intelligence Versus Superintelligence

Let us finally address the question whether one could eventually create something akin to an omniscient superintelligence, if we used all the data of the Internet? In fact, the *Google Brain* project probably seeks to create such a superintelligence (and other projects, too). However, I expect a distributed, networked system of multiple brains interacting with Artificially Intelligences to be superior to a superbrain. Remember that the "wisdom of crowds" often exceeds that of the best experts.[25] This implies that collective intelligence can outperform superintelligence. In a complex world the consideration of multiple different, independent perspectives and goals is key to success. Therefore, to cope with the challenges of the future, we need a pluralistic and participatory approach, as I will discuss it later.

It's probably not by accident that our society is based on distributed intelligence, characterized by many moderately sized brains which think differently, with relatively small information flows among them. If this wouldn't be useful, biological evolution might have created different solutions already.[26] Smaller brains learn more quickly. Thus, collective intelligence can probably better adjust to the variability and combinatorial complexity of our world.[27]

---

[25]While *Google* can easily implement many different algorithms, the very nature of a large corporation with its self-regarding goals, uniform standards, hiring practices and communications means that the teams developing these approaches will be inherently more prone to observe and follow each other's successes. Similar ways of thinking produces less diverse opinions.

[26]High bandwidth communication could, for example, make it impossible to stop destructive ideas from spreading. One could certainly not be interested in fostering the "madness of crowd" (or "social flash crashes"), which can lead to violence or war, as known from the psychology of masses.

[27]As we know, intellectual discourse can be a very effective way of producing new insights and knowledge.

9.17   There Is More to Come: New Rewards, Virtual Worlds

Diversity is often seen as an obstacle to efficient socio-economic solutions. However, it is not by chance that the most advanced species on Earth is so diverse, in contrast to social insects, for example. Diversity is and always has been a major driving force of evolution. Over millions of years, diversity has hugely increased, creating a growing number of different species. Diversity drove (and still drives) differentiation and innovation, creating new characteristics of life. Once humans became social and intelligent beings, cultural evolution set in.[28] The slow process of genetic evolution was then complemented by a rapid evolution of ideas. Since then, humans have increasingly emancipated themselves from the limitations of matter and nature.

Nowadays, spreading ideas or memes (in the original sense of this word) has become more important than spreading genes. Now, besides the real world, digital virtual worlds also exist, such as massive multi-player on-line games. Thus, humans have learned to create new worlds out of nothing but information. Multi-player online games such as Second Life, World of Warcraft, Farmville, and Minecraft are just a few examples.

It is equally fascinating that new incentive systems have evolved out of these digital worlds, which are now influencing our behaviors as well. While it is perhaps unsurprising that some people care about their position in the Fortune 500 list of richest people, people are not only competitive about money and don't just respond to material payoffs. There are many other drives as well. Tennis players and soccer teams worry about their rankings. Scientists care about the number of citations to their work, and actors live for the applause they receive.

Some people even care about the scores they reach in gaming worlds. Although such rankings usually don't imply any immediate material or other real-world benefits, they motivate people to make an effort. It's also pretty surprising how much time people spend to increase their number of *Facebook* friends or *Twitter* followers, or their *Klout* score. It is obvious, therefore, that virtual environments offer the potential to create multi-dimensional reward systems, as it is required to enable self-organizing socio-economic systems.

There can be little doubt that we are now living in a cyber-social world, in which much of our social interaction takes place online. The evolution of global information systems will drive the next phase of human social evolution. Information systems support "networked minds" and enable "collective intelligence".[29] Humans, computers, algorithms and robots increasingly weave a network that can be characterized as an "information ecosystem". Therefore, one question becomes absolutely crucial—how will this change our socio-economic system?

---

[28]Therefore, for the survival of the human species, socio-cultural diversity is as important as genetic diversity. Our planet may be exposed to a variety of disruptive events, and we can't tell in advance what culture will cope with these challenges best. For example, if a major solar storm destroys the functionality of our electrical infrastructure, those cultures that don't use electricity may be best prepared to survive.

[29]Rifkin [20].

9.18 Appendix 1: How Selfish Are People Really?

Our daily experience tells us that many people do unpaid jobs for the benefit of others. A lot of volunteers work for free and some organize themselves in non-profit organizations. In addition, we often leave tips on the restaurant table, even if nobody is watching and even if we'll never return to the same place (and that's also the case in countries where tips are not as socially obligatory as in the USA). Furthermore, billionaires, millionaires and normal people make donations to promote science, education, and medical assistance, often in other continents. Some of them do it even anonymously, meaning that they will never get anything in return, not even recognition.

Indeed, this has puzzled economists for quite some time. To fix the classical paradigm of rational choice based on selfish decision-making, they eventually assumed that everyone would have an individual utility function, which reflects their personal preferences. However, as long as there is no theory to predict personal preferences, the concept of utility maximization does not explain much. If we are to take rational choice theory seriously, we need to believe that people who help others must derive utility from it, otherwise they wouldn't do it. But this appears to be pretty circular reasoning.[30]

9.18.1 Ultimatum and Dictator Games

In order to test economic theories and understand personal preferences better, scientists have performed a large number of decision experiments with people in laboratories. The findings were quite surprising and totally overturned conventional economic theory at the time.[31] In 1982, Werner Güth used the "Ultimatum Game" to study stylized negotiations. In these experiments, a person (the "proposer") was given a sum of money (say, $50) and asked to decide how much of this money he/she would offer to a second person (the "responder"). If the responder accepted the amount offered by the proposer, both got their respective share. If the responder rejected the offer, however, both received nothing.

According to the concept of the self-regarding "homo economicus", the proposer should offer no more than $1 and the responder should accept any amount. After all, surely it is better to get a little money rather than nothing! However, it turns out that responders tend to reject small amounts, and proposers tend to offer about 40% of the money on average. A further surprise is that, all over the world, proposers tend to share with others. Similar experimental outcomes are found when playing for amounts as high as a monthly salary. To reflect these findings, Ernst Fehr (*1956) and his colleagues proposed that humans hold inherent principles of fairness and

---

[30]Now, however, the model for the emergence of "homo socialis" presented in this chapter can be used to understand the evolution and variation of individual preferences and utility functions.

[31]Henrich et al. [21]

have an aversion to inequality.[32] Others, such as Herbert Gintis (*1940), assumed that cooperation has a genetic basis ("strong reciprocity").[33]

There is also a simpler game, known as the "Dictator Game", which is even more stunning in some respects. In this game, one person receives some money and is asked to decide how much he/she would give to another person—it could be nothing! Although the potential recipient does not have any influence on the outcome, many people tend to share nevertheless. On average, people give away about 20% of the money they receive from the experimenter. Of course, there are always exceptions and some people don't share.

Why do we have this tendency of people to share? In principle, this could result from the feeling of being watched by others, which might trigger behavior that complies with social norms. If this were the case, sharing wouldn't occur when decisions are taken anonymously. To test this, we made a Web experiment with strangers who never met in person.[34] Both the proposer and the responder received a fixed sum of money to participate in the experiment. However, rather than sharing money, they had to decide how to share a workload involving several hundred calculations. In the worst case, one of them would have to do all the calculations, while the other would get money without having to work! To our great surprise, even in this anonymous setting, the participants tended to share the workload quite fairly. Thus, there is no doubt that many people behave other-regarding when making decisions. They clearly have a preference for fair behavior. As observing people ("surveillance") doesn't largely increase the level of fairness, people apparently behave according to inborn principles or internalized norms.

9.19   Appendix 2: A Smarter Way of Interacting, not Socialism

Many governments currently try to counter inequality by redistributing wealth using centralized social welfare systems. In contrast, the concept of "homo socialis" has nothing to do with socialism or with wealth redistribution. Therefore, "homo socialis" should not be considered to be some form of tamed "homo economicus", who shares some of his/her payoff with others. As we have discussed before, "homo economicus" tends to suffer as a result of "tragedies of the commons", whereas "homo socialis" can overcome them by considering the externalities of decisions. Thus, "homo socialis" can generate higher profits on average. The consideration of externalities by "homo socialis" creates more favorable outcomes, at least in the long term.

In contrast, when everyone behaves like "homo economicus", a few individuals generate high profits in social dilemma situations by exploiting others, while the vast majority of people lose out. Wealth redistribution does not overcome such "tragedies

---

[32]Fehr and Schmidt [22].

[33]Gintis [23].

[34]Ciampaglia et al. [24, 25]; see also Berger et al. [26].

of the commons" and cannot produce average levels of profit comparable to those earned by "homo socialis".[35] In conclusion, an economy in which externalities are considered and "homo socialis" can thrive is expected to produce better and more sustainable outcomes. From a socio-economic and individual perspective, deciding like "homo socialis" is a smarter way of decision-making. It is far superior to an economy in which "homo economicus" dominates and social policies try to fix the damage afterwards (if at all).

Finally, let me address the question of whether friendly, other-regarding behavior is more likely if people have a lot of resources and can "afford" to consider the interests of others, or whether it also occurs under particularly bad conditions. In fact, in the desert and other high-risk environments, people can only survive through cooperative behavior. However, other-regarding behavior can also create benefits in low-risk environments, in which people could survive by themselves. By considering the externalities of our own behavior, our socio-economic system can continuously improve so that it produces increasingly better payoffs.[36] In other words, "homo socialis" improves the performance of a socio-economic system, creating on average greater advantages for everyone. Even in a world with large cultural differences, it seems that those countries and cities achieve a particularly high quality of life, which manage to foster other-regarding behavior and consider externalities. As I said before, friendly, other-regarding behavior is to everyone's benefit, if enough people (inter)act in this way. Nevertheless, it would be desirable to develop institutions that protect "homo socialis" from exploitation by "homo economicus". Reputation systems and merit-based matching are such institutions, which can help to produce better outcomes in a globalized world.

References

1. F. Galton, Vox populi, Nature 75, 450–451 (1907).
2. J. Surowiecki, The Wisdom of Crowds (Anchor, 2005).
3. S.E. Page, The Difference: How the Power of Diversity Creates Better Groups, Firms, Schools, and Societies (Princeton University, 2008).
4. A.W. Woolley, C.F. Chabris, A. Pentland, N. Hashimi, Evidence for a collective intelligence factor in the performance of human groups, Science 330, 686–688 (2010).
5. D.N. Thompson, Oracles: How Prediction Markets Turn Employees into Visionaries (Harvard Business, 2012).
6. E. Siegel, Predictive Analytics: The Power to Predict Who Will Click, Buy, Lie, or Die (Wiley, 2013).
7. N. Silver, The Signal and the Noise: Why So Many Predictions Fail - but Some Don't (Penguin, 2015).
8. E. Bonabeau and G. Theraulaz, Smarm Intelligence: From Natural to Artificial Systems (Santa Fe Institute, 1999).
9. S.E. Asch (1951) Effects of group pressure on the modification and distortion of judgments. In H. Guetzkow (Ed.) Groups, Leadership and Men, pp. 177–190 (Carnegie Press, Pittsburgh).

---

[35]When the latter interact among each other.
[36]Wright [27].

10. J. Lorenz, H. Rauhut, F. Schweitzer, and D. Helbing (2011) How social influence can undermine the wisdom of crowd effect. *Proceedings of the National Academy of Sciences USA (PNAS)* **108**(28), 9020–9025.
11. P. Ekman, W. V. Friesen and P. Ellsworth, Emotion in the Human Face: Guidelines for Research and a Review of Findings (Pergamon, New York, 1972).
12. P. Ekman and W. Friesen, Facial Action Coding System: A Technique for the Measurement of Facial Movement (Consulting Psychologists Press, Palo Alto, 1978).
13. Paul Ekman, Wallace V. Friesen, and Joseph C. Hager, Facial Action Coding System: The Manual on CD ROM (A Human Face, Salt Lake City, 2002.
14. T. Grund, C. Waloszek and D. Helbing (2013) How natural selection can create both self- and other-regarding preferences, and networked minds. *Scientific Reports*, **3**, 1480.
15. D. Helbing (2015) Homo Socialis—The road ahead, Review of Behavioral Economics 2 (1–2), 239–253.
16. D. Helbing (2013) Economics 2.0: The natural step towards a self-regulating, participatory market society. *Evolutionary and Institutional Economics Review* **10**, 3–41.
17. E. Brockfeld, R. Kühne, and P. Wagner, Calibration and validation of microscopic traffic flow models, Transportation Research Record 1876, 62–70 (2004).
18. D. Helbing, Pluralistic modeling of complex systems, Science and Culture 76, 315–329 (2010).
19. J.E. Kelly III and S. Hamm, Smart Machines: IBM's Watson and the Era of Cognitive Computing (Columbia University Press, 2013).
20. J. Rifkin, The Empathic Civilization: The Race to Global Consciousness in a World in Crisis (Penguin, New York, 2009).
21. J. Henrich, R. Boyd, S. Bowles, C. Camerer, E. Fehr, and H. Gintis (eds.) Foundations of Human Sociality: Economic Experiments and Ethnographic Evidence from Fifteen Small-Scale Societies (Oxford University, 2004).
22. E. Fehr and K.M. Schmidt, A theory of fairness, competition, and cooperation, *The Quarterly Journal of Economics* 114, 817–868 (1999).
23. H. Gintis, Strong reciprocity and human sociality, Journal of Theoretical Biology 206, 169–179 (2000).
24. G.L. Ciampaglia, S. Lozano, and D. Helbing, Power and fairness in a generalized ultimatum game, PLoS ONE 9(6): e99039.
25. G.L. Ciampaglia, S. Lozano, and D. Helbing, Anonymous sharing behavior in Web experiments with different balance of power, see http://papers.ssrn.com/sol3/papers.cfm?abstract_id=255 2202.
26. R. Berger, H. Rauhut, S. Prade, and D. Helbing, Bargaining over waiting time in ultimatum game experiments, Social Science Research 41, 372–379 (2012).
27. R. Wright, Nonzero: The Logic of Human Destiny (Vintage, 2001).

Chapter 10
The Economy 4.0

A Participatory Market Society Is Born

> It is human nature to exchange not only goods but also ideas,
> assistance, and favors out of sympathy. It is these exchanges that
> guide men to create solutions for the good of the community.
> —Adam Smith

The invention of the steam engine turned the agricultural society – the „Economy 1.0"
– into an industrial society – the „Economy 2.0". Later, the spread of education enabled the
service society – the „Economy 3.0". Nowadays, the pervasiveness of digital technologies is
driving another technological revolution, which is creating the „Economy 4.0", which I call
the Participatory Market Society. This society is characterized by the ubiquity of information,
bottom-up participation, „co-creation", self-organization and collective intelligence as new
organizational principles. We will also see more personalized products and services and an
increased engagement in a serious and fair partnership with citizens, users, and customers.
Finally, the spread of „projects", empowered by social collaboration platforms, will enable
more flexible and efficient forms of production and services.

Our economy is in the middle of a once-in-a-century transformation. Big Data,
Artificial Intelligence, and the Internet of Things will fundamentally change many
of our current procedures and transform the institutions on which our economies and
societies are based.

The invention of the steam engine was a crucial technological catalyst in the
transition from the agricultural society ("Economy 1.0") to the industrial society
("Economy 2.0"). This new industrial society was driven from the bottom-up by
entrepreneurs, whose self-interested "rational" mindset is often reflected by the
theoretical concept of the "homo economicus". Consequently, very little attention
was paid to social and environmental "externalities" such as mass unemployment,
poverty, malnutrition, child labor, pollution, and the exploitation of resources.

In response, many countries established health insurances, social security systems,
and environmental laws. By increasingly complicated regulations, new jobs were
created. A new service sector (the "Economy 3.0") was grown. Affordable mass
education played a key role in creating the societal conditions in which this shift
could take place. In service societies, administrations based on top-down planning
and optimization became the new organizational basis. Now, driven by the digital
revolution, a new kind of economy ("Economy 4.0") is on the rise. How will it look
like?

© Springer Nature Switzerland AG 2021
D. Helbing, *Next Civilization*,
https://doi.org/10.1007/978-3-030-62330-2_10

10.1   We Can, We Must Re-Invent Everything

Nowadays, however, the world is often too complex for timely or proactive top-down governance. In particular, neither governments nor multi-national companies have so far been able to overcome complex problems such as climate change, overfishing, unsustainable use of resources, international conflicts, and global financial instability. Moreover, the problems of overregulation and high government debts are basically unsolved. Conventional modes of governance, which are based on standardized procedures, are also not very good at satisfying diverse local needs.

Therefore, the creation of new institutions and approaches is inevitable. In particular, we need information platforms which can harness the diverse knowledge in our society and support more effective decision-making. Before I describe how I think the Economy 4.0 will work, I would like to provide a quick overview of the transformation which is already underway. This will help to illustrate the fundamental changes to our socio-economic systems that we should expect. Indeed, these changes will be of a magnitude greater than anything we have witnessed in the last hundred years.[1]

10.2   Personalized Education

In the past, the academic system did a remarkable job in compressing increasing quantities of knowledge into surprisingly few lectures. Often, centuries of knowledge were squeezed into a single course. However, our world is now changing faster than ever. We are probably in a situation where most of our personal knowledge is already outdated and where it is impossible to follow all of the news that might be relevant to us. Even scientists and doctors are struggling to cope with the exponential increase in knowledge. Politicians and business leaders are confronted with similar problems, too.

In response to this challenge, the education system has already started to change. We are now seeing an explosion of new bachelor and master degrees, which go beyond the classical fields of study such as mathematics, physics, chemistry, biology and history. This increasing diversity means that people can now find study directions which better fit their interests and talents. Moreover, massive open online courses (MOOCS) are quickly spreading through platforms such as *Coursera, Khan Academy, edX* or *Udacity*, which enable people to learn about almost any subject from some of the best professors in the world. Such virtual courses are often taken by huge numbers of visitors. Up-to-date, high-quality knowledge can now be shared with millions of people, who can learn a subject whenever they want.

---

[1]To some extent, we might compare the effect of the digital revolution with that of a tsunami.

The next step in the evolution of personalized education might be the use of particular multi-player, interactive online games. This would allow participants to explore the laws of nature, to experience different historical ages or cultures, and to interact with fellow students. Professors may act as coaches or "chatroom masters", providing guidance, correcting misunderstandings and answering questions that cannot be readily addressed with knowledge from the Internet. Furthermore, virtual, three-dimensional video meetings may become standard, once the necessary technology and bandwidth are in place.[2]

Of course, schools must change dramatically as well. Teachers complain about burnout, aggression in the classroom, and worsening test results. Are schools still preparing children well enough for a successful future? Pupils complain that the knowledge they learn is outdated and not very useful to them. They feel that the Internet now provides faster and more accurate answers than their teachers. It might be true that pupils are losing the ability to memorize facts and focus, and that they are over-stimulated. But they certainly want to learn things that fit their personal interests and needs. Unfortunately, the standardization inherent in today's education systems inhibits the motivation and creativity of many pupils.

In future we will need less standardized education but more personalized learning focused on imagination, creativity, innovation, and collaboration. Rather than presenting children with a predetermined curriculum of things that they "should know", greater emphasis must be placed on the ability to learn individually and in teams. This would involve teaching pupils where they can find reliable information, and how they can critically assess it and use it responsibly. Rather than forcing children to put their smartphones and tablets aside, we must reach out to them on their own territory—the Internet. We need to help them to use information technology to develop their ideas and successfully engage with others.

In conclusion, the schools of today belong to a bygone era. They are the manifestation of an outdated concept of education which is no longer fit for purpose. Drumming standardized knowledge into pupils' heads simply doesn't work anymore. Education must become an exercise in systematically exploring the wider world and critically assessing information.

10.3 Science and Health, Fueled by Big Data

We are also at the beginning of a dramatic change in the way we analyze scientific data and manage our health systems. Although I disagree with Chris Anderson, who posits that Big Data will bring about the "end of theory", I do believe that it has the potential to change research methods and scientific disciplines. Traditionally, we have

---

[2]For this, hologram or virtual reality technology could be used. Facebook's recent acquisition of Oculus Rift for $2 billion and Microsoft's development of a similar product called HoloLens show that some of the biggest players in the technology sector envisage that virtual reality will soon become viable.

had three pillars of knowledge creation, namely theoretical analyses, experiments, and computer simulations. But the analysis of Big Data now provides a fourth pillar.

Big Data is not an oracle which will answer all the world's questions, but the availability of huge data sets will empower us to gain insights in a way that was impossible before. The analysis of Big Data helps us to find initial evidence quickly, even though the correlations and patterns discovered by this kind of analysis will often not imply reliable conclusions. Nevertheless, if combined with good theory, computer simulations, and/or targeted experiments to verify or falsify certain hypotheses, Big Data can help us to generate better knowledge faster and more effectively than ever before. For this reason, Big Data will certainly transform research and innovation.

The health system is another institution which will benefit a lot from Big Data. For example, *IBM*'s Watson computer will soon support doctors in diagnosing and treating diseases by evaluating a much larger body of medical literature and evidence than doctors could do on their own. In addition, it will become possible to correlate our health status (or diseases) with our diet, genetic predisposition, and data about our socio-economic environment. Rather than having to undergo general medical treatments with considerable side effects, as was the case with antibiotics and other medication in the past, the health systems of the future will enable personalized medicine and treatments which will be designed and calibrated to meet the needs of individuals. These treatments will be more effective while reducing undesirable side effects: what is good for one person might be bad for someone else. Furthermore, we might be able to diagnose emerging diseases early on, before they break out, thereby helping us to prevent them. As a result, disease prevention might become more important than treatment. Obviously, this will cause a major paradigm shift in the health system.

A further interesting opportunity is opened up by a recent study conducted by Olivia Woolley Meza, Dirk Brockmann and myself.[3] To fight pandemics, it is important to immunize people in order to reduce the spread of an infection from one person to another. But how can we ensure that a sufficient number of people gets immunized (particularly when supply is limited)? Immunizing everyone is typically not possible because there are usually not enough immunization doses. Informing the public about the infection through mass media is also not effective because many people will either not respond or overreact. Surprisingly, however, if people knew about the number of infections in their social circle, this would let those who are most likely to be infected seek immunization. In other words, local information can be more effective than global information. Information platforms such as *Flu Near You* or *Influenzanet*[4] are now making this type of local information flow possible.

To benefit from the opportunities mentioned above, we must carefully establish a trustful relationship with the patient. Before sensitive personal and health data can be used, trustworthy and reliable information technologies and governance frameworks need to be developed so that patients have a sufficient level of control over the use of their treatment and data and how this affects their lives. Most likely, the

---

[3] Woolley-Meza et al. [1].

[4] See https://flunearyou.org/ and http://influenzanet.info.

health system can only be sustainably improved, if there is a reasonable involvement of the patients.

10.4 Banking and Finance

The financial sector has been in turmoil since a long time. This is probably best illustrated by the latest financial crisis and by new phenomena such as flash crashes, which can wipe out almost $1 trillion of stock market value in a couple of minutes, as it occurred on May 6, 2010.[5] These flash crashes are considered to be a side effect of algorithmic trading. In fact, about 70% of all financial transactions are now autonomously performed by computers, whereby markets are continuously and automatically monitored for potential opportunities. An entirely new financial business model based on high-frequency trading has emerged, which has undermined the foundations of traditional financial investments.

There has also been an explosive expansion of shadow banking and the market of derivatives.[6] In many cases, derivatives have replaced insurance contracts. They have entirely transformed the insurance business and the way real estate is financed.[7]

Furthermore, payment processes and money are also undergoing a transformation. While most financial transactions were based on cash or bank transfers in the past, they are increasingly being replaced by credit card payments and systems such as *Paypal, Google Wallet* or *Apple Pay*.

Furthermore, we see a trend towards microcredit, peer-to-peer lending and peer-to-peer money transactions, be it through *BitCoin, P-Mesa* or other means. Hence, the trend is towards decentralized approaches, where banks are not needed as intermediaries anymore. Monetary transactions are directly executed between people. This may be seen as a response to the failure of banks to provide a good service for everyone, evidenced by the unaffordability of loans and private homes for a broad range of companies and people. Some people even think the latest financial crisis, by far the biggest ever, was a direct result of the digital revolution. This is, because interactions based on trust were replaced by credit default swaps and other financial derivatives, which sought to insure traders against high losses. In the future, banks and other companies will certainly have to pay more attention to trustable products, procedures, and systems to be successful.

---

[5] See https://en.wikipedia.org/wiki/2010_Flash_Crash.

[6] In finance, a derivative is a financial product. Its worth is derived from its underlying assets, indices or interest rates.

[7] And ultimately triggered the financial crisis.

10.5   In the Wake of Big Data, the Pillars of Democracies Are Shaking

Democracies are often said to rest on four pillars: the legislative body, the executive body, the judiciary and the media. Personally, I think that science should actually be included as a fifth pillar of democracy. How would one otherwise want to make well-informed decisions? If governments want to fulfill their democratic mandate well, evidence-based decision-making is certainly essential. All of these pillars, however, will soon undergo a fundamental transformation and the signs of change are already there.

We have seen perhaps the most dramatic changes so far in the media. The Internet has undermined the traditional business model of printed newspapers because news is often provided for free in an attempt to attract more readers and online advertisers. A similar dynamic is at play in the music and film business. However, these businesses were the architects of their own downfall: they entered the digital arena before they had developed a viable online business model.

In addition, people are now turning away from mass media in favor of TV on demand and engage with personalized information sources such as *tumblr*. In this field too, there is clearly a trend towards individualization and decentralization, where people play a more active role in choosing and curating the information they consume. In an internal report on the digital strategy of the *New York Times*, for example, the paper's readers were identified as its most "underutilized resource".[8] By writing comments and blogs, and by sharing news stories on social media, readers are reshaping the media landscape. They determine which stories and which news sources reach a wide audience online. Grassroots journalism which builds upon local expertise is another interesting development.[9]

Beyond the media world, however, the executive branch of governments and the judiciary are also quietly changing. Police, secret services and other authorities now routinely use surveillance and Big Data to identify crime hotspots, terrorist suspects, traffic violations, tax evaders, and corruption. To accelerate trials, plea bargaining is becoming increasingly common. In addition, the judicial process is being shortened, which often means that defendants have fewer opportunities to defend themselves. The international trade agreements currently under negotiation[10] even envisage that conflicts of interest will be settled outside of the existing court system. Again, these changes are driven by the desire to increase efficiency by removing regulatory obstacles. Complementary, conflicts of interest may also be resolved through voluntary settlements negotiated in a community-based mediation process.

In the area of administration, many routine jobs will be taken over by computers. I believe it is just a matter of time until the legislative process will undergo a fundamental transformation, too. First, the concept of centralized decision-making is

---

[8] See http://www.scottmonty.com/2014/05/what-brands-can-learn-from-bombshell.html.

[9] Williams [2]; see also https://en.wikipedia.org/wiki/Citizen_journalism and https://krautreporter.de/.

[10] Such as TTP, CETA, TTIP, and TISA.

increasingly being questioned by countries and citizens who feel that diversity is not valued highly enough. Second, we have to overcome the problem of over-regulation by adopting new approaches which allow for more innovation and for solutions which are tailored to local needs. Therefore, I believe that long-term planning and administration will increasingly be complemented or replaced by more flexible approaches. I am convinced that the information systems of the future, particularly the Internet of Things, will enable self-organization, (co-)evolution, and collective intelligence. These approaches will probably become the new organizational principles of the digital society and the Economy 4.0 to come.

Instead of trying to control innovation through complicated regulations, it might be better (particularly for job creation) to transfer the responsibility for their externalities (i.e. the positive and negative consequences) to the beneficiaries. The simple rule that the originator of externalities would have to compensate others for damage created might replace thousands of complicated regulations by simple measurement standards and unleash a wellspring of creativity which is currently stifled by red tape.

In fact, over-regulation, too low innovation rates, and unemployment levels are currently among our greatest worries. Recently, there were about 25 million unemployed people in the European Union (the EU-28 states), and close to 20 million in the Eurozone.[11] However, this does not even begin to account for the vastly greater number of people who are not officially seeking employment (and thus are not counted as unemployed), while they would actually like to have a job if one were available for them. Unfortunately, things will get even worse as many employees will be replaced by intelligent machines. Experts predict that the number of jobs in the industrial and service sectors will drop by 50% in the next 20 years.[12] Unfortunately, large technology companies are unlikely to create enough new employment in the digital sector to make up for the difference. As a consequence, we may see an unemployment rate of 30 to 50% or even more—a number that probably no country can sustain based on the current socio-economic framework. Faced with such challenges, it is clear that we need to reinvent everything—our economy, the way we innovate, the way we do business, and the way we run our societies (see also Appendix 10.1).

10.6 Industry 4.0

Recently, many newspaper columns have been devoted to the rise of the "Industry 4.0". To explain what this is about, let us begin with the "Industry 1.0", which represents the first stage of industrial automation, involving innovations such as the steam engine and the mechanical weaving loom. In turn, the "Industry 2.0" was the age of the conveyor belt, which heralded the dawn of mass production. This process was further refined by the "Industry 3.0", whereby many production tasks

---

[11]In Japan, too, about half of young adults don't have a permanent position, see http://www.zeit.de/2015/18/japanjugend-lebensstandard-oecd.

[12]Frey and Osborne [3], Rifkin [4], Brynjolfsson and McAfee [5], Kurz and Rieger [6].

were carried out in a computerized way or even by robots. Finally, the "Industry 4.0" marries mechanization with communication technology to make it possible for intelligent machines (or robots) to directly communicate with each other or with the few remaining production workers. The "Industry 4.0" represents the next step of automation, which will eventually lead to a largely self-organizing production system. For example, in modern car factories, there are almost no workers. Most of the tasks are performed by robots, which are remotely controlled by a few skilled engineers.

The key communication technology driving this development is the "Internet of Things", which uses networked sensors to generate information that can be used to manage production in real-time. At home, the Internet of Things plays a similar role, allowing us to control our Bluetooth stereos or TV sets with our smartphones or tablets. But it would be naive to believe that the only consequences of digital technologies will be more efficient production processes and smarter gadgets. The digital revolution will transform our entire economy, and our societies as well. We will see a trend towards selforganizing systems everywhere. Concepts from complexity science will be empowered by Internet of Things technologies and vice versa. This will provide us with the key to a better future.

10.7   New Avenues in Production, Transportation, and Marketing

Let us now look at disruptive innovations in the areas of business and transportation. For 100 years, vehicles have looked more or less the same. They have had four wheels, a petrol or Diesel engine, a steering wheel and a driver. The production of vehicles was centered around a few companies who could benefit from "economies of scale" and advertise their products to a mass market.[13]

But now, it has suddenly become fashionable to drive electric vehicles produced by companies such as *Tesla*, while *Google* and other firms are developing driverless cars. Furthermore, *Uber* is challenging the traditional taxi business model by connecting passengers directly with freelance drivers in their vicinity. *Amazon* is even experimenting with automatic delivery by drones.

Rather than creating catch-all messages for a mass audience, advertisers are increasingly using personal data to deliver tailored messages to specific target markets. The customers don't anymore have to search for products that might interest them—the products and services find their way directly onto the screens of potential customers. Online shopping platforms know our desires and suggest which book we should read, which product we should buy, and which hotel we should book. Rather than visiting a shop and hunting for a product, as we would have done a few years

---

[13] Henry Ford famously stated that Ford customers could order their vehicles "in any color, as long as it's black".

ago, we now increasingly shop online and get a far greater array of products delivered to our homes than what we can find in a shopping mall.

But that is not all. Companies like *eBay* allow everyone to sell products. This creates a peer-to-peer market, whereby items can be reused. Rather than throwing a used product away, we can now sell it, donate it, or share it with others. In fact, we are currently witnessing the emergence of a "sharing economy". Car sharing is just one of the earlier examples for this. Now we enjoy new services such as *Couch-surfing* or *Airbnb*, which offer strangers the opportunity to stay as guests in other people's homes—something that would have been unimaginable just a few years ago (before the emergence of online reputation systems). In the meantime, neighbors share drilling machines and other tools, books and bikes, and offer personalized services.

The emergent sharing economy seems to be the result of the recent economic crisis. While earning less, people try to maintain a high quality of life by sharing goods. As a positive side effect, resources are being used in a more sustainable way. In future, the sharing economy is a great opportunity to further increase living standards despite the competition of a growing world population for the limited resources of our planet. The sharing economy will certainly be an integral and important part of the circular economy, which we need to build in order to reduce scarcities and waste.

The sudden move towards shared use (which may be seen as a special case of "recycling") is enabled by novel online platforms that can directly match supply and demand at a local level. It is now possible to coordinate the socio-economic activities between many more individuals, even though they might have very diverse interests, skills, resources, and needs. The companies facilitating this coordination enjoy remarkable growth rates of around 20%, which benefits from the recent trend towards sharing rather than exclusively owning goods.

However, we can also see a trend to buy and sell personalized products, for example, tailor-made jeans. In fact, entire shops have been established to customize mass products to individual tastes. Individualized production is becoming a trend. This is fueled, for example, by 3D printers and other new production methods.[14] After the "democratization of consumption" in the twentieth century,[15] we see a "democratization of production" in the twenty-first century. In fact, the separation between producers and consumers is already eroding. We are becoming a mixture of producers and consumers, so-called "prosumers", who play an active role in the conception and production of the goods we buy. On the Internet, and even more so on social media, this has already been true for some time.

Eventually, it will not make much sense anymore to manipulate opinions through advertising. Instead, companies will simply give us what we really want, if we are willing to let them know. In the future, a symbiotic relationship between producers and consumers will be key to economic success. Companies that don't care about the wishes and opinions of their customers will have little chance in the marketplace,

---

[14]"Additive" production methods now make it easy to generate anything from pizzas to artificial skin to houses, using particular 3D printers.

[15]Where ordinary people have more comfort than kings used to have some hundred years ago.

while those which engage in a fair partnership with their customers will thrive. Therefore, we will see a shift from a company-centric market to a user-centric market.[16] Cooperative networks between consumers and producers will also play an increasingly important role for providing the information that is necessary to create better services and products.

This trend towards more personalized and individually customized products will create a "hyper-variety market". In fact, the digital economy will open up a vast array of business opportunities and new product categories, as the information age unleashes a wave of creativity. The possibilities to produce creative products such as music, news, blogs, and videos will be endless. The current growth in the market for smartphone apps and virtual goods and services is a harbinger of this potential. For example, every minute, more than 500,000 posts are published on *Facebook*. Moreover, there are now more than half a million smart phone apps in existence and about as many app developers in Europe. In 2013, *Apple* alone generated more than $10 billion dollars in revenue for developers through its *AppStore*. In December 2013, more than three billion apps were downloaded.

This has very interesting implications for the future structure of markets. While today, we have a few monopolies or core businesses and some peripheral business activities, it is likely that peripheral products will become increasingly important in future.[17] This doesn't mean that large companies wouldn't grow any further. We should rather imagine this transition like a digital desert turning into a digital rain forest, where species of all kinds and sizes coexist.[18] This transition will benefit from a number of facts: First, in the emerging "zero marginal cost society",[19] producing copies of a digital service or product is cheap, and creating more copies doesn't cause much higher costs. Second, it is important to realize that the digital economy isn't a zero-sum game, where one can only gain when others lose.[20] Therefore, besides getting rich by cutting costs and rationalization, where immaterial (information-based and cultural) goods are produced, co-creation and social synergy effects offer a second way of creating value. Third, participatory information platforms can act like "catalysts". In some sense, they are the "humus" fueling the evolution of a thriving digital information, innovation, production and service ecosystem. Such digital ecosystems enable exponential innovation (rather than producing a new version of a certain product every few years, as we often have it today). To allow this exponential innovation to happen, we need interoperability and suitable intellectual property rights (IPR) (see Appendix 10.1). Even the G8 Open Data report points

---

[16]This also implies a transition from a push economy to a pull or "on demand" economy.

[17]One may visualize this by representing today's core businesses by the core of a sphere and peripheral businesses at its surface. Interestingly, the relationship between the surface area A of an n-dimensional sphere of radius r and its volume V is $V = rA/n$. Therefore, the higher the dimensionality n of a market, the more economic activity happens at the surface, i.e. in peripheral businesses!

[18]See https://horizon-magazine.eu/article/open-data-could-turn-europe-s-digital-desert-digital-rainforest-prof-dirkhelbing.html.

[19]Rifkin [7].

[20]Wright [8].

out that the competitiveness of countries will depend on the sharing and reuse of data.[21] This will be crucial, in particular, to avoid mass unemployment by creating opportunities for small and medium-sized companies, and for self-employment, too.

10.8 Where Are We Heading? New Forms of Work

The digital economy will profoundly affect the world of work. In the past, long-term employment at one company was common (at least in large German companies such as Daimler Benz, BMW, or Bosch). In the meantime, however, companies increasingly opt to employ their staff on rolling, short-term contracts. A growing percentage of the workforce already works on a temporary basis, and we will probably see a further rise in short-term employment. *Amazon Mechanical Turk*, for instance, enables companies to outsource certain tasks to a casually employed workforce, where the "working relationship" between employer and employee often lasts for a few minutes only![22] In this way it is, for example, possible to translate a 1000-page document within minutes, by splitting the task into 1000 subtasks, done by 1000 people.

Such short-term commitments may come at the cost of significant increases in the stress levels of employees. In view of the growing insecurity and existential risk created by these new forms of employment, an increasing number people advocate the introduction of a basic income enabling everyone to survive, even though not very comfortably.[23] This is a hotly debated topic: Why should people be entitled to get an unconditional payment? Would the state be able to pay for it? Would people still work hard, or would they get lazy? Probably, even if a basic salary were introduced, our economy and society should continue to build on merit-based principles such that most people would try to increase their incomes through paid work in order to reach a higher quality of life.

---

[21] Open Data in the G8: A Review of Progress on the G8 Open Data Charter, see http://www2.dat ainnovation.org/2015-open-data-g8.pdf.

[22] In some countries such as Great Britain, many people work on so-called "zero hour contracts". Besides Mechanical Turk, there are now a number of other platforms such as oDesk, guru, freelance, Elance, online outsourcing, 99 designs, or crowd flower.

[23] Some institutions have already calculated whether this would be affordable. If everyone got an unconditional "minisalary", a complicated and expensive social welfare system would not be necessary. As a result, the overall public budget would not change significantly. See also Konrad Adenauer Stiftung, Die Idee des Grundeinkommens – Ein Weg zu mehr Beteiligungsgerechtigkeit?, http://www.kas.de/wf/doc/kas_20403-544-1-30.pdf? 100831182341, http://www.sueddeutsche.de/wirtschaft/bedingungsloses-grundeinkommen-was-waere-wenn--1.1785811; Finland has recently decided to commit to a basic income, see https://basicincome.org/news/2015/06/finland-new-government-commits-to-a-basic-income-experiment/. The Netherlands is currently making experiments: https://web.archive.org/web/20150626120829/, http://www.independent.co.uk/news/world/europe/dutch-city-ofutrecht-to-exp eriment-with-a-universal-unconditional-income-10345595.html.

10.9   Everyone Can Be an Entrepreneur

Whatever your opinion on the issue of a basic income may be, it's likely that people will increasingly set up "projects" to earn money with own products and services. Such projects will be task-driven, supported by digital assistants, short-lived, and very flexible.[24] The entrepreneurs of the future will launch and coordinate such projects and organize the necessary support, as discussed below. Once completed, a project will end and the workforce will look for new projects to coordinate or participate in. Many people will probably coordinate a project and simultaneously participate in others.

Such a project-based socio-economic organization has a number of advantages. The projects will provide their participants with opportunities to influence the issues they care about. Project participants will also be able to work with a greater degree of independence in fields that genuinely excite them. As the projects will be short-lived, this will further overcome the "Peter principle", which posits that, in a conventional working environment, employees tend to be promoted until their position is beyond their ability.

Due to the various advantages, companies, political parties and other established institutions will increasingly use the organizational form of projects to complement their existing activities in a flexible manner. This will also help to overcome another crucial problem: many institutions today are too slow to adapt to our quickly changing world in a timely way.

10.10   Prosumers—Co-Producing Consumers

Digital technologies are now enabling entirely new and more flexible ways of organizing our economy and society. People are starting to use social media platforms to organize their own interests and are establishing their own "projects". In principle, everyone could do this now, given the required technical and social skills. The World Wide Web, social media, and homemade products created with 3D printers are just three examples of this trend. In fact, 3D printing technology is now enabling local producers or individuals to sell products to friends, colleagues and the rest of the world. Rather than merely specifying the color and individual features of a product when we order it, we can now design its components or composition to produce bespoke products, tailored to our needs. It is even possible to set up a team of designers, engineers, marketing professionals, and other specialists to design a customized smartphone with components produced by multiple companies or with new, tailor-made components. Furthermore, conventional factories are

---

[24]Michael Bernstein, a professor at Stanford University and leading crowdsourcing expert, speaks of "flash teams".

increasingly working hand in hand with collaborative projects brought about by the newly emerging digital economy.[25]

Note that "projects" as I have described them above are already in existence. Open-source software projects, for example, are largely run by people who want to use open-source components to develop own software more quickly, which they then make available to everyone in exchange. Such "open innovation" is typically built on a framework of "viral" open-source licenses, which encourage users to contribute something in return for what they get for free. These software licenses (such as the GNU General Public License) reward a culture of collaborative creation ("co-creation"), reciprocity, and fair sharing.[26]

This is fueled by the principle of "co-epetition"—a special form of competition that goes along with cooperation. In the context of open-source development, the *GitHub* platform has become particularly popular among software developers. The platform shows who has contributed what, whereby it creates incentives to contribute. Thus, everyone can benefit from the growing "ecosystem" of open-source software.

We must further pay attention to another fundamental change: as the Internet connects citizens, customers and users, we see a trend towards more bottom-up participation in our socio-economic system. Apparently, it can be counterproductive to engage in too much centralization and standardization without providing sufficiently diverse opportunities for an array of countries, regions, and local communities with widely differing interests, needs, weaknesses and strengths.[27] As systems become more complex, it takes more local knowledge to satisfy the increasing diversity of needs, which calls for more participation. Otherwise, complex socio-economic systems will perform poorly or may even become unstable over time. Therefore, let us now discuss the advantages and disadvantages of top-down and bottom-up organization in more detail.

10.11 Top-Down Versus Bottom-Up Organization

Top-down organization is common in military organizations, administrations, and many companies. The aim of this approach is to "command and control" so that leaders are able to exert their power and get their wishes and decisions implemented. The hierarchy inherent in such systems also makes it easier to establish a system of accountability. Top-down organization, furthermore, enables rapid decision-making

---

[25]Leadbeater [9], Ormerod [10].

[26]This reflects the spirit of the other-regarding behavior discussed in a previous chapter and benefits the entire system.

[27]The current problems with the EURO currency in a large economic area with very different growth rates, GDP per capita, and public debts in different European countries demonstrate this well.

and timely coordination of resources over large distances. The collection and evaluation of information required to make decisions, however, is often quite slow, because
it takes a long time for information to flow up the chain of command[28].

Using a top-down approach, it is possible to reach optimal results in a system, if
the goal is well defined and the outcomes can be well determined, i.e. if the system
isn't too complex, if it doesn't vary much, if changes are reasonably predictable,
and if problems can be well detected and quickly solved. Under such conditions,
top-down control can increase the performance of a system. Often, however, at least
one of these conditions is not fulfilled.

While the top-down approach can harness the individual intelligence and expertise of exceptional leaders, the large accumulation of power implies that incidental
mistakes can be disastrous.[29] In other words, the centralized leadership structure of
top-down systems can help to solve problems, but it often creates new (and sometimes even bigger) problems. Moreover, although top-down control enables faster
change, it often delays necessary change.

In sum, a top-down approach might be best if a system is sufficiently simple
and deterministic, and its variability is low. Therefore, it benefits from standardization.[30] A top-down approach works well in situations where it is more important
to be decisive than to wait for consensus. A medical emergency might be a good
example. But systems which are controlled from the top-down are vulnerable, and
the concentration of power makes them easily corruptible.

By contrast, bottom-up organization performs better under complex and highly
variable conditions, if suitable coordination mechanisms are in place. Good examples
are our immune system, markets and ecosystems. Bottom-up organization supports
flexibility, local adaptation, diversity, creativity, and exploration. It also tends to be
more resilient to disruptive events.

Bottom-up organization fosters democratic processes, but may also cause herding
behavior. To work well, good education and a willingness to take responsibility at
the bottom is required. Using a decentralized approach, more information can be
processed, and collective intelligence is enabled. However, collating and integrating
this information to make it useful tends to be challenging.

On the whole, top-down organization is based on the principles of power and
control, whereas bottom-up organization empowers people to help themselves and
each other. Top-down governance results from the accumulation of resources and
asymmetry in the system, while bottom-up governance benefits from distributed
control and tools that support peer-to-peer coordination. However, given that top-
down and bottom-up approaches both have strengths and weaknesses, neither

---

[28]And often also for commands to reach the lowest level from the top-down.

[29]Remember that 40% of TOP 500 companies disappear within a 10 years time
period; see, for example, https://www.aei.org/publication/charts-of-the-day-creative-destruction-
in-the-sp500-index/ and http://www.cnbc.com/2014/06/04/15-years-to-extinction-sp-500-compan
ies.html. Moreover, many of the most powerful historical figures have caused some of the greatest
tragedies in the history of humankind.

[30]Standardization has a number of undesirable side effects, however, including low innovation and
growing inequality, see Footnote xvi of Chap. 10.

approach will work best in all situations. In fact, these approaches can complement each other, and both are necessary. They must be applied in the right circumstances or suitably combined.

Currently, Big Data is being used to strengthen the top-down governance of our socio-economic systems, but a number of factors promote the spread of bottom-up organization, too. High levels of education as well as access to reliable, high-quality information and to information systems supporting evidence-based decision-making are important for this trend. In addition, Social Technologies are beginning to foster collaboration and trust in circumstances where it has been lacking before. For example, social media platforms are helping people to coordinate resources, while the growth of reputation systems lets people and companies behave more accountably and responsibly. Bottom-up approaches have further been strengthened by the spread of Open Data, citizen science, moderated Internet communities, and the "maker" movement.[31]

To cope with the increasing variability and complexity of socio-economic systems, more autonomy at lower organizational levels and distributed control approaches are often necessary. This is fundamentally changing the landscape for policymakers and citizens alike. In fact, more bottom-up participation is required for a number of reasons: there is a need for more capacity ("bandwidth"), more creativity, better solutions, and more resilience. Further advantages of bottom-up participation are related to diversity (such as innovation and collective intelligence), but we must certainly learn to master the challenges of diversity better. For this, we need suitable information systems, which support a differentiated kind of interoperability. Digital assistants and new kinds of Social Technologies (as I have discussed them in previous chapters) are examples of such information systems.

For all of the above reasons, decentralized, bottom-up organization is currently spreading. The rise of *BitCoin* and peer-to-peer lending may be taken as examples. Moreover, using smart grids, electricity can now be (co-)produced by citizens. "Swarm intelligence" is another powerful, decentralized approach inspired by the self-organized behavior of socially behaving animals. Taking this principle several steps further, citizen science and crowd sourcing have given rise to a number of credible information platforms and community services. Here, I have not just *Wikipedia* and *OpenStreetMap* in mind. In California, for example, citizens are collectively detecting earthquakes using a distributed network of sensors,[32] while in Japan a crowd-based sensing system has been developed to monitor nuclear radiation.[33]

---

[31]Anderson [11].

[32]Minson et al. [12].

[33]After Fukushima: Crowd-Sourcing Initiative Sets Radiation Data Free, see https://newsroom.cisco.com/feature-content?type=webcontent&articleId=1360403.

Fig. 10.1 Illustration of
classical hierarchical
top-down organization

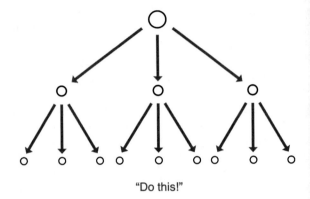

"Do this!"

10.12   Allowing Diverse Resources to Come Together Quickly

Let us now explore how modern information systems can allow top-down and bottom-up organization to come together in order to create new and superior forms of organization. In this context, it is instructive to examine the example of disaster response management, which has traditionally been executed from the top-down. But recently my perspective on the efficacy of the traditional approach has dramatically changed, namely when I helped organize a hackathon in San Francisco together with Thomas Maillart, Alexei Pozdnoukhov and *Swissnex*.[34] The aim was to discover new ways to recover from earthquakes. Even though the event took place on the National Day of Civic Hacking and was therefore competing with a lot of other hackathons, it attracted about 80 people. Obviously, the event hit a nerve. Participants formed nine teams, each of which dealt with a specific project idea. The results established a new paradigm of disaster response management, facilitated by modern information systems.

First, let us discuss the traditional approach: there, a commander instructs his chiefs of staff what to do, and they relay these instructions to their subordinates, who carry out the orders (see Fig. 10.1). However, during a disaster, the flow of information is often not smooth and the chain of command is often interrupted. This might be caused by a lack of information, disturbances, delays, or resource constraints. In fact, it often takes 72 h until disaster response units can work at full capacity. Sadly, within that time many people will have already died from injuries or from dehydration.

In contrast, when using modern information systems, some autonomy on lower levels can achieve better results. My conclusion from the hackathon is that it is better if commanders draw up and publicize a set of specific macro goals ("we need *x*

---

[34]What if the Big One Hits? Hacking Earthquake Resilience, see https://www.swissnexsanfran cisco.org/event/earthquakeresilience/, https://www.swissnexsanfrancisco.org/press/latest-news/ear thquakehackathon/.

Fig. 10.2 Illustration of a superior combination of top-down and bottom-up organization

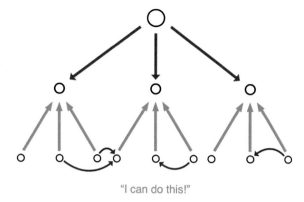

"I can do this!"

and y" or "we must do z"). In response, people, companies and non-governmental organizations can find the best way to use their individual and collective skills and resources to contribute to reaching these objectives (see Fig. 10.2).

For this approach to be efficient and effective, it is important to have suitable tools to coordinate the various activities, and to match resources with needs. In fact, new information and communication tools can play a crucial role for this. Such platforms were often either absent or not publicly accessible in the past. Nowadays, however, the technology necessary to enable good coordination is widespread. For example, information platforms such as Ushahidi play a valuable role in mapping crises and responding to disasters. Let us now discuss how new information technologies can serve as survival toolkit in future disasters.

10.13 Towards a More Resilient Society

One of the three teams which won the hackathon, *Amigocloud*,[35] came up with a smartphone app that allows everyone to take and annotate pictures of broken infrastructure and other problems, which are automatically uploaded to a public webpage whenever the mobile communication system is operational. When the central communication infrastructure is down, emergency connectivity can now also be provided by "meshnets", whereby offline mobile devices form adhoc peer-to-peer networks to transfer information. This is possible using smartphone communication protocols such as *FireChat*.

Of course, for this we have to assume that the mobile phones' batteries aren't empty. However, the second winning team proposed *Charge Beacons*,[36] i.e. autonomous local charging stations using solar panels, which allow citizens to

---

[35] See https://www.amigocloud.com/homepage/index.html.

[36] See https://www.youtube.com/watch?v=LVGwHAtLwVQ.

recharge their smartphones even during a blackout. As a positive side effect, the *Charge Beacons* also serve as community hotspots where people meet.

The third winning team, *Helping Hands*, developed a smartphone app that allowed everyone to ask for and offer help. For example, one could request baby food, fresh water, or support for someone in need at a specific location. Others could offer clothes, water, food, or various kinds of support. The smartphone app would locally match resources with needs. This empowers the community to help itself and to harness the generosity witnessed during crises. As a consequence, the public disaster response teams will be able to focus more on providing aid in areas where people can't help each other well enough. Thus, the above-described approach frees up resources for urgent and strategic matters. Consequently, both the citizens on the bottom and the governance system on the top will benefit. Remarkably, all of the above concepts emanated from a single hackathon, on a single day!

The observations of Yossi Sheffi in his book *The Resilient Enterprise* are also relevant in this context.[37] He describes what it takes to keep a business running when struck by disaster. Sheffi (*1948) contends that the boss must provide a framework that allows the company's experts to find solutions. However, he/she shouldn't interfere with the details of this process, as attempts to micro-manage are often counterproductive. Instead, the aim should be to empower the staff to find solutions for themselves.

But given that we may never experience a disaster in our lifetime, why should we all care about disaster response management? Because it offers guidance how to successfully deal with complex problems. In fact, this example is highly relevant to many of the challenges faced in politics and business today. It sometimes seems that, by the time one problem is solved, there are already one or two more problems waiting to be fixed. In a rapidly changing, highly variable world which is difficult to predict, we need flexible governance systems that can quickly adapt to new events and local needs (see Fig. 10.3). This will determine the resilience, viability and success of a complex system under pressure, which may be a company challenged by competitors, a city struck by disaster, or a country exposed to the pressures of globalization.

10.14   A New Kind of Economy Is Born

As discussed above, the optimal organizational approach to disaster response management combines elements of both, centralized top-down control and distributed bottom-up organization. Could we adopt this approach not just in crisis situations, but every day? In fact, the creation of goods, services and knowledge in the emerging Economy 4.0 works exactly like this.

A critic might argue that I am "reinventing the wheel", because some companies have already been managed this way. In fact, many companies subcontract work

---

[37]Sheffi [13].

Government 1.0   Government 2.0   Government 3.0   Government 4.0

 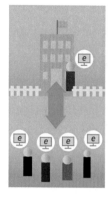  

Fig. 10.3 Illustration of the likely evolution of government models. Government 1.0 tells people what to do and is separated from its citizens. Government 2.0 uses electronic means to engage in a two-way interaction with its citizens. Government 3.0 engages in individually fitting interactions, services, and solutions. This is currently being implemented in South Korea. (You may want to watch this related movie list: https://www.youtube.com/watch?v=KgVBob5HIm8&list= PLDmlT_+Ptfv0fcMD29kNKOIhV0yK9AhyQl) Government 4.0 also builds on the coordination among its citizens and on mutual help, which establishes participatory resilience

to third parties who can carry out tasks more efficiently. This works similarly to the suggested mode of organization illustrated in Figs. 10.2 and 10.3. However, most companies and institutions still seem to be governed predominantly in a top-down way.

While outsourcing avoids certain undesirable consequences, such as inefficiencies due to duplicate efforts and some other problems, too, it also prevents favorable things from happening. The same applies to hierarchical forms of organization. So, where do we have potentials for improvement?

Suppose a company experiences an economic downturn because it sells fewer products than expected. Traditionally, the firm would lay off employees to improve its balance sheet. However, one obvious reason why the company does not sell enough products is that it doesn't offer enough interesting products that people would buy. Therefore, suppose the company decided to call for new product ideas from within the firm to make use of the knowledge, skills, and machinery at its disposal. This would probably generate several interesting ideas for new products, which could help to overcome the stagnation and the lack of innovation that large companies often experience. In other words, a company could offer its employees a platform ("playground") that allows them to set up their own "projects" from the bottom-up. This would enable groups of staff to work autonomously on the development of new products for a limited period of time. Depending on the results, the company would then decide whether to close these units down or turn them into spin-off companies.

10.15 Emergence of a Participatory Market Society

While, in the past, most of us couldn't participate in improving the man-made systems around us (because the tools necessary to coordinate the knowledge and skills of many people weren't widely available), this is now changing. New information systems and organizational principles will allow individuals to actively engage as citizens in the public arena, as employees within firms, and as consumers and users within the production process. Therefore, if set up well, involving users, customers, and citizens will naturally lead to better services, better products, better businesses, better neighborhoods, smarter cities, and smarter societies. By building suitable platforms to coordinate information and action, we can create more opportunities, enable people, companies and institutions to make better decisions, and encourage people to act responsibly.

So far, I have discussed the growth of short-term projects, co-creation, home production, sharing, personalization, hyper-variety markets and the importance of bottom-up participation. Another crucial trend is the increasing relevance of "networked thinking". In this connection, studying the Silicon Valley provides some interesting insights.

The Silicon Valley exhibits a surprisingly fluid exchange of people between companies. If a company goes bankrupt, which is quite common, people usually find a new job quickly. In a sense, it is not unreasonable to think of people as being employed on a long-term basis by the Silicon Valley rather than being employed on a short-term basis by its companies. In other words, the Silicon Valley is like a "super-company", in which there is an invisible flow of knowledge and people that connects all of the companies located there. The companies themselves are the niches, in which experimentation can take place and among which a lot of diversity can exist. In effect, therefore, the Silicon Valley is nourished and sustained by the co-evolution of the companies and ideas within it.

Furthermore, it is instructive to view the interaction of companies within the Silicon Valley as forming a huge "information, innovation, production and service ecosystem", where the various economic stakeholders may be compared to different biological species. For the economy to be in good shape, it is important that all of the sectors, which consume and produce, do well. If a significant part of the production ecosystem disappears, this is as if some species disappear, which can disrupt the entire economic ecosystem and cause a recession. In fact, in an interesting study, Cesar Hidalgo, Laszlo Barabasi, Ricardo Hausmann and others[38] showed that economic development and prosperity largely depend on the economic diversity of this "ecosystem", which mirrors the concept of biodiversity in nature. Similarly to the natural world, a greater diversity of economic production leads to more innovation and economic prosperity.

However, an "information, innovation, production and service" ecosystem can be organized in different ways. Fueled by almost unlimited amounts of venture capital, many innovators in the Silicon Valley aspire to build global monopolies.

---

[38]Hidalgo et al. [14].

This creates "walled gardens",[39] between which very limited information exchange and cooperation exists. To build complex products, it is common to acquire other companies, but this reduces their usefulness for third parties.

To catch up and be competitive, Europe may take another route. After all, the organization of our economic system should serve our society best. Considering that data will be cheap (since we produce as much data in one year as in all the years before), it would make sense to engage in an Open Data approach and in open innovation, where "walled gardens" are avoided, but information exchange is built on merit-based principles. Money would be made by distilling useful information from raw data and by distilling applicable knowledge from information. By requiring differentiated interoperability, complex products could be created in a modular way by organizing "projects" into networks of projects. For such "super-projects" to grow in a self-organized way, the interaction must be mutually beneficial and will often involve a multi-dimensional value exchange (see Appendix 10.2).

As I explained in the previous chapter, evolutionary principles will anyway sooner or later cause a paradigm change from self-regarding to other-regarding forms of organization, since this produces better outcomes on average. For companies, this means that they need to communicate and cooperate more with their customers and suppliers. Next-generation social media will provide suitable tools for this. Companies that manage to offer individually tailored, customized products and services will have a competitive advantage. Obviously, this requires more information to be shared, and in order for this to be viable in the long term, a trustworthy and fair system of bidirectional communication and collaboration is crucial. As a consequence, the business leaders of tomorrow will have to be well acquainted with "systems thinking", an approach which integrates and balances different interests and perspectives. Companies like *Porsche*, for example, are already aware that they can only produce and sell top-quality cars by engaging in a partnership with both their employees and their customers.[40]

10.16 Supporting Collective Intelligence

The challenges posed by the increasing complexity of our anthropogenic systems also call for more intelligence. Interestingly, collective intelligence can surpass the intelligence of smart people and powerful computers. It thrives in environments, where diverse knowledge and skills can come together. But collective intelligence isn't possible, if decisions are taken from the top-down (as this is restricted to the intelligence of a single mind or a small circle of people). In comparison, majority decisions are often better than those of single individuals or centralized authorities, because they take more perspectives and knowledge on board. However, in our

---

[39] In fact, most smartphone apps are "walled gardens", and they are usually not interoperable with other apps.

[40] See http://www.zeit.de/2014/34/porsche-betriebsrat-uwe-hueck.

increasingly complex world, the outcomes of majority votes are also too limited, and the contributions of minorities become increasingly important. This is one of the reasons why more participatory ways of organizing our economy and society are emerging.

To foster collective intelligence we must build platforms allowing people to coordinate and integrate information so that they can learn, innovate, and produce collaboratively. The goal must be to create a participatory "information, innovation production, and service ecosystem". This must provide sufficient niches for diverse, competitive problem-solving approaches, but also incentives for interoperability and cooperation. Furthermore, it must be sufficiently easy to integrate partial solutions in a modular way to reach higher-level goals. In fact, when thinking about the development of a modern car or plane, we immediately understand that some systems are so complex and require so many different skills that no one person can fully grasp all the details.

Importantly, by growing a diverse ecosystem in which many different approaches can co-exist and co-evolve, we will also make our society more resilient. The flexibility of this approach, furthermore, means that whatever happens, we will have a rich arsenal of options to respond. In contrast, the international trade and service agreements which have recently been negotiated may homogenize and standardize the world too much, and thereby eliminate the niches in which new ideas emerge and thrive.[41] Niches provide necessary opportunities for experimentation. Indeed, this is why evolution is so effective.[42] If we homogenize our socio-economic systems too much, they will be more vulnerable to disruption—with potentially disastrous consequences.

10.17   Preparing for the Future

What can we do to prepare ourselves for the Economy 4.0? In the past, we built public roads to facilitate the industrial society and public schools to provide the workforce for the service society. Now, we will need to build the public institutions for the digital society. This will include open and participatory information platforms, which will support people to make better decisions, to act more effectively, and to coordinate and help each other. In addition, participatory platforms should support creative projects and collaborative production.

In order to overcome the challenges ahead of us successfully, it is important to acknowledge that citizens, consumers, and users are essential for our society's success and must be treated as partners—in a fair way and with respect. If we want to create new jobs, we must also increase opportunities for small and medium-size

---

[41] Large companies are often rather bad at innovating, and that's why many of them need to acquire other companies to survive.

[42] Remember also that local interactions are important for cooperation to occur when collective goods shall be created, as I have shown in a previous chapter.

companies, and self-employed people. This can be reached by avoiding too much standardization while demanding differentiated interoperability, particularly from large companies. In this respect, incentivizing information exchange and Open Data is of key importance.[43]

Well-designed, participatory information platforms will help everyone to engage in the Economy 4.0 more effectively. They will support people in communicating more easily, identifying suitable partners to work with, coordinating activities, and collaboratively creating goods and services. Furthermore, such platforms could be used to do financial and project planning, to manage supply chains, to schedule processes, to carry out accounting, and to perform many other management tasks such as health insurance and tax administration. Such a platform could largely reduce the barriers of entry to the market and provide everyone with the tools needed to easily set up collaborative projects. In fact, future job and work platforms should have all of these features. Why don't we build them as public infrastructures to foster the Economy 4.0?

10.18   Appendix 1: Re-Inventing Innovation

Compared to material goods, information is a special resource. While material goods are limited, which can lead to resource conflicts, information can be reproduced cheaply and as often as we like. Nevertheless, current intellectual property rights treat digital goods more or less like material goods. I believe, a different kind of intellectual property right (IPR) would dramatically accelerate innovation and create many more jobs. While we have to catch up with the pace at which our world is changing, the current IPR regime creates major obstacles. Therefore, we need a new paradigm which will allow collaborative creation ("co-creation") to flourish.

In fact, we could fundamentally change the way we foster innovation. Currently, many people don't like to share their best ideas, because they don't want other people to become rich in the wake of their research, while getting very little compensation. As a result, it often takes years until an idea is shared with the world through a publication or patent. But what if we innovated cooperatively from the very first moment? Let us assume an idea is born in America, and it is shared with others through a public portal such as *GitHub*. Afterwards, experts from Asia could work on these ideas for a couple of hours, then experts from Europe could build on their results, and so on. In this way, we could create a research and development paradigm that never sleeps, that overcomes the limits of a single team, and that embraces "collective intelligence".

---

[43] A report by McKinsey & Co. estimates the additional value of Open Data to be of the order of $3–5 trillion per year: https://www.mckinsey.com/business-functions/mckinsey-digital/our-ins ights/open-data-unlocking-innovationand-performance-with-liquid-information.

Such an approach would produce considerable synergy effects. My colleagues Didier Sornette and Thomas Maillart recently demonstrated that, by open collaboration, two people can produce software that would otherwise have required 2.5 developers ("1 + 1=2.5").[44] Geoffrey West, Luis Bettencourt and I, together with some others, discovered a similar pattern in cities: productivity that depends on social interactions tends to disproportionately increase with population size.[45] For example, a city with two million inhabitants would be about 20% more productive per 1 million inhabitants than two cities of one million. This is probably the main reason for the rapid and on-going urbanization of the world.

Interestingly, Internet forums of all kinds have nowadays created something akin to virtual cities. Many citizen science projects (and also the famous *Polymath* project on collaborative mathematics) underline that a crowd-based approach can complement or even outperform classical research and development approaches.[46]

Given the great advantages of collaboration, what are the main obstacles? A central problem is the lack of incentives to share. Currently, researchers are motivated by two kinds of rewards: they receive a basic salary and they earn the recognition by their peers in the form of citations of their published work. For this reason, many scientists do not share their ideas until they have been published.

Patents are a further obstacle to the sharing and widespread implementation of good ideas. While patents are actually intended to stimulate research and development by protecting the commercial value of ideas, in the digital economy patents seem to hinder innovation more than they foster it. It is as if everyone would own a certain number of words and could charge others for using them—this would certainly obstruct the exchange of ideas considerably.

However, it has recently become difficult to legally enforce hardware and software patents, and there have been an increasing number of patent deals between competing companies. The electric car company *Tesla* has even decided to allow others to use their patents.[47] All this might indicate that a paradigm shift in terms of intellectual property rights is just around the corner.

Moreover, it has become increasingly difficult to earn large amounts of money by publishing music, movies or news. This is not just a problem of illegal downloads. In contrast to material resources, information is becoming an abundant resource. Given that every year, we produce as much data as in the entire history of humankind, information will become increasingly cheaper.

---

[44]Sornette et al. [15].

[45]Bettencourt et al. [16].

[46]See https://en.wikipedia.org/wiki/Polymath_Project, http://polymathprojects.org/, http://dl.acm.org/citation.cfm?id=1979213.

[47]Why Elon Musk just opened Tesla's patents to his biggest rivals, see http://www.bloomberg.com/bw/articles/2014-06-12/why-elon-musk-just-opened-teslas-patents-to-his-biggest-rivals.

10.18.1 Micropayments Would Be Better

So why not pursue an entirely different IPR approach, perhaps in parallel to the current intellectual property regime? By it's very nature, information "wants" to be free and to be shared. Every culture is based on this. Information is a virtually unlimited resource, which in principle can be reproduced almost for free. In contrast to material resources, this allows us to overcome scarcity, poverty and conflict. Nevertheless, we currently try to prevent people from copying digital products. What if we simply allowed copying, but introduced a micropayment system to ensure that every copy generates revenue for the content creator (and those who help to spread content)? Under such circumstances, we would probably love it when others copy our work!

Rather than complaining about people who copy digital products, we should make it easier to pay for the fruits of creativity and innovation. Remember that, some time back, *Apple's iTunes* made it simple to download and buy songs, for 99 cents each. It would be great to have a similarly simple, automatic compensation scheme for digital products, ideas and innovations. Modern text-mining algorithms could form the basis of a system, where content creators and companies would be automatically paid whenever their ideas are used. This payment could be calibrated according to the scale of the initial investment, the age of the invention and its "innovativeness", i.e. the degree to which it made advances over already existing solutions. This would encourage cooperative innovation without providing a disincentive for new research.

Establishing a micropayment system would also allow companies and citizens to earn money on the data they generate and exchange. Then, everyone could benefit from contributing to the global information ecosystem. This would create an incentive system that would reward the sharing of data. But to get paid for every copy, one would need a particular file format. Copies ("offspring") of data would have to be linked with their respective source ("parent") via a kind of "data cord", so that micro-payments between the owners and users of the data can be processed.[48] In fact, something like a "Personal Data Store" would be needed to execute these payments.[49]

---

[48]Lanier [17].

[49]Another function of this Personal Data Store would be to give each user control over his or her own personal data. Whenever personal data would be (intentionally or accidentally) produced about someone, it would have to be sent to that person's data store (which would be like a mailbox for data). The person could then determine what kind of data they are willing to share, with whom, for what period of time, and for what purposes.

10.19   Appendix 2: Multi-Dimensional Value Exchange

So far, we haven't addressed the kind of financial system we will need for the Economy 4.0. In its current form, money has a serious short-coming: it is one-dimensional.[50] This makes it unfit to manage complex dynamical systems. From the perspective of control theory, this problem is completely obvious. For example, complex chemical production processes cannot be governed using a single control variable such as the concentration of a particular chemical ingredient. In a complex production process, one must be able to control multiple variables, such as the temperature, pressure and concentration of numerous ingredients.

It is also instructive to compare this with ecosystems. The plant and animal life existing in a particular place is not solely determined by a single factor such as the amount of water. Temperature, humidity and various kinds of nutrients such as oxygen, nitrogen and phosphor play a role, too. Our bodies also require many kinds of vitamins and nutrients to be healthy. So why should our economic system be any different? Why shouldn't a healthy financial system need several unrelated kinds of value exchange?[51]

It is helpful to recognize that the financial system is primarily a system to coordinate the use of resources. Therefore, we shouldn't hesitate to invent better coordination mechanisms. For example, besides conventional money, our future value exchange system could consider environmental factors and other material and non-material externalities such as social capital (for example, trust and reputation).[52] In fact, the Finance 4.0 system introduced in a later chapter does this.

10.19.1   We Could All Be Doing Well

The circumstance that people respond to many different kinds of rewards, as we have seen in the previous chapter, allows us to establish a multi-dimensional incentive and exchange system (which is very different from a nudging approach[53]). This system would support the self-organization of individuals and companies, which is highly important to successfully manage complex socio-economic systems in future. However, compared to the financial system we have today, such new forms of value would not necessarily be easily convertible. As a result—depending on how many dimensions of value we would distinguish—everyone could be doing well, when

---

[50]Even though there are many different currencies in the world, we can convert them in an almost frictionless way, which makes money effectively one-dimensional.

[51]This would lead us to "multi-dimensional finance", and entirely new era in the history of money.

[52]Further details of this idea can be found in the International Application No. PCT/IB2015/050830.

[53]In contrast to nudging, this approach is not trying to manipulate the decisions of people by exploiting their cognitive biases, but it builds on active decisions and trustable information systems, which are minimizing data biases.

measured in the kind of value which best suits the personal strengths, skills and expertise[54]. That in itself is an interesting perspective worth pursuing!

References

1. O. Woolley-Meza, D. Helbing and D. Brockmann (2015) Limited information activates resonant epidemic control, preprint, https://www.researchgate.net/profile/Dirk_Helbing/public ation/329963800.
2. E. Williams, Grassroots Journalism (Dollars & Sense, 2007).
3. C.B. Frey and M.A. Osborne, The future of employment: how susceptible are jobs to computerisation? Available at: http://www.oxfordmartin.ox.ac.uk/downloads/academic/The_Future_ of_Employment.pdf (2013).
4. J. Rifkin, The End of Work (Tarcher, 1996).
5. E. Brynjolfsson and A. McAfee, Race Against the Machine (Digital Frontier Press, 2011).
6. C. Kurz and F. Rieger, Arbeitsfrei: Eine Reise zu den Maschinen, die uns ersetzen (Riemann, 2013).
7. J. Rifkin, The Zero Marginal Cost Society: The Internet of Things, the Collaborative Commons, and the Eclipse of Capitalism (St. Martin's Griffin, 2015).
8. R. Wright, Nonzero: The Logic of Human Destiny (Vintage, 2001).
9. C. Leadbeater, We-Think: Mass innovation, not mass production (Profile Books, London, 2008).
10. P. Ormerod, Positive Linking: How Networks Can Revolutionise the World (Faber & Faber, 2012).
11. C. Anderson, Makers: The New Industrial Revolution (Crown Business, 2014).
12. S.E. Minson et al., Crowdsourced earthquake early warning, Science Advances 1(3), e1500036.
13. Y. Sheffi, The Resilient Enterprise: Overcoming Vulnerability for Competitive Advantage (MIT Press, 2007).
14. C.A. Hidalgo, B. Klinger, A.-L. Barabasi, and R. Hausmann, The product space conditions the development of nations, Science 317, 482–487 (2007).
15. D. Sornette, T. Maillart, and and G. Ghezzi, How much is the whole really more than the sum of its parts? 1 + 1=2.5: Superlinear productivity in collective group actions, PLoS ONE 9(8): e103023.
16. L. M. A. Bettencourt, J. Lobo, D. Helbing, C. Kühnert, and G. B. West (2007) Growth, innovation, scaling and the pace of life in cities. Proceedings of the National Academy of Sciences USA (PNAS) 104, 7301–7306.
17. J. Lanier, Who Owns the Future? (Simon & Schuster, 2014).
18. D. Helbing (2018) Breaking Free: Freedom, Peace and Prosperity, TEDxVarese. https://www.youtube.com/watch?v=PJGZpV4PUwY

---

[54]Helbing [18]

Chapter 11
The Self-Organizing Society

Taking the Future in Our Hands

We are faced with the growing complexity and diversity of an increasingly interdependent world. But Big Data and Artificial Intelligence, while potentially powerful and useful, are not a panacea for our problems. The idea of super-governments or multi-national companies running the world like a perfect clockwork is doomed to fail. Therefore, we must learn to turn complexity and diversity into our advantage. This requires a distributed governance approach. Now, the „Internet of Things" enables self-organizing systems, which can create socio-economic order and many benefits from the bottom-up. While solving the problem of over-regulation, this can harness diversity and foster innovation, collective intelligence, societal resilience, and individual happiness.

In the course of this book, we have made a number of unexpected discoveries[1]:

- Having and using more data is not always better (e.g. due to the problem of "over-fitting", "spurious correlations", or classification errors, which can make conclusions meaningless or wrong).[2]
- Even if individual decisions can be correctly predicted in more than 95% of all cases, this does not mean that the macro-level socio-economic outcome would be predicted well.[3]
- In complex dynamical systems with many interacting components, even the perfect knowledge of all individual component properties does not necessarily allow one to predict what happens if components interact.[4] In fact, interactions may cause new, "emergent" system properties.

---

[1] The following points have first been presented in a contribution on Societal, Economic, Ethical and Legal Challenges of the Digital Revolution to Jusletter IT (May 21, 2015), see http://jusletter-it.weblaw.ch/issues/2015/21-Mai-2015.html and http://papers.ssrn.com/soL3/papers.cfm?abstract_id=2594352.

[2] Having a greater haystack does not make it easier to find a needle in it.

[3] Maes and Helbing [1].

[4] Assume we exactly know the psychology of two persons, but then they accidentally meet and fall in love with each other. This incident can change their entire lives, and in some cases it can change history too (think of Julius Caesar and Cleopatra, for example, but there are many similar cases). A similar problem is known from car electronics: even if all electronic components have been tested well, their interaction often produces unexpected outcomes. In complex systems, such unexpected, "emergent" system properties are quite common.

© Springer Nature Switzerland AG 2021
D. Helbing, *Next Civilization*,
https://doi.org/10.1007/978-3-030-62330-2_11

- The most important issue is whether a system is stable or unstable. In case of stability, variations in individual behavior do not make a significant difference, i.e. we don't need to know what the individuals do. In contrast, in case of instability the system is often not predictable or controllable due to amplification and cascading effects.[5]
- In complex socio-economic systems, surprises will sooner or later happen. Therefore, our economy and society should be organized in a way that can flexibly respond to disruptions. Socio-economic systems should be able to resist shocks and recover from them quickly and well. This is best ensured by a modular, "resilient" system design.[6]
- In complex dynamical systems, which vary a lot, are hard to predict and cannot be optimized in real-time, distributed control can outperform top-down control attempts by flexibly adapting to local conditions and needs.
- While distributed control may be emulated by centralized control, a centralized approach might fail to identify the variables that matter.[7] Depending on the problem, centralized control is also considerably more expensive, and it may be less efficient and effective.[8]
- Filtering out information that matters is a great challenge. Explanatory models that are combined with little, but suitable kinds of data are best to inform decision-makers. Such models also indicate what kind of data is needed.[9]
- Diversity and complexity are not our problem. They come along with innovation, socio-economic differentiation and cultural evolution. However, we have to learn how to use diversity and complexity to our advantage. This requires us to understand the hidden forces behind socio-economic change as well as to use self-organization and digital assistants to support the coordination of actors with

---

[5]In case of cascade effects, a local problem will cause other problems before the system recovers from the initial disruption. Those problems trigger further ones, etc. Even hundreds of policemen could not avoid phantom traffic jams from happening, and in the past even large numbers of security forces have often failed to prevent crowd disasters (they have sometimes even triggered or deteriorated them while trying to avoid them), see Helbing and Mukerji [2].

[6]Helbing et al. [3].

[7]Due to the data deluge, the existing amounts of data increasingly exceed the processing capacities, which creates a "flashlight effect": while we might look at anything, we need to decide what data to look at, and other data will be ignored. As a consequence, we often overlook things that matter. While the world was busy fighting terrorism in the aftermath of September 11, it did not see the financial crisis coming. While it was focused on this, it did not see the Arab Spring coming. The crisis in Ukraine came also as a surprise, and the response to Ebola came half a year late. Of course, the possibility or likelihood of all these events was reflected by some existing data, but we failed to pay attention to them.

[8]The classical telematics solutions based on a control center approach haven't improved traffic much. Today's solutions to improve traffic flows are mainly based on distributed control approaches: self-driving cars, intervehicle communication, car-to-infrastructure communication etc.

[9]This approach corresponds exactly how Big Data are used at the elementary particle accelerator CERN; 99.9% of measured data are sorted out immediately. One only keeps data that are required to answer a certain question, e.g. to validate or falsify implications of a certain theory.

diverse interests and goals. These digital assistants could harness the potentials of artificial and collective intelligence.

- Diversity is also crucial for distributed collective intelligence, which is better suited to respond to the combinatorial complexity of our world than an artificial superintelligent system.[10]

11.1 Cybernetic Society Versus Synergetic Society: Why Top-Down Control Will Fail

We have seen that, after the automation of production and the invention of self-driving cars, the automation of society is next. However, there are two kinds of automation: centralized top-down control and bottom-up self-organization based on distributed control. The first option, corresponding to a technocratic solution, might be called a cybernetic society, while the second one might be called a synergetic society, as it builds on the local coordination of autonomous processes and on catalyzing mutual benefits.

New kinds of information systems will certainly allow us to manage the world more successfully, but we must be careful about the approach we take. Would a cybernetic society work, or would a synergetic society be better? It turns out that, for a number of reasons, it would be impossible for a supercomputer to optimize the world in real-time. The attempt to create a data-driven "crystal ball" to predict the future and a "magic wand" to impose the desired changes would sooner or later fail. While digital tools, which collect a large fraction of the world's data, can certainly be powerful, they would not work reliably.

Therefore, a centrally controlled cybernetic society would sometimes make mistakes. But as a powerful information tool might have a large-scale systemic influence, a single mistake could be highly destructive. For example, imagine powerful digital tools to get in the hands of a misguided group of individuals or a criminal organization. This could easily create a despotic regime. Therefore, the more powerful an information systems is, the greater are the safety measures needed to protect citizens and companies from potential harm.[11] Today, however, no information system seems to be a hundred percent secure. Most companies and public institutions have been hacked already.

Surprisingly, not even a "benevolent dictator" or "wise king" with the very best intentions and all the data and technology in the world could make optimal decisions. First, although computational power grows exponentially, the data volume grows even faster. Therefore, no single person, company or institution will ever be able to optimize our rapidly changing world in real-time. Supercomputers cannot even

---

[10]As we know, intellectual discourse can be a very effective way of producing new insights and knowledge.

[11]This calls for a suitable combination of encryption, decentralization, transparency, participation, reputation systems, community moderation mechanisms, and legal protection.

perfectly optimize the traffic lights in a big city. This is because the computational effort required becomes insurmountable when the traffic system is large.

Second, our ability to optimize systems in real-time from the top-down even decreases, as the complexity of man-made systems grows faster than the data volume. As a result, our relative lack of computational power will increase rather than decrease over time. Despite this, business leaders and policymakers have so far focused their efforts on attempting to control complexity from the top-down through many regulations, laws and enforcement institutions. While this approach has served us reasonably well for a long time, it is eventually coming to its limits. It has led to over-regulation and high debt, while many ills of the world are still not cured. In fact, we seem to have more problems than ever. To solve them, we don't need more power, but more wisdom, and more citizen participation.

11.2   Omnibenevolence Doesn't Exist—It's an Illusion, Despotism

For all the above reasons, I question the usefulness of mass surveillance with the aim to enable a top-down controlled "cybernetic society" ruled by a "benevolent dictator" or "wise king". Such a technocratic approach is dangerous and totalitarian in nature. We must also realize that it is impossible to be omnibenevolent,[12] i.e. to make decisions that benefit everyone. This is because people pursue different goals. Such pluralism, however, is essential for the survival and success of our species, i.e. for our ability to master challenges of various kinds.

The Big Data dream—which promises almost infinite knowledge and power to governments and a select elite of companies—turns out to be a dangerous illusion. Big Data is far from being a universal panacea for all the world's ills. Big Data has not even solved Silicon Valley's problems, and the rest of the world is much larger, more diverse, and more complex. Therefore, it is time to wake up from this dream before it becomes a nightmare and stop clinging to the flawed logic of our current data-driven approach to Big Data, which is based on mass surveillance. The terrible terror attacks in Boston and Paris, for example, have shown that mass surveillance

---

[12] Immanuel Kant, one of the masterminds of the age of enlightenment, argued that a state that decides how its citizens should be happy is a despot. He wrote: "It thus becomes evident that the principle of Happiness, which is properly incapable of any definite determination as a principle, may be the occasion of much evil in the sphere of political Right, just as it is in the sphere of morals. And this will hold good even with the best intentions on the part of those who teach and inculcate it. The sovereign acting on this principle determines to make the people happy according to his notions, and he becomes a despot", see http://oll.libertyfund.org/titles/kant-kants-principles-of-pol itics-including-his-essay-on-perpetual-peace/simple. In fact, the greatest humanitarian disasters in our history were caused by people who wanted to impose a better world order on a large number of people, see also https://www.volkswagenstiftung.de/aktuelles-presse/aktuelles/mächtige-werkze uge-der-abstraktion. Socio-economic misery or war were frequent medium-term outcomes. These resulted from reduced diversity and damage to the (eco-)systemic organization and self-organized functionality of society.

can't guarantee safety. The same applies to security. Organized crime using digital backdoors (such as "zero day exploits") causes an exponential increase of cybercrime, which currently produces losses of $3 trillion per year.[13]

The attempt to control individuals may have even caused an increasing loss of control. Indeed, studies show that extremism and crime often result from a failure to integrate communities of minorities or migrants into society (i.e. from socio-economic marginalization and a failure to create a culture of mutual respect).[14] Control is certainly not a good substitute for trust.[15] Whoever wields power must carefully avoid violating widely accepted moral, cultural or legal values, as this can seriously undermine trust, legitimacy, and power. In the long run, this can substantially weaken the credibility of companies and governments and their core interests. It might even produce a legitimacy crisis and loss of control.

I have further shown that, in a multi-cultural world, the use of coercive means tends to be counter-productive. Therefore, power based on force tends to be unstable in the long run. Constructive power, in contrast, requires citizen consent and a trustful, symbiotic relationship, in which all parties, including citizens, benefit. Thus, we need institutions that can help us to establish and maintain a proper balance between various interests and to foster the self-organization of our society and economy.

11.3 Time for a New Approach

Where will the digital revolution take us? (see Appendix 11.1) Due to many instances of misuse, public trust in conventional Big Data uses has been undermined. But the digital revolution does not necessarily mean that we must forfeit our human rights, decision-making autonomy, dignity, and democracy. There are better ways to create social order and socio-economic well-being than by amassing vast quantities of sensitive personal data and establishing surveillance of all kinds—from speed control to Internet control and, one day, perhaps even mind control.[16]

To understand the complexity of our world and turn it into our advantage we need collective intelligence, which requires diversity rather than conformity.[17] In order to

---

[13]See   http://www.cybersecurityintelligence.com/blog/cybhttps://www.cybersecurityintelligence.com/blog/cybercrime-more-profitable-than-drug-trading--212.html;   https://www.europol.europa.eu/content/eu-serious-and-organised-crime-threat-assessment-socta;   https://web.archive.org/web/20190707082247/, http://www.infosecisland.com/blogview/24439-Cybercrime-Is-Now-More-Profitable-Than-The-Drug-Trade.html; in the year 2009 the damage was $1 trillion http://www.cnet.com/news/study-cybercrime-cost-firms-1-trillion-globally/.

[14]Social networks and cultural norms can be very effective in creating social order and resilience.

[15]And "trusting" means "not knowing". In this connection, I recommend to watch the TEDx talk by Detlef Fetchenhauer explaining "Six reasons why you should be more trustful": https://www.youtube.com/watch?v=gZlzCc57qX4.

[16]"Brain hacking" has recently become a scientific field.

[17]I therefore, propose to establish a "Plurality University" to provide people with "digital literacy" and the various skills they will need for the digital society to come.

manage our future in an increasingly complex society, it is important to encourage and consider multiple perspectives. A symbiotic relationship with digitally literate citizens, customers and users is key to success. Our society can only live up to its capacity, if it makes best use of the skills, ideas and resources of its citizens. It will be of strategic importance to offer participatory social, economic, and political opportunities. In future, those societies will be leading, which manage to create a win-win-win situation between businesses, citizens, and state.

Generally, to reap the benefits of the digital revolution, I recommend engaging more in *distributed* storage, processing, and control. In complex systems, a decentralized kind of organization can be superior to a centralized system.[18] This is surprising, but a centralized approach often ignores local knowledge (since it is usually not possible to centrally process all local information). Bottlenecks such as insufficient processing power and data transmission rates are limiting factors, also in future. The use of local knowledge, in contrast, allows decentralized approaches to thrive.

The local interactions between the many components of a complex dynamical system can produce emergent structures, properties, or functionalities based on self-organization. However, as traffic jams, crowd disasters, financial crises and "tragedies of the commons" show, self-organization does not always create desirable outcomes. Nevertheless, these phenomena are now reasonably well understood and can be replicated using mathematical models and computer simulations. Those simulations tell us that the negative outcomes of self-organization can often be avoided by changing the interaction rules, i.e. the mechanisms by which the components of the system interact. In some cases such as traffic flows or the financial system, simply altering the system's parameters (such as the vehicle density or interest rate) can avoid or reduce undesirable consequences. In fact, while the "invisible hand" (which may be seen as another term for "self-organization") often fails if network effects or externalities matter,[19] we can now overcome such failure. 300 years after the concept of the "invisible hand" was invented, we can let it work for us! The sensor networks behind the "Internet of Things" enables us to realize Adam Smith's brilliant vision of self-organizing systems for the first time in human history. While this is an unprecedented opportunity to make our increasingly complex world manageable again, we need to fundamentally change the way we think about global governance.

---

[18]Couldn't we emulate a decentralized system by a centralized one? This seems plausible, but besides being more expensive, it is not obvious which data is relevant and what data obfuscates the truth. The relevant signal in one application might be noise in another application and vice versa. Therefore, in contrast to the conventional wisdom on Big Data, less information can be better. Let me give an example. As I previously discussed, it is possible to understand the spread of epidemics combining small datasets of infection cases and airplane trips. This approach performs better than a large centralized system powered by Big Data such as *Google Flu Trends*. The reason is that too much data produces problems such as "spurious correlations" and "over-fitting", i.e. the fitting of random, meaningless patterns. In other words, results of Big Data analytics may be irrelevant and misleading, which can cause bad decisions.

[19]Note that eliminating network effects by world markets and free trade is not a perfect solution, as it eliminates the niches which are needed for innovation. Moreover, externalities will always play a role when production takes place in densely populated areas or in vulnerable ecosystems.

The emergence of the "Internet of Things" means that we will soon be able to measure almost anything in real time, using networks of sensors that can communicate with each other in a wireless way. Interestingly, to support self-organization, it is not necessary to store the measured data for long. Therefore, the collection of as much data as possible, which is at the core of today's Big-Data paradigm, can be replaced by a superior Smart Data approach, where tailored measurements are made to produce temporary data for specific uses. Such real-time measurements would be sufficient to provide the information needed for the self-organized structures, properties and functions which we may want to produce. Moreover, as I have underlined before, it is anyway impossible to process all of the data currently available. Storing more data does not necessarily mean better results—it's just more expensive. So, why should we retain more data in the first place?

Thus, to fully unleash the power of information, we will need to go beyond the brute-force machine learning approach of contemporary Big Data analytics. We must learn to combine knowledge from computer science, complexity science and the social sciences to get the measurements and interactions right. So far, we have rarely combined the knowledge and skills from these disparate disciplines effectively. Silicon Valley is probably too technology-driven, while the social sciences tend to underutilize technology. Both don't pay enough attention to complexity science, but it's value for solving real-world problems will soon be obvious. The golden age of complexity science is near.

11.4 How to Make the "Invisible Hand" Work

Interactions between the components of a complex dynamical system produce "externalities", i.e. external effects such as reputation, happiness, or wealth, emissions, waste, or noise, or other consequences that affect the environment or others in a positive or negative way. These externalities can be altered by introducing or modifying *feedback loops* in the system, for example, by introducing value exchange. Such feedbacks allow the system components to adapt to the local conditions in ways that produce or restore the desired functionality. In economic systems, feedback mechanisms are often produced by financial costs or rewards, while in social systems it is common to use incentives or sanctions. However, certain kinds of information exchange and coordination mechanisms can be even more efficient ("altruistic signaling", for instance). It is also important to consider that the use of a single feedback mechanism (such as money) is usually too restricted to let a complex socio-economic system self-organize successfully, and therefore we need a multi-dimensional incentive and value exchange system, as I have proposed before.

To allow for real-time measurements of our world, my collaborators and I have started to work on a distributed Digital Nervous System as a participatory citizen web. With this enabling technology, called *Nervousnet*, one could measure externalities and feed them back to the decision-making entities[20] in such a way that efficient *and* desirable outcomes are produced. For example, one could build assistant systems to dissolve traffic jams or produce fluent traffic flows in cities. One could also build an assistant system to stabilize global supply chains and thereby reduce the "bull-whip effect" that would otherwise produce booms and recessions. Furthermore, one could build digital assistants to support cooperation and avoid conflict. These "Social Technologies" would help one to ensure favorable outcomes of interactions for all sides.

In fact, interactions between two entities (be it people, companies, or institutions) can basically have four possible outcomes:

1. If an interaction would be unfavorable for both entities, as it is often the case in conflicts and wars, the interaction should be avoided.
2. If the interaction would be favorable for one side, but bad for the other and negative overall, the interaction should also be avoided. To ensure this, the second entity should be protected from exploitation by the first one.
3. If the interaction would be favorable for one side and bad for the other, but positive overall, it can be turned into a win-win situation by means of a value transfer.
4. Finally, if the interaction would be beneficial for both sides, one should engage in it, but one might still decide to share the overall benefits in a fairer way by means of a value exchange.

Digital assistants could support us in all these situations. They could help us to create situational awareness, including the potential side effects and risks implied by certain decisions and (inter)actions. Without such assistants, we would certainly overlook many opportunities for beneficial interactions we could actually engage in. Digital assistants could also help us to organize protection against exploitation, which would otherwise deteriorate the overall state of the system. And finally, Social Technologies could support us with multi-dimensional value exchange, as I discussed it before. Social Technologies can assist us particularly in avoiding the systemic instabilities which are the main source of our unsolved problems.

11.5   The Secrets of Self-Organization

At times, self-organization seems to be almost magic. So, how does it work? Surprisingly, it is often based on simple local interactions which enable mutual adaptation. Social norms, for example, are akin to the physical forces governing the universe. They determine our everyday lives based on compliance mechanisms such as sanctions and rewards. In contrast to physics, however, the socio-economic forces governing the structure, dynamics, and functions of our society may change due to innovation.

---

[20]Which can be people, institutions, companies, or even algorithms.

Besides negative compliance mechanisms such as peer punishment, money is an important reward mechanism in our society, but not the only one. Indeed, social reward mechanisms can be even more effective than money. The weakness of today's financial system is that money is de facto one-dimensional. In future, we will need a more diverse, multi-dimensional incentive and exchange system to manage complex dynamical systems. These can now be created, because the virtual world offers novel ways to create incentive mechanisms. Rating and reputation systems are good examples.

Finally, self-organization requires suitable sets of rules to work well. But how to foster self-organization and the emergence of societally beneficial interaction rules? Over time, top-down regulation has produced over-regulation and inequality.[21] A self-organization approach, in contrast, may overcome such problems (at least to some extent), as it aims at maximizing opportunities rather than hampering them through standardization and unsatisfactory compromises (see Appendix 11.2). Proper self-rule achieves the goals set while fostering socio-economic diversity, innovation, happiness and the resilience of the overall system. Local experimentation supports socio-economic and cultural evolution. However, favorable self-organization requires an active endeavor to find and implement suitable sets of rules. This is not trivial and its importance should not be underestimated.

11.6 Cultures as Collections of Invisible Success Principles

In the past, humans haven't been very good at identifying suitable interaction rules, which has impeded self-organized and decentralized governance approaches. Fortunately, recently developed tools can help us to identify suitable institutional settings and interaction rules ("rules of the game") that can produce favorable self-organization. For example, one can perform experiments more easily than ever before. In fact, we can test out different permutations and combinations of various new rules in advance with the aid of computer simulations, lab or web experiments, interactive multi-player online games, or Virtual Worlds.[22] We can also try to identify the hidden mechanisms on which the cultures of the world are based. These cultural mechanisms, in fact, are highly important for the success of well-functioning societies and their resilience to disruptions. Surprisingly, most of these success principles are not explicitly known, but are "internalized" subconsciously while we grow up. This situation may be compared with the time when we didn't have alphabets to express our knowledge in writing. However, if we managed to explicate and formalize the success principles of the world's cultures, we could combine them in entirely new

---

[21]The issue is that each new rule implies costs to adjust, but these are very uneven. For some, the costs are lower than average and they benefit from the new rule. In the case of many rules, however, only a few players benefit while others are placed at a relative disadvantage. As a consequence, overregulation implies a large degree of inequality.

[22]Helbing and Yu [4].

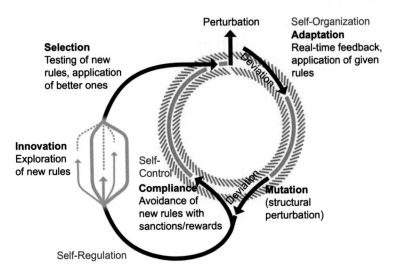

Fig. 11.1 Illustration of innovation in a framework of self-organizing, self-controlling and self-regulating systems, as it may be realized in a so-called Policy Wind Tunnel

ways. The project which I propose to achieve this might be called "Culturepedia" or "Cultural Genome Project".

The above implies three important differences as compared to the conventional policy-making of today. First, computer simulations and interactive Virtual Worlds can be used as a kind of "policy wind tunnel" to explore the implications of different sets of rules in advance. Second, alternative sets of rules can be continuously generated and tested. Third, the most promising set of rules would be implemented on a large scale only after prior testing (Fig. 11.1).

11.7   Locality as Success Principle of the Universe

Albert Einstein (1879–1955) pointed out that "we cannot solve our problems with the same kind of thinking that created them". But information and communication technologies now enable new approaches. Distributed bottom-up governance considering externalities might be used to manage complex dynamical systems such as our society more efficiently.[23] It is a promising alternative to classical top-down control, which is based on thousands of complicated laws and regulations that inhibit innovation and produce high costs.

Distributed systems are based on real-time interactions at a local level (where "local" does not necessarily refer to geographic space). Locality is very important in our universe. Most physical forces are extremely short-range. Locality is also crucial

---

[23]Colander and Kupers [5].

to self-organizing processes in socio-economic systems. For example, niches that support diversity and innovation can only exist locally. Moreover, local interactions foster cooperation, as we have seen it in a previous chapter. One might even say that the most interesting socio-economic phenomena are based on co-evolutionary processes that happen on the meso-level, which is a layer between individual system components and the entire system.

So we need to be aware of the importance of locality in complex systems and use it to our benefit. This leads us to the concepts of "glocality" and "glocalization"— approaches where people aim to reach global goals based on local activities, using locally adapted solutions. What practical implications may this have for the way we should manage complex and global systems in future?

11.8 Cities as Agents of Change

Elaborating further on the thoughts above, how would a decentralized, bottom-up organization change the way we govern our increasingly complex world? It would mean that, besides trying to find global solutions through institutions like the United Nations, we would have to build complementary institutions focused on local challenges and opportunities, namely cities and regions. For many years, we have failed to negotiate binding global agreements to reduce climate change. As a consequence, climate action has been dramatically delayed. It has also proved impossible to solve a number of other problems. Maybe a bottom-up approach would be more effective at times?

In fact, more than half of the world's inhabitants live in cities now, and the percentage is steadily growing. Cities are the places where the problems occur and where the solutions are developed and applied. Urban areas are often centers of pollution and crime, but also of production and innovation.[24] Cities are most threatened by disasters, too. Thus, our efforts to increase societal resilience need to focus on them.

In this context, it is worth listening to the insights of the former chief city planner of New York City, Alexandros Washburn, as expressed in his book on "The Nature of Urban Design".[25] Interestingly, although New York is the leading metropolis of the twentieth century, it has no master plan. Instead, it gradually adapts to the local needs of its neighborhoods. Washburn emphasizes how important it was that he could influence everything, while he underlines that he could control nothing. He contends that the first and foremost priority of any urban planner should be to listen, and that the function of public space should be to build public trust by bringing a wide variety of people together. He argues that in order to make the city more resilient and simultaneously meet quantitative, qualitative, and natural needs, it must be managed through an intricate interplay of top-down and bottom-up governance

---

[24]Bettencourt et al. [6].

[25]Washburn [7].

processes, pretty much as I have discussed it in the previous chapters. The same can be said about the "virtual cities" on the Internet, i.e. the social communities that have formed in the digital world, where transparency is also important to foster trust.

11.9   City Olympics to Improve the World

Going a step further, in full agreement with Nobel prize winner Elinor Ostrom's "polycentric" approach to solving global problems,[26] I believe that cities and social communities can be important agents of global change. A suitable combination of competition and collaboration among cities can advance us in our efforts to solve the challenges of the twenty-first century. If we manage to find ways to make our cities smarter, this will make our planet smarter.[27] In this way, acting locally will cause a global change for the better. For example, I recently proposed that we might establish something like a "City Olympics" to address global problems such as climate change.[28]

As we know, calls to combat climate change are often met with skepticism by companies and citizens, who see it as a threat to their preferred ways of business and life, and that's why these attempts receive so little support. However, doing something for our climate could be rewarding and even fun, if we ran a climate-oriented City Olympics every few years. These events would have a sporting spirit, whereby cities all over the world would engage in a friendly competition to develop the best science, technology, and architecture to counter climate change. They would also compete to achieve the highest degree of citizen engagement (in terms of environmental-friendly mobility, investments in renewable energy technology, better thermal insulation, and more). These events could be presented by the public media in pretty exciting ways. Furthermore, after each Climate Olympics, there would be a cooperative phase, where the best ideas, technologies and urban governance concepts would be exchanged among the participating cities, thereby allowing them to make faster progress. Which city or country can reach its climate goals first? Let's be ambitious! While we may dislike regulations that tell us what to do, we love competitions, we love winners, and we love cooperation, too!

We could address other global challenges in a similar way. This would simply mean a change of the disciplines in which cities and regions compete. It also seems natural that cities would form global networks with other cities that struggle with similar problems. Exchanging knowledge, ideas, technology and experts, or supporting each other when disaster strikes would give such global networks of cities an advantage. Why shouldn't we have an alliance of cities that takes a lead in supporting better, climate-friendly technologies? Just suppose that cities next to rising oceans, such as New York City, Singapore, London, Hamburg, Sydney, and

---

[26] Ostrom [8].

[27] Barber [9].

[28] See https://www.youtube.com/watch?v=TaRghSuzBYM.

a few others would start this together. Wouldn't that create a first-mover advantage, which others would soon seek to copy?

11.10 Just a Thought: Regions Rather Than Nations?

The "glocality" principle "think global, act local" can be implemented in various ways. For example, it might be beneficial to establish governance structures based on representatives of regions. Global negotiations between nation states have often failed, because nations have acted selfishly and often wielded veto powers. But what if we built institutions that could make decisions from the bottom-up in parallel to the top-down institutions we already have today? An institution such as a council of regions, for example, might help to reach agreements that are better adjusted to local needs and would provide more space for local cultures and diversity. We might even have institutions working towards the same goals in parallel from the top-down and the bottom-up. Wouldn't such a competition between two institutional frameworks accelerate progress?

To have a strong degree of legitimacy, regional representatives should be directly elected. In order to avoid political casts, all adult citizens should be eligible to be a candidate, regardless of whether they belong to a political party or not. Moreover, it would promote integration if all adult residents of a region (including foreigners) would have the right to vote. Remember that lack of participation is one of the most important factors causing crime, extremism and conflict.

To solve problems that have trans-regional relevance, the corresponding regional parliaments could send representatives for a limited time into trans-regional and global councils, which would be established to address specific problems. After all, these representatives would know best how to serve the needs of the people they are representing. To ensure flexibility and avoid accumulation of power and corruption, the global representatives of the regional parliaments should rotate every few months, or have a mandate which is restricted to certain subjects, or both.

11.11 How to Manage Our Future: Some Proposals for Immediate Action

In recent years, I have spoken to a lot of people, many of whom expect that we will soon see major changes. There are many signs that our world has become unstable and that global conflict or war might result due to power shifts, but also due to systemic instabilities that have internal rather than external reasons. If we want to manage a smooth transition into a better future, we must change not only what we are doing, but also the way we are doing it, and how we think about the world. In particular, we need to learn how information, stronger interactions and increased interdependencies are changing our systems.

People expect that governments act on their behalf, but this doesn't mean that they want governments to micromanage their lives. In fact, citizens are calling for more opportunities to participate in decisions about matters which concern them. In this regard, new opportunities are now emerging. New information and communication systems can support participatory decision-making and coordination.

Given that we will probably face a major change in the way our economy and society is organized, how can we support the transition from where we are today into our digital future? Below, I detail some proposals for action which we could begin with.

1. **Improve systemic resilience.** Most global or large-scale networks (and to an even greater degree, networks of networks) are prone to highly damaging cascading effects. Therefore, the basic functionality of our critical infrastructure is vulnerable. To make such systems more resilient, it is important to apply modular design principles, as they are common in management science. As a consequence, we must do at least two things. First, we need to build "shock absorbers" or "engineered breaking points" into our systems, which can effectively stop cascades by decoupling different parts of the network.[29] Second, we must learn to use diversity as an asset. For example, to achieve sustainable and good systemic solutions, it's important to combine diverse solutions in ways that create "collective intelligence" and resilient systems. This requires a joint effort of all stakeholders, which will typical involve independent representatives from politics, business, science, and the citizenry. It would be useful if, besides professional politicians, independent, qualified citizens would be represented in decision-making bodies as well.[30]

2. **Reduce laws and regulations in order to support diversity and its many positive side effects.** Diversity is not only the basis of societal resilience, but also of cultural evolution and individual happiness. Furthermore, diversity drives innovation, collective intelligence and economic well-being. Thus, the complaints of companies about over-regulation and the reservations of citizens about attempts to standardize their cultures, lives, and cities must be taken seriously. Otherwise, great projects such as the European Union may fail in the long run. We should try to combine the strengths of different cultures rather than making them all the same. Copying the leading economic system is not the best solution.[31] Therefore, we might proceed as follows: every law (apart from constitutional principles) could have a limited term of validity. Over-standardization should be avoided. Instead, it is important to create opportunities. Therefore, we should allow different self-organizing systems to coexist and compete with each

---

[29]Note, however, that the specific design of shock absorbers and engineered breaking points strongly depends on the particular kind of system. Therefore, an interdisciplinary approach is needed.

[30]Usually for a short time period and for a specific task.

[31]Remember the section on the Netflix challenge in my chapter on the Economy 4.0.

other. Importantly, when trying to reach high-level social or environmental standards or similar goals, countries, cities and companies shouldn't be compelled to implement a single, "one-size-fits-all" solution. In the very best sense of pluralism, one should have a choice of at least two or three options, which are based on best practice. Then, a culturally and locally fitting solution can be found. This will increase diversity and resilience, as there is probably not just one good solution, but several. It will also increase societal support.[32] Finally, in many cases, compulsory regulations can be replaced by best practice guidelines, thereby helping everyone to improve established practices.

3. **Build a reputation system to promote awareness, quality and responsible action.** If we reduce the number of laws and regulations, we need to replace them with something else. More freedom can be given to decision-makers if they behave more responsibly. Merit-based and reputation systems can be used to promote considerate and responsible action. They can foster cooperation and social order in efficient and effective ways. In fact, reputation systems are rapidly spreading across the Internet precisely because they are so useful. They help to give customers better services, and allow sellers to get a higher price for better quality products and services. However, reputation systems could be improved in a numbers of ways. Attempts to manipulate rankings or to spam the system should be discouraged. Facts, advertisements and opinions should be clearly distinguished from each other. It should be possible to post ratings in an anonymous, pseudonymous and personal capacity, but these distinct forms of engagement should be assigned different weights. Reputation and recommendation systems should be community-specific, and based on multiple quality criteria. Users should be able to choose, configure, create and share information filters and recommendation algorithms, in order to support pluralism and create an evolving ecosystem of increasingly better information filters.

4. **Rebalance top-down and bottom-up decision-making according to the well-established principle of subsidiarity.**[33] In order to enable everyone to make better-informed decisions and act more effectively, we need to build open and participatory information platforms. This will empower people to contribute to the management of our systems from the bottom-up, producing outcomes which are more attuned to the diverse local needs, considering local knowledge that matters. Altogether, we will increasingly see a change away from hierarchical decision-making ("you should do this!") towards autonomous but other-regarding activities ("I can do something that needs to be done!"). This obviously needs differentiated multi-dimensional reward systems and information platforms that help to coordinate local activities and facilitate the self-regulation of communities. Such systems could also resolve many conflicts of

---

[32] A broader support for laws can be often achieved by increasing the number of options.

[33] According to the subsidiarity principle, higher-level organizational units should only manage tasks, which cannot be efficiently performed on lower levels. Otherwise higher-level units should interfere as little as possible. They should rather provide suitable incentives or institutional settings that support the self-organization of lower-level activities.

interest through a self-organized system of community moderators, who would consider the externalities associated with decisions and actions. These community moderators will judge and foster compliance with local rules, while staying within the framework of the fundamental, constitutional principles. They should be instated for a limited time period, based on their previous record of respecting both, fundamental principles and local (community) rules.

5. **Establish a new data format based on the data cord principle to enable informational self-determination and micro-payments.** I have pointed out that some of the current problems associated with the Internet extend beyond issues of security and cybercrime. These problems mainly result from a lack of user control over their personal data, a lack of accountability, and an inability to easily reward companies and people for the data, ideas and cultural goods they produce. I think that all of these problems could be solved by a combination of a Personal Data Store[34] (i.e. a personal mailbox for data) with special encryption techniques and a new kind of data format based on the concept of a "data cord". This system would connect the contents of each data store with the respective producer or owner and allow them to control access to their data.[35] In case of personal data, the subject of the data should be considered the owner, and he or she should be able to control the rights of third parties to use it. Furthermore, a micro-payment system should enable a multi-dimensional value exchange. The more often data is copied or used, the more (material or immaterial) profit would be automatically distributed to the different beneficiaries on the value-generating chain. Such a micro-payment approach would be superior to our current system of intellectual property rights (IPR), such as the software patenting system. Current IPR approaches tend to inhibit the efficient co-evolution of ideas, which is pivotal to the success of human culture.[36]

6. **Create a multi-dimensional financial system to have a backup system and make our financial system more functional and resilient.** We have seen that our financial system is more fragile than we thought, and we cannot rule out that it might collapse one day. It is essential, therefore, to establish a backup financial system, which could facilitate economic exchange if our current system fails. Therefore, I am calling for a complementary, multi-dimensional exchange system. This would create welcome competition with our current financial system, and thus help it to improve. In fact, we currently see peer-to-peer payment and lending systems emerging. If they meet certain quality standards and serve public interests (such as providing loans so that companies can invest), governments could support the development of such systems. For example, such payment systems could be subject to a special tax status and fewer regulations

---

[34]World Economic Forum, Personal Data: The Emergence of a New Asset Class, see http://www. weforum.org/reports/personal-data-emergence-new-asset-class and also de Montjoye et al. [10].

[35]See Big data, privacy, and trusted web: What needs to be done, http://papers.ssrn.com/sol3/pap ers.cfm?abstract_id=2322082.

[36]Just suppose we would all own a few words and would have to negotiate about their use with others. This would obstruct and limit our language and culture immensely!

(as long as these systems are not "too big to fail"). The current payment systems (including *BitCoin*) are not yet perfect, but competition[37] will lead to further innovations. In particular, I have pointed out that a one-dimensional incentive system does not allow our complex socio-economic systems to self-organize and function well. For this reason, a multi-dimensional reward and exchange system is needed (which I call "multi-dimensional finance"). This would be akin to having several bank accounts for different kinds of use.

7. **Use information systems and other measurement methods to determine externalities and compensate for them.** For self-organization to work well, it is important to quantify the externalities of decisions and actions. For example, people and companies could be rewarded for external benefits created by them. Similarly, if everyone had to pay for the damage produced, this would strongly reduce the frequency and size of such damage. An important step, therefore, is to build an infrastructure that is able to measure and quantify benefits and damage to our physical and biological environment, and also to our socio-economic system ("social capital", for example). The sensor networks underlying the emerging "Internet of Things" can now serve this goal, but it will also be important to increase awareness and create incentives for responsible behavior.

8. **Tax systemic risks and provide rewards for transparency, responsibility, data access, informational self-determination, and open innovation.**[38] Besides charging for damage which has already occurred, it would also make sense to charge for *likely* socio-economic damage ("systemic risks"), similarly to how insurance companies calculate the risks posed by individuals. In the past, we have often had business models that lead to "tragedies of the commons" or that undermine privacy, pollute the Web with spam, or advertise products and services in ways that are barely distinguishable from user ratings and facts. For the time being, until we figure out a better approach, taxation might be a relatively simple and straight-forward way to improve our techno-socio-economic systems, including information systems. Rather than taxing labor more than profits from financial investments or robotic production, a tax on systemic risks would make sense and reduce risks. This would encourage the modularization or simplification of complex systems which would increase their resilience. It would also encourage firms to collect "Smart Data" rather than Big Data (i.e. discourage the collection of huge quantities of data that are of limited use and are often quite problematic in many ways). So, it might be worth considering to introduce a progressive tax on the number of network links, and to promote openness, transparency, interoperability, participatory opportunities, and informational self-determination through tax incentives. Such an approach could reward local interactions and the provision of high-quality data, while encouraging the deletion of old and irrelevant data. Moreover, the development of

---

[37]Maybe jointly with some insurance system to cover damage.

[38]On the long run, it would be interesting to figure out how to tax in a distributed, adaptive and bottom-up way.

participatory information systems that would benefit everyone could be incentivized, too. The money generated from the taxation measures mentioned above could be used to pay for public information infrastructures and other necessary institutions for the digital age to come. This would help to quickly build a mutually beneficial information ecosystem. In other words, suitable kinds of taxation could reward desirable and responsible innovation and the private activities which contribute to this process. Finally, let me stress that such taxation schemes should not stand in the way of Open Data and open innovation, and they should not be based on surveillance. Free, open, high-quality data should be tax exempt. In this context, one should also remember that the additional economic value accrued from Open Data has been estimated by McKinsey to be of the order of $3–5 trillion globally per year.[39] It would be great if everyone could get a share of this cake!

9. **Build the infrastructure and institutions for the digital society.** I believe that, so far, no country in the world is well prepared for the digital era to come and the new principles governing it. Therefore, it would make sense to engage in an Apollo-like program. The equivalent of a Space Agency in the field of Information and Communication Technology (ICT) could produce an *Innovation Alliance* with a mission to develop institutions and informational infrastructures for the emerging digital era. This is crucial in order to respond to the challenges of the twenty-first century in a smart way and to release the full potential of information for our society. For illustration, it is instructive to recall the numerous factors that enabled the success of the automotive age. The first prerequisite was the invention of cars and the emergence of mass production. Public roads, gas stations, and parking lots were then constructed to provide the infrastructure to make these vehicles useful. The establishment of driving schools and driver licenses enabled the population to develop the skillset necessary to benefit from this new transport technology. Traffic rules, traffic signs, speed controls, and traffic police were used to ensure that the traffic system ran smoothly. Finally, the invention of new technologies such as guardrails, anti-lock braking systems (ABS) and airbags greatly improved the safety of vehicles. All of this requires many billions of dollars of investment each year. In fact, we invest a lot of resources into the agricultural sector, the industrial sector, and the service sector. But are we investing enough in the emerging digital sector? While the digital revolution certainly creates new challenges for our societies, it also opens up many promising opportunities to address the challenges we are faced with.

---

[39]McKinsey & Company, Open Data: Unlocking Innovation and Performance with Liquid Information (2013), see https://www.mckinsey.com/business-functions/mckinsey-digital/our-insights/open-data-unlocking-innovationand-performance-with-liquid-information; see also The Boston Consulting Group, The Value of Our Digital Identity (2012), s. https://www.bcg.com/publications/2012/digital-economy-consumer-insight-value-of-our-digital-identity, https://web.archive.org/web/20140327104517/http://www.libertyglobal.com/PDF/public-pol icy/The-Value-of- Our-Digital-Identity.pdf.

What do we need to do to make the digital age a great success? First of all, we need to build *trustworthy, transparent, open, and participatory information and communication systems, which are compatible with our cultural values.* For example, it would make sense to establish a *citizen web* to manage a joint "Internet of Things" and enable complex self-organizing systems by means of real-time measurements and feedbacks. I call the intelligent information platform needed for this the "Planetary Nervous System". This platform could also fuel a new kind of search engine. To protect privacy, all data collected about individuals should be saved in a *Personal Data Store.* Subject to the consent of the individual in question, this data could be processed in a decentralized way by third-party *trustworthy information brokers*, which would allow everyone to control the use of their potentially sensitive personal data. A *micro-payment system* would allow data providers, intellectual property right holders, and innovators to be rewarded for their services. It would also encourage the exploration of much-needed new intellectual property right paradigms. A pluralistic, *user-centric reputation system* would promote responsible behavior in the virtual (and real) world. It would even enable the establishment of a new, multi-dimensional value exchange system, which would overcome the weaknesses of the current financial system by providing additional adaptability. A *global participatory platform* would empower everyone to contribute data, computer algorithms and related ratings, and to benefit from the contributions of others (either for free or for a fee). It would also use next-generation *social media* to measure, produce and protect *social capital* (which encompasses network-based social attributes such as trust, reputation, and cooperativeness). A *job and project platform* would support crowdsourcing, collaboration, and socio-economic co-creation. Altogether, this would build a rapidly growing *information and innovation ecosystem*, which would unleash the potential of data for business, politics, science and citizens alike. We could also create a *digital mirror world* which would use sophisticated computer simulations to assess the likely risks and opportunities associated with decisions we might make. Furthermore, *digital assistants* and *Social Technologies* could help us to cope with the diversity of our world and benefit from it. Finally, *Interactive Virtual Worlds*, potentially based on different economic systems, decision-making institutions, and intellectual property rights, would allow us to unleash the full creative potential. They could also help us to identify suitable institutional settings and interaction rules for self-organizing systems, before we deploy them in the real world.

10. **Build a new educational system that prepares people for the digital age to come and for creative work.** It becomes increasingly clear that most of our current institutions and jobs will fundamentally change. Much of the work, which has been performed by people in the past, will be done by computers, algorithms, or robots in the future. This particularly applies to procedural and rule-based work. Hence, many people will have to find other work, which will revolve around their ability to create information and knowledge, including cultural products. Rather than standardized education, we will need more personalized education and training to foster creativity and imagination. I propose that

the fundamental skills should encompass languages, mathematics, computer science, and the ability to find relevant information and critically evaluate it. This would empower people to curate information and to produce new knowledge. In addition, skills enabling people to share knowledge, collaborate with others, and create services and products collaboratively, while bearing in mind the externalities, will become increasingly important. In future, those seeking paid work must be able to concentrate on tasks, but also to flexibly adapt to new opportunities. Furthermore, they will have to apply an interaction- and systems-oriented way of thinking in order to understand and manage the complex dynamical systems around us. In conclusion, digital literacy and good education will be more important than ever.

With these preparations, the emerging "Internet of Things" and participatory information platforms could unleash the power of information and turn the digital society into a great opportunity for everyone. All it takes to make the digital age a great success is the will to establish the necessary institutions. Are we ready for this?

11.12   Let's Get Started!

Of course, governments can help to set this in motion, and they should! Spending on wars in the past 15 years has cost the world many lives and many trillions of dollars. Instead, we could have used this money to build the basis for the digital society of the future. Why shouldn't we assist people in making better decisions, by providing good information? For this, access to high-quality information is key, which in turn requires openness and transparency. Participatory opportunities can create added value and trust. Citizens have become part of our global information system. They should now be able to contribute to the collective intelligence needed to solve the increasingly complex problems of our world. A "new deal on data" should treat citizens as first-class partners in exploring the opportunities of the future and in mastering our challenges.(see Footnote 34)

Regardless of whether politicians support self-organization or not, companies will increasingly use the underlying success principles to create more efficient systems and make money with them. That's simply the logic of automation implied by the digital revolution. As self-organizing systems spread, they will also change the way our world is governed. Advances in Information and Communication Technologies will drive this process. But the citizens can drive it, too!

Given that *Instagram* was initially built by 13 people and *WhatsApp* by around 50, it is clear that a few people can now have impact on a global scale. Moreover, note that *Wikipedia* has a lot of contributors, and *OpenStreetMap* is now supported by 1.5 million volunteers. Thus, citizens don't have to wait. They can take action themselves. The Information and Communication Technologies of the future will enable us to change the world for the better! We can build a user-controlled Internet of Things ourselves as a citizen web. We can measure externalities of socio-economic activities. We can create a "Culturepedia" to collect information about the various success principles on which our cultures are based. We can build digital assistants and other Social Technologies to understand each other better and interact successfully. We can establish and run pluralistic information platforms to share data, algorithms, and information filters. And we can create a global maker community to produce our own products.

Thanks to the digital revolution, incredible inventions have become possible. Many utopian dreams are not science fiction anymore. We are only limited by our own imagination and our will to shape our future together. Do you want to be part of it? Then, follow the *FuturICT* blog[40] and social media channels, join the *Nervousnet* community,[41] and contribute to a trustworthy and respectful, participatory society, using the power of information.

Now, everything is finally coming together: science, politics, business, and social affairs. We can create self-organizing and self-improving systems with massively increased efficiency in a way that is perfectly compatible with participation and democratic principles. This approach respects the autonomy of decision-making and supports free entrepreneurship. However, the consideration of externalities will also create benefits for our environment and society. So, what are we waiting for? Let's build this all together! The next two chapters will explicate in more detail how.

11.13 Appendix 1: Where Might the Digital Revolution Take Us?

In this book, it became obvious that we need to see the world in a different light, as entirely new principles will apply. In future, the world won't be well characterized by political categories such as "left" or "right." It will have its own logic: "future-oriented" is probably the best way of putting it. Even though the digital era will be unprecedentedly different, we can already see it on the horizon. We can analyze the new trends that will underpin the digital revolution and draw conclusions by studying the transformative "forces" at work.

It is entirely possible that we experience a phase of super-governance driven by Big Data. However, I predict that societies will eventually use the "Internet of Things" to build decentralized and self-organized governance structures. This will happen because such systems have the potential to be more effective and efficient in promoting innovation, flexibility, adaptiveness and resilience. In short, they are superior.

The self-organization approach described above has nothing to do with anarchism. It is consistent with human rights and constitutional principles, and it combines the best elements of democracies and market systems. It is also very different from

---

[40]See http://futurict.blogspot.com, in particular this one: http://futurict.blogspot.ch/2015/04/societal-economic-ethical-and-legal.html.

[41]Contact us at nervousnet@ethz.ch.

communism and socialism. First, rather than involving a command economy, this approach implies as little top-down planning and control as possible. Second, rather than redistributing wealth, it creates socio-economic opportunities for everyone. It enables individuals to help themselves and to cooperate more effectively. Self-organization is built on individual self-determination ("self-control") within a framework that promotes collective intelligence and helps everyone to make better decisions. This framework is also designed to encourage responsible, other-regarding behavior.

Suitably designed reputation and merit-based systems can be a powerful catalyst for cooperation and socio-economic progress in a globalized world. If properly implemented, the economic system of the future will be more efficient and effective. Today, we still have an unbalanced and dysfunctional struggle between top-down regulation and bottom-up self-organization. This causes frictional losses, conflict and high costs. The expense of maintaining this over-regulated system has become unsustainable. Most industrialized countries have reached historical heights of public debt levels in the order of 100–200% of gross domestic product (GDP) or more. Nobody knows how we will ever pay for this, never mind the cost of even more regulation.

In the coming decades, however, I expect a superior digital society to emerge in most of the world. In the previous chapter, I have given examples showing that a participatory market society is already on its way (for example, the "sharing economy" and the quickly growing "makers community" reflect this well). The participatory market society will build on the new opportunities created by modern information and communication systems. To get a better idea of how this society might approximately look like, it is useful to discuss the Swiss system, which comes closest to my imagination of how the Participatory Market Society might work.

The Swiss system works pretty well and is based on a number of important features. It is a federally organized system, which harnesses great science and good education. It exhibits a form of direct democracy, where people can vote on many issues of public concern (remarkably, Swiss voters also decided in recent referenda not to increase holidays and not to reduce taxes!). Swiss society is multilingual and multicultural and based on a consensus-oriented tradition of decision-making. A political culture of multi-party power-sharing at all levels of government including the executive ensures that no one person or party accumulates too much power. Consistent investments have resulted in a well-maintained public infrastructure and an excellent public transportation system. Switzerland has low levels of debt in comparison to other industrialized countries. Nevertheless, I expect this kind of system to further evolve consistently with its socio-economic and cultural traditions, by taking advantage of the new opportunities created by modern Information and Communication Technology and by exploring better mechanisms to create collective intelligence.

Furthermore, note that self-organization is a conservative approach in that it builds on principles that have already been proven to work in our societies—and on core cultural and ethical values (see also Appendix 14.1). By accounting for externalities, it helps us to create more sustainable systems, to preserve our environment, and to make our society more resilient. This is achieved by enabling our society to better adapt to new conditions, including technological, environmental, and demographic changes. But what if we prefer our society to stay as it is? Can we preserve our current society, or go back to how it was before? Many of us wish that we could—we had a good time in the past! But this is a romantic wish and dangerous dream, because we can't stop our economy and our societies from progressing. Indeed we shouldn't really want to stop this progress, because we would miss out on great opportunities. Other countries would surely use these to gain a competitive advantage. Why would we want to fall back, if we could lead this development?

11.14 Appendix 2: Future Governance: Options Rather Than Compromises

People might be more satisfied with our governance outcomes if decisions would involve or consider all those that are affected by a decision, regardless of whether it is about a local, regional, national, supranational, global, commercial, or community issue. In principle, we could now establish such decision-making processes using electronic participatory voting platforms. Individual points of view could be collected in an argument map, such as a *debate graph,* to work out a reasonable number of perspectives on the problem. Then, representatives of these different perspectives should talk to each other in a round-table-like setting and try to integrate different perspectives as much as they can. This process would eventually elaborate a small number of good alternatives (say, two or three). These should ideally be interoperable. A democratic vote would then be used to decide between these alternatives. The relative number of votes of everyone might depend on his or her respective externalities. However, with the exception of a few fundamental principles, we should not strive to implement just one solution everywhere in the world. A suitable degree of diversity and culturally fitting solutions are important. Moreover, solutions should be regularly evaluated to further improve them over time according to the evolutionary principles of innovation and spreading of superior solutions.

I would like to suggest that the more diverse a community is the smaller the decision areas should be chosen (e.g. cities or regions). Problems that can only be addressed by a homogenous solution (i.e. by global standardization) require sufficiently large and diverse committees to reflect the various perspectives sufficiently well. Otherwise collective intelligence cannot be successfully built. Let's assume we have various perspectives i. Then, each could be represented by $a*ln\, W_i$ people, rounded down to integer numbers, where ln denotes the natural logarithm and W_i stands for the expected externalities of the respective decision on those that are represented by perspective i (for example, the contribution that would have to be made to create a new collective good). Finally, a is a constant that determines the overall size of the committee. I also think that a supermajority (ideally about two thirds of all votes) should be required to establish laws which are intended to be permanent. Usually, such a high level of support can only be established by allowing for a number of different options, from which companies or regions could choose the most suitable one.[42] This decision procedure would enable some degree of standardization while also creating opportunities that fit local culture and needs. In other words the self-organization approach aims to create options rather than compromises for everyone. This can harness the power of diversity, collective intelligence, and combinatorial innovation, which will be the basis of successful digital societies in the twenty-first century.

References

1. M. Maes and D. Helbing (2017) Random deviations improve micro-macro predictions: An empirical test. Sociological Methods & Research 49(2), 387–417. Title of preprint: "Noise in behavioral models can improve macro-predictions when micro-theories fail".
2. D. Helbing and P. Mukerji, Crowd disasters as systemic failures: Analysis of the Love Parade disaster, EPJ Data Science 1:7 (2012).
3. D. Helbing et al. FuturICT: Participatory computing to understand and manage our complex world in a more sustainable and resilient way. Eur. Phys. J. Special Topics 214, 11–39 (2012).
4. D. Helbing and W. Yu (2010) The future of social experimenting. Proceedings of the National Academy of Sciences of the USA 107(12), 5265–5266.
5. D. Colander and R. Kupers, Complexity and the Art of Public Policy: Solving Society's Problems from the Bottom Up (Princeton University, 2014).
6. L. M. A. Bettencourt, J. Lobo, D. Helbing, C. Kühnert, and G. B. West (2007) Growth, innovation, scaling and the pace of life in cities. Proceedings of the National Academy of Sciences USA (PNAS) 104, 7301–7306.
7. A. Washburn: The Nature of Urban Design (Island Press, Washington, 2013).
8. E. Ostrom, Beyond Markets and States: Polycentric Governance of Complex Economic Systems, The American Economic Review 100(3), 641–672 (2010).
9. B.R. Barber, If Mayors Ruled the World: Dysfunctional Nations, Rising Cities (Yale, 2014).
10. Y.-A. de Montjoye, E. Shmueli, S.S. Wang, and A.S. Pentland, openPDS: Protecting the privacy of metadata through SafeAnswers, PLoS ONE 9(7): e98790.

---

[42]Today, it's still common to impose one single standard on everyone, which creates locally unfit solutions, inequality, and dissatisfaction.

Chapter 12
Digital Democracy (Democracy 2.0, 3.0, 4.0)

How to Make It Work?

Is democracy outdated, is it broken? Many people feel the current political system will not work much longer. Suddenly, we are faced with unsustainability, mass migration, terror, climate emergency, a financial system at the verge of collapse, and „Corona emergency". Some have suggested it is time for a data-driven digital state, and China would lead the way. Eventually, however, people have realized that this might establish a global technological totalitarianism. They understand that a data-driven and AI-controlled society could easily end in a data dictatorship, where optimization will overrule more and more freedoms. So, how to prevent that the world would eventually be run like a digitally optimized „Animal Farm"? How to upgrade democracies with digital means? Here, I suggest to build participatory platforms that support collective intelligence, and to engage in open formats such as „City Olympics", which combine competition with cooperation and bottom-up engagement.

12.1 The "Benevolent Dictator" Is Dead

When the book "Limits to Growth" was published in the early 1970*ies*,[1] the world started to worry about its future. In the twenty-first century, it was suggested, some of the planetary resources needed for survival would fall short, for example, oil or water. Moreover, all simulation scenarios indicated that the world would fail to get on a sustainable path, and the economy and world's population would collapse. We would see sky-rocketing death rates as never before.

At this time, people started to change their behavior. They avoided plastic bags, had car-free Sundays, and established environmental movements. Politically, it should have been required that, whatever we do from then on, would reduce the overall consumption of resources every year. In fact, had we increased resource efficiency at a rate of only 3% per year, the world would be sustainable by now. However, the industry had different plans and demanded: "Put all the obstacles out of our way, and we will fix the problems for you." Neoliberalism, globalism, and free trade were proposed as solutions. From the perspective of the year 2020, however, one can surely

---

[1]Meadows [1].

© Springer Nature Switzerland AG 2021
D. Helbing, *Next Civilization*,
https://doi.org/10.1007/978-3-030-62330-2_12

say that we are not on a good course to achieve the UN Agenda 2030 sustainable development goals (SDGs) on time.

It turns out that neoliberalism, globalization and free trade have multiplied global resource consumption.[2] The lack of sustainability of the industrialized word was exported to the entire planet, and so the consumption of resources increased dramatically even in industrialized states and per capita. The global competition of everyone against everyone had created a "race to the bottom", a "tragedy of the commons",[3] and the outcome could be billions of early deaths.

So, what should we do? It appeared natural to measure the remaining resources in the world, where they were located, and who consumed them. It also seemed plausible to control consumption. For this, however, it would be necessary to control individual behavior. Nevertheless, some people found it justified to do this. It seemed necessary to "save the planet". Humans were "enemies" of the very environment they needed to survive—and effectively an enemy of humanity, some people claimed.[4] And so it appeared justified to create a system to control humanity altogether.

In the meantime, mass surveillance made its way.[5] The Big Data collected about the world and us would be fed into giant AI systems, which—some day in the future—would potentially be smarter than any human on this planet. Consequently, some argued, people would have to do what this superintelligence recommends or demands, as if it was a "digital God".

Moreover, tools were developed to simulate the entire world.[6] In these simulations, everyone would be represented by a digital double.[7] You may imagine it as a "black box"[8] learning to behave like you, when fed with a lot of surveillance data.

The world simulator would simulate different scenarios. The best one for the planet would be implemented. For this, however, everyone would have to contribute to the realization of the plan. Everyone would have to submit to it and behave as suggested. We would be manipulated by personalized information (called "big nudging"[9]). If we would not follow the recommendations, i.e. "mess with the plan", we would be punished by negative points in their citizen score (no matter whether this is a "social credit score"[10] as in China, a "customer lifetime value"[11] as companies in the

---

[2] Why Malthus Got His Forecast Wrong (December 12, 2012) https://ourfiniteworld.com/2012/12/12/why-malthus-got-his-forecast-wrong/ (accessed August 6, 2020).

[3] See https://en.wikipedia.org/wiki/Tragedy_of_the_commons.

[4] King and Schneider [2].

[5] See https://en.wikipedia.org/wiki/Mass_surveillance.

[6] Sentient world: war games on the grandest scale, The Register (June 23, 2007) https://www.theregister.com/2007/06/23/sentient_worlds/.

[7] See https://en.wikipedia.org/wiki/Digital_twin.

[8] Pasquale [3].

[9] I.e. nudging based on Big Data, namely personal data gained through mass surveillance, see https://www.spektrum.de/kolumne/big-nudging-zur-problemloesung-wenig-geeignet/1375930.

[10] See https://en.wikipedia.org/wiki/Social_Credit_System.

[11] See https://en.wikipedia.org/wiki/Customer_lifetime_value.

West use it, or some other "super-score"[12]—a one-dimensional score created from a multitude of measurements). This score would determine our value for society, and it would determine our rights and opportunities.

According to its engineers, the superintelligent system would act like a "benevolent dictator".[13] Even though the above storyline sounds kind of plausible, it contains a couple of serious mistakes:

1. If you want to optimize the world, you would need to choose a goal function. Unfortunately, we don't know, what is the right goal function for the world. There is not even a science I know of, which would tell us—on scientific grounds— what goal function to choose. Should it be profit (GDP per capita)? Should it be sustainability (if yes, how to define it)? Should it be life expectancy? Should it be happiness? Or what else should it be?

2. In the past years, the economic players tried to maximize profit. This, however, has brought our planet to the brink of disaster—so much so that we are now talking about a "climate emergency". Suppose we would now tell a superintelligent system to maximize sustainability. Then, it could easily happen that the system would suggest to end the lives of many people—or if it was equipped with tools for this (such as Skynet seems to be[14]), it might even put some people to death.

3. I actually doubt that optimization is the right approach, even though "optimization" sounds like a good thing. The reason is as follows: In order to optimize, one needs a one-dimensional goal function, otherwise one cannot decide what solution is better and which one is worse. (Such decisions need ">" and "<" operations.) Hence, the complexity of the system that should be optimized needs to be projected on one dimension. In many cases, this will lead to terrible over-simplifications. In any case, once you have decided for the goal, you would push all other goals into the background, and after some time of neglecting these goals, say, 50 years later, one of them will have become an even bigger problem to solve. Therefore, in our society, we cannot afford to focus on one goal and push back all others, in contrast to what a company might do. For a society to thrive, one needs to pursue various goals at the same time, and find a suitable balance between these goals. That, however, requires pluralism and diversity, not global centralized optimization. In the next chapter, we will discuss how this may be done.

4. Even if it was the right thing to maximize a particular goal function, one would need to rely on the correctness of the outcome. However, algorithms (machine learning and others) do not always converge.[15] They may imply biases and

---

[12]Superscoring: Wie wertvoll sind Sie für die Gesellschaft? PC Welt (February 5, 2020) https://www. pcwelt.de/ratgeber/Superscoring-Wie-wertvoll-sind-Sie-fuer-die-Gesellschaft-10633488.html.

[13]See https://en.wikipedia.org/wiki/Benevolent_dictatorship.

[14]The NSA's SKYNET program may be killing thousands of innocent people, Ars Technica (February 16, 2016) https://arstechnica.com/information-technology/2016/02/the-nsas-skynet-program-may-be-killing-thousands-of-innocent-people/.

[15]Particularly if the underlying data have scale-free characteristics.

discrimination.[16] They may also be sensitive[17] to the data set, to the algorithm, or even the hardware used. In many cases, there will be classification errors[18] and spurious correlations.[19] So, Chris Anderson's dream,[20] according to which Big Data would just tell us what is true or false, and what had to be done, has NOT come true. Big Data has NOT made science obsolete. In the data deluge, where there is ever more "dark" data[21] than ever before,[22] science is needed to decide what data to look at and in what ways. Note that these problems do not go away with more data.

5. When there is no solution for the world's existential problems within the current system (which, therefore, cannot be found by optimization), creativity and innovation is the right approach. That is, we need to think out of the box in order to expand the solution space and find solutions outside the current system.[23] Note, however, that innovation always challenges established ways of doing things—it challenges the system. This may not happen, if people are punished for deviations from the grand plan of a superintelligent system. Hence, controlling people and manipulating their behavior may create an even bigger problem. Solutions derived from data of the past may reinforce a system that is not serving us well. This can cause disasters, where there would otherwise be solutions (not contained in the data of the past).

6. Tragically, if we do not have one centralized system optimizing the world, but different companies trying to optimize it in parallel, with different goal functions, this is not making things better. The problem is that each optimization implies constraints—it reduces freedoms in order to reach the optimal solution. So, when many optimization processes are happening in parallel, a lot of freedoms will be sacrificed. Your car insurance, your health insurance, your doctor, your dentist, your employer, your electricity provider, your sustainability guide...—all of them will demand something, and some of them will probably ask you for contradictory actions. You will become "a slave of many masters", so to say. Compared to this, it is easier to follow the demands of one master, as in the Chinese credit score system, but this is totalitarian in nature. It always claims to be right. It cannot be questioned (while in a system with competitive demands, one can at least say "but X demanded something else—please sort it out with X").

Given the above problems, I would say "the (idea of the) benevolent dictator is dead". Therefore, in the following, I will propose systems that are based on

---

[16] Helbing [4].

[17] See https://en.wikipedia.org/wiki/Sensitivity_analysis.

[18] See https://en.wikipedia.org/wiki/False_positives_and_false_negatives.

[19] Vigen [5].

[20] The End of Theory: The Data Deluge Makes the Scientific Method Obsolete, Wired (June 23, 2008) https://www.wired.com/2008/06/pb-theory/.

[21] Helbing [6].

[22] The data volume generated doubles in less than a year, i.e. in one year we are producing more data than in all the years before in human history.

[23] Helbing [7].

empowerment and coordination rather than control and manipulation, and capable of promoting mass innovation.

12.2 The Concept of Digital Democracy

The question is, how can we upgrade democracy with digital means—in a way that is competitive with the Chinese system? Such an upgrade would have to be built on (1) digitally unleashing creativity, (2) combinatorial innovation, and (3) better decision-making. The first and second might be promoted by participatory approaches, such as Open Innovation,[24] enabling people—as we will see—to do things by themselves that they could not do in the past. The second and third bring us to the subject of "collective intelligence".[25]

"Collective intelligence" is also often called "swarm intelligence",[26] which is an impressive phenomenon known from the animal world, as I mentioned before. Well-known examples are flocks of birds, fish swarms, bee hives or ant colonies. "The fable of the bees," published in 1714,[27] suggests that the economy should work like a beehive.[28] Even though the queen bee does not give commands and even though the activities of the hive are self-organized in a distributed way, bees maintain a highly differentiated animal society, including different kinds of "jobs". Ant colonies[29] are impressive as well. Even though a single ant has a brain with 250.000 neurons only,[30] the number of neurons in an ant colony could be as big as the number of neurons in a human brain. Nevertheless, ants run an entire society in a distributed way, using an interesting communication system.[31] Can we learn something from the way these social animals are organized?

When it comes to bees, it is known that they send out "scouts" to scan the surrounding for food sources. They fly into different directions, explore the environment, and return. Then, they communicate their findings by means of a "bee dance".[32] The average direction indicates the direction of the food source they found, while the excitedness of the dance indicates the amount of food. The bystanders will evaluate the dances of many bees and then decide for the food source to exploit. The "scouts" don't have an interest to exaggerate, because the food returned will benefit the entire bee hive. If they would lie, it would harm all bees. Therefore, bees have

---

[24]Chesbrough [8].

[25]Page [9].

[26]Bonabeau et al. [10]

[27]See https://en.wikipedia.org/wiki/The_Fable_of_the_Bees.

[28]See https://en.wikipedia.org/wiki/Beehive.

[29]See https://en.wikipedia.org/wiki/Ant_colony.

[30]See https://en.wikipedia.org/wiki/List_of_animals_by_number_of_neurons.

[31]Dussutour et al. [11].

[32]See https://en.wikipedia.org/wiki/Waggle_dance.

no incentives to trick their fellow bees. Most importantly, however, there are three stages of the process:

1. *Exploration:* The "scouts" explore the food sources independently of each other.
2. *Sharing:* Information is shared with others.
3. *Integration:* The information of several "scouts" is evaluated, compared and integrated. This is the basis of the collective decision taken by the swarm.

In a sense, one could say that democracies are built on the principle of "swarm intelligence" as well. Alternatively, one speaks of the "wisdom of crowds".[33] When one needs to solve a problem where nobody knows the exact answer, a collective of people may often outperform the judgment of experts. The "Netflix Challenge"[34] described in a previous chapter is a famous example for this, but there are in fact many other examples. A study at the prestigious MIT (Massachussetts Institute of Technology) has clearly demonstrated the existence of social intelligence, if a group is diverse and communication is balanced.[35] In fact, publications on "collective intelligence" are abundant.

On a population level, today's democracies use the principle of "collective intelligence" only every few years (during elections), while we could benefit from it every day. In parliament, the "collective intelligence" principle may be used on a more regular basis, but it is often overruled by coalition agreements and party-based voting discipline.[36] In the end, all that remains is a "yes/no" decision in parliament, where the majority wins over the minority. In the worst case, this could end with something like a "dictatorship of the majority", as it occurs in some populist systems.[37] (By the way, if we had an AI system, which figures out our opinions based on mass surveillance and always does whatever the majority wants, we would end in a society of "digital populism" or "cyper-populism".) Under such conditions, minorities would be systematically marginalized, even though most functions of our society are based on minorities: intellectuals, inventors, entrepreneurs, politicians, judges, artists, etc. One can easily imagine that such a system might turn against the people on which societies depend.

A few months after we published our paper "Build Digital Democracy",[38] it was claimed that Brexit was the first major casualty of "Digital Democracy".[39] This is, of course, not true. That article was using the word "Digital Democracy" in a very different, misleading way. There, it meant something like "*Facebook* democracy".[40]

---

[33] Surowiecki [12].

[34] See https://en.wikipedia.org/wiki/Netflix_Prize.

[35] Woolley et al. [13].

[36] Called "whip" in English and „Fraktionszwang" in German.

[37] See https://en.wikipedia.org/wiki/Populism.

[38] Helbing and Pournaras [14].

[39] Brexit: The first major casualty of digital democracy, Brookings (June 29, 2016) https://www.brookings.edu/blog/order-from-chaos/2016/06/29/brexit-the-first-major-casualty-of-digital-democracy/.

[40] Marichal [15].

As many people realized after the Cambridge Analytical scandal,[41] *Facebook* can be used in very manipulative ways. Algorithms determine which opinions will spread and which ideas will never get noticed. Some people would say, it is a propaganda and censorship tool, which is disguised as a platform for the "freedom of speech". A similar thing might apply to other Social Media as well.

In recent years, many Social Media have been criticized for becoming platforms that promote flame wars and hate speech, misinformation and fake news. Hence, they are often presented as proofs for the "madness of crowds".[42] It is then typically concluded that one should not offer The People a stronger participation in political decision-making processes. However, this conclusion is short-sighted, as the Social Media of today are not designed to promote the "Wisdom of Crowds" and even undermine it.[43] They are built on the principle of the "attention economy".[44] That is, those who get more attention will have more influence. No wonder discussions are getting ever more noisy, more fake, and more emotional. This is what maximizes attention. It is also no wonder that Social Media have become battlefields for our minds.[45] Often enough, one gets the impression we are in the middle of an information war.

The "Digital Democracy" I have in mind is of a very different nature. It is not about one fraction of people winning over another fraction of people. It is about learning to combine the best ideas of many minds which each other. The goal is to find better solutions—solutions that work for many, solutions that empower us. But how to do this? Similar to what we discussed before, this goes in several steps:

1. *Exploration:* Information search and search for solutions.
2. *Sharing:* Information exchange to give the "big picture".
3. *Integration:* Development of integrated solution approaches through deliberation.
4. *Voting:* Selection of the best integrated solution.

This process could be realized with suitable platforms for "Massive Open Online Deliberation" (MOODs).[46] Let me now explain the different steps in more detail.

First, it is important that various individuals will search for relevant information on a given problem and for possible solution approaches. These would be analogous to the "scouts" mentioned before. In this step, it is crucial that the scouts pursue diverse approaches. It is also important that these individuals will not be manipulated, because this could reduce the solution space explored, which might prevent finding the best possible solution. Consequently, during the first step, the information platforms used should not influence them by recommendations, and the scouts

---

[41]Kaiser [16].

[42]MacKay [17].

[43]Lorenz et al. [18].

[44]See https://en.wikipedia.org/wiki/Attention_economy; Davenport and Beck [19].

[45]Wylie [20].

[46]How to make democracy work again in the digital age, Huffington Post Blog (August 4, 2016) http://www.huffingtonpost.com/entry/how-to-make-democracy-work-in-the-digital-age_us_57a2f488e4b0456cb7e17e0f.

should not communicate with each other. Otherwise, the "wisdom of crowds" effect would be undermined.[47]

Second, the information found needs to be shared with others. In this stage, it is not important to "win against others". Rather, the purpose of this step is to add to a "bigger picture". When a complex problem needs to be solved, it will require the combination of many different perspectives in order to get a more or less complete picture. For this, the information of each contributing individual does not have to be complete. It is important, however, that the partial views will complement each other well.

In the second step, the various bits of information should be well structured and put into a logical order. What follows from what? Which arguments are adding further details to a particular perspective, and which arguments establish new perspectives? Such different perspectives could be due to different interest groups, but not necessarily so. (Just think of a beautiful building, which cannot be captured by a single photograph, but only by a collection of photographs from different perspectives.)

In other words, the arguments should be mapped out on a virtual table. For this, one might use so-called "argument graphs". In the end, everyone should be able to see all relevant arguments in a well-structured way, such that different perspectives become visible for all.

In the third step, well-versed representatives of the different perspectives should be invited to a round table—either a virtual or real one. In this step, the goal again is not to win against the others. Rather, the representatives of the different perspectives would have the task to work out integrated solution approaches that take on board many perspectives. If the round table does not succeed with representatives of the different perspectives, it is worth trying to work with people who are to represent perspectives that are not their own. In many cases, of course, it will be difficult to find a solution that satisfies all of the different perspectives at the same time. However, as a result of step three, one should have several integrated solutions that satisfy various perspectives simultaneously.

Hence, in step four, a decision needs to be taken for one of the integrated solutions. This would happen by voting, where typically those people would be the voters, who are affected by the solution (i.e. people with "skin in the game"[48]).

However, rather than deciding by majority vote, one may consider to use different voting rules, for example, "Quadratic Voting".[49] This allows every voter to give a certain number of points to each solution, representing the "pain" the solution would produce to the respective voter. The solution with the smallest number of points would be chosen, corresponding to the "minimization of pain". A similar procedure could be applied for the "maximization of gain". It is clear that our society urgently needs to collect experiences with various voting rules in order to see which one works best in what kind of situation.

---

[47]Mann and Helbing [21].

[48]Taleb [22].

[49]Posner and Weyl [23, 24].

Note that, even in step four, the overall goal is not to "win against others", but to find a solution that works for everyone, or at least for many. The more people benefit from a solution, the greater will be the chance that society benefits as well. Of course, it is unlikely that everyone can benefit from every decision taken, but it is expected that the resulting decisions would be better in the sense that more people would benefit than when majority voting is applied. It would, moreover, be desirable to ensure that it is not always the same group of people that benefits from the decisions made, but that benefits are distributed over different interest groups in a fair way.

All in all, the above four-step process is expected to deliver solutions that will benefit more people. When averaged over many decisions and fairness considerations are taken into account, it might be even reached that everyone benefits (some from one decision, others from other decisions). Hence, the resulting system is expected to be superior to one that is based on classical majority voting or dictatorial decisions. It could raise the quality of the outcome for everyone, thereby promoting prosperity and peace.

Note that the above approach is kind of similar to the way the Swiss basic democracy[50] works, which is geared towards consensual decision-making and is using rotation principles[51] to ensure that everyone can raise their most important issues. Also, The People can interfere with the process at any time by means of referenda.[52] Digital technology would now allow us to implement these democratic principles in a digital platform. Such a platform should be based on a "democracy by design" approach.[53] It could increase the efficiency of democratic processes, and it would allow to export this successful system to other countries—and to companies and institutions, too. Finally, it might be possible to scale up the system in order to solve some global problems as well. Hence, you can see that "Digital Democracy" is not primarily about electronic voting[54] (which many people find pretty concerning, because it may create possibilities to manipulate democratic elections electronically). Instead, Digital Democracy is about unleashing "collective intelligence", which can be supported by digital means.[55]

12.3 Participatory Resilience

When the world is in trouble, we would want to have a "resilient system". This means, whatever disaster, crisis, shock or surprise society experiences, it will be able to flexibly adapt and recover, and in the best case even get to a better system

---

[50]See "direct democracy", https://en.wikipedia.org/wiki/Direct_democracy, https://www.eda. admin.ch/aboutswitzerland/en/home/politik/uebersicht/direkte-demokratie.html.

[51]See https://de.wikipedia.org/wiki/Rotationsprinzip.

[52]See https://en.wikipedia.org/wiki/Referendum.

[53]Metcalf [25], Thomas [26].

[54]See https://en.wikipedia.org/wiki/Electronic_voting.

[55]Mann and Helbing [21].

performance afterwards (as the principle of "anti-fragility"[56] suggests). For such flexibility, we need a special system design and operation.

The good news is that we know some of the principles that can make systems more resilient. Among them are[57]

1. redundancy,
2. decentralization and modular design,
3. local autonomy,
4. solidarity,
5. diversity and pluralism,
6. distributed control,
7. participatory approaches,
8. local digital assistance.

Redundancy ensures that, if one system element is broken, there is still a backup system that one can rely on.

A **modular system design** makes sure that, if one part of the system gets in trouble, other parts can be decoupled and saved, particularly if the modules can operate **autonomously** (for some time at least). Behind the concept of modular design is the idea of creating "firewalls". This will be able to keep a problem from spreading, which would happen in a densely connected system, (known as "domino effect" or "cascading effect"). Note that autonomy (or sovereignty, as some people like to call it) typically comes with **sustainability**. Interestingly, "autonomy" is also one of the most important factors that matter for the **happiness** of people.[58] Another one is "having good relationships with others".[59] This means people like to show empathy, responsibility, and **solidarity**, particularly in situations when other people need help.

Diversity makes sure that, if one mechanism or approach fails, there are still others, which work under adverse conditions.

Participatory approaches (such as "participatory resilience" or "participatory sustainability") allow people to take action locally, while first aid units are still not there. After a natural disaster, it often takes 72 h until public help is fully operational. However, most people die within 3 days after a disaster strikes, such that public help often comes late.

Digital assistance can keep up communication by creating an ad hoc network and empower people to help themselves, coordinate, and support each other.[60]

Interestingly, the above discussed resilience principles question the usefulness of centralized information and control systems to master future disasters, existential

---

[56]Talib [27].

[57]Helbing [28].

[58]The No. 1 Contributor to Happiness, Psychology Today (Jun 30, 2011), https://www.psycholog ytoday.com/us/blog/bouncing-back/201106/the-no-1-contributor-happiness.

[59]The Secrets of Happiness Revealed by Harvard Study, Forbes (May 27, 2015) https://www.for bes.com/sites/georgebradt/2015/05/27/the-secret-of-happiness-revealed-by-harvard-study/.

[60]Banerjee et al. [29].

threats and crises. Centralized systems tend to roll out one solution (the supposedly "best" one) everywhere, which however undermines diversity. Centralized systems also violate the modular design principle. They often fail during crises, when help is needed most. For example, when disasters strike, the regular communication network and other critical infrastructures often break down.[61]

Given the recommended distributed organization, cities (and the regions around them) are suitable organizational units for a resilient world. In fact, if issues are regulated locally rather than globally, one will have more degrees of freedom to find a fitting solution. It is easy to imagine that it largely restricts the freedom of decision-making and the capacity to act, if people, who live hundreds or thousands of kilometers away, are trying to interfere with local decisions. Rather than focusing on regulation, which restricts possibilities, one should focus on responsible empowerment, i.e. empowerment that cares about the impact on others and on nature. So, how to do this? How can we activate the full potential of cities and the regions around them?

12.4 Beyond Smart Cities[62]

How can we combine Smart Cities with collective intelligence?[63] How can we empower cities and citizens? How can we turn cities and regions into innovation motors? How can cities help to make the world more sustainable and resilient? These are some of the questions discussed in the following.

The dream of building "good cities" is old.[64] Since the twentieth century, there have been many attempts to create, develop or shape cities, sometimes even from scratch. Examples range from gigantic modernistic approaches known from Brasilia and Chandigarh, to more radical, but theoretical concepts aimed at changing society and engineering social order, such as Ecotopia or the Venus project. Recent developments are driven by the planetary trend towards urbanization, mass migration, and the need for sustainability. New visions of a global urban future were developed, such as "Sustainable", "Eco", or "Resilient" Cities, typically based on a top-down approach to the design of urban habitats.

Cities created from scratch heavily depend on massive private investments, for example, Songdo in South Korea or Lavasa in India. Despite ambitious goals and many technological innovations, their long-term success cannot be taken for granted, as they are often conceived by urban planners without the participation of people

---

[61]Helbing et al. [30].

[62]This and the following sections contain materials from the joint manuscript "Open Source Urbanism: Beyond Smart Cities" that Sergei Zhilin and I have written some time ago, see https://futurict.blogspot.com/2018/11/open-source-urbanism-beyond-smart-cities.html.

[63]Using the Wisdom of Crowds to Make Cities Smarter, FuturICT Blog (March 28, 2019) https://futurict.blogspot.com/2019/03/using-wisdom-of-crowds-to-make-cities.html.

[64]Sennett [31].

who later live in these cities. Such projects are typically implemented without much feedback from citizens. This makes it difficult to meet their needs. In fact, some of these cities have ended as "ghost cities".[65]

In the wake of the digital revolution, data-driven approaches promised to overcome these problems. "Smart Cities",[66] "Smart Nations,"[67] and even a "Smarter Planet"[68] were proposed. Various big IT companies decided to invest huge amounts of money into platforms designed to run the "cities of the future". Fueled by the upcoming Internet of Things, cities would be covered with plenty of sensors to automate them and thereby turn them into a technology-driven "paradise."[69] So far, however, these expectations have not been met.[70] Why?

Geoffrey West (*1940) points out that cities cannot be run like companies.[71] A company is oriented at maximizing profit, i.e. a single quantity, while a city must balance a lot of different goals and interests. This tends to make companies efficient, but vulnerable to mistakes. Cities are often less efficient, but more resilient. Driven by diverse interests, cities naturally do not put all eggs in one basket. This is why cities typically live longer than businesses, kingdoms, empires, and nation states.[72]

Importantly, cities are not just giant supply chains. They are also not huge entertainment parks, in which citizens consume premanufactured experiences. Instead, they are places of experimentation, learning, social interaction, creativity, innovation, and participation. Cities are places, in which diverse talents and perspectives come together, and collective intelligence emerges. Quality of life results, when many kinds of people can pursue their interests and unfold their talents while these activities inspire and catalyse each other. In other words, cities partly self-organize, based on a co-evolutionary dynamics.[73]

While rapid urbanization comes with many problems, such as the overuse of resources, climate change and inequality, cities become ever more important, as they are motors of innovation. Presently, more than half of humanity lives in cities, and the urban population is expected to increase to 68% by 2050. To meet the social, economic, and ecological challenges, innovation must be further accelerated, as the UN Agenda 2030 Sustainable Development Goals[74] stress.

Given the digital revolution and the sustainability challenges, we now have to re-invent the way cities and human settlements are built and operated, and how cities

---

[65] See https://en.wikipedia.org/wiki/List_of_ghost_towns_by_country, https://en.wikipedia.org/wiki/Under-occupied_developments_in_China.

[66] See https://en.wikipedia.org/wiki/Smart_city.

[67] See https://en.wikipedia.org/wiki/Smart_Nation.

[68] See https://en.wikipedia.org/wiki/Smarter_Planet.

[69] A Paradise, But Freezing Cold, Wissenschaftskolleg zu Berlin (February 2020) https://www.wiko-berlin.de/en/wikotheque/koepfe-und-ideen/issue/15/das-kalte-paradies.

[70] Hugel and Hoare [32].

[71] West [33].

[72] Sassen [34].

[73] Bettencourt and West [35], Batty [36].

[74] See https://en.wikipedia.org/wiki/Sustainable_Development_Goals.

can contribute to the solutions of humanity's present and future existential problems. In the past, we had primarily two ways of addressing such issues:

(1) nation-states (and their organization in the United Nations) and
(2) global corporations.

Both have not managed to deliver the necessary solutions on time, for example, to problems such as climate change and lack of sustainability. Therefore, we propose a third, complementary way of addressing global problems: through networks of Smarter Cities, which enhance the classical, technology-driven Smart Cities concept **(Smart Cities 1.0)**

- by collective intelligence **(Smart Cities 2.0)**,
- by co-creation **(Smart Cities 3.0)**, and
- by design for values (namely, constitutional and cultural ones) **(Smart Cities 4.0)**.[7]

Accordingly, one can define the concepts of Democracy 2.0, Democracy 3.0, and Democracy 4.0. But how to unleash the urban innovation engine to benefit citizens, societies, and the world?

12.5 "City Olympics"

"City Olympics," "City Challenges" or "City Cups" could boost innovation on the level of cities and regions and across cities, involving all stakeholders. They would be national, international or even global competitions to find innovative solutions to important challenges. Competitive disciplines could, for example, be

- to reduce climate change,
- to increase energy efficiency,
- to reduce the consumption of resources,
- to improve sustainability,
- to enhance resilience,
- to promote fairness, solidarity and peace, and
- to develop organizational frameworks that empower cities and citizens to be innovative, take collective action, and make effective contributions to achieving local and global goals.

Increasing the role of cities and regions as drivers of innovation would allow innovative solutions and initiatives to be taken in a bottom-up way. All stakeholders and interested circles would be encouraged to contribute to City Challenges. Politicians would mobilize the society and call for everyone's engagement. Scientists and engineers would invent new solutions. Of course, citizens would also be invited to participate, e.g. through Citizen Science.[75] Media would continuously feature the various

---

[75]See https://en.wikipedia.org/wiki/Citizen_science.

projects, the efforts, and progress made. Companies would try to sell better products and services, thereby promoting the practical implementation of better solutions.

Overall, this effort would create a positive, playful and forward-looking spirit and collective action, which could largely promote the transformation towards a sustainable digital society. In the short time available (remember that the UN wants to accomplish the sustainability goals by around 2030), the ecological transformation of our society can only succeed if the majority of our society is taken on board, and if everyone can participate, contribute, and benefit.

The resulting solutions would then be evaluated and "best of" lists created for the different disciplines of the City Challenge. Accordingly, prizes would be handed out to the winners. However, as the creation of these solutions would be publicly funded, they would be a public good, i.e. Open Source[76] (for example, under a Creative Commons[77] license). This will allow that *any* city can take and implement *any* of the solutions developed. In other words, *any* city can potentially benefit from *all* the innovations made by other cities. Moreover, big business, small and medium-size enterprises, spin-off companies, scientific institutions, non-government organizations (NGOs), and citizen initiatives could take *any* of these solutions, combine them with each other and develop them further. This would enable **combinatorial innovation** and create a lively, participatory **information and innovation ecosystem.** Hence, City Challenges would combine competitive and cooperative aspects. Everyone with good ideas and solutions could add something to the public city platform, thereby benefiting cities, citizens, and society altogether.

We are actually not that far from such solutions. Berlin, for example, has recently organized a "Make City" festival.[78] Suppose, such festivals would take place in many cities in a synchronized way, and that there would be more reporting and more participation. Further assume that the solutions would be evaluated and shared, and that there would be an alternation of competitive and cooperative phases, as suggested above. Then, we could truly identify the best ideas in the world, and combine them with each other, thereby promoting the emergence of a **global collective intelligence.**

The proposed approach combines some of the greatest success principles we know of:

- competition (capitalism),
- collaboration (social systems),
- collective intelligence (democracies),
- experimentation and selection of superior solutions (evolution, meritocratic cultures), and
- intelligent organization and design (using AI and other suitable methods).

The proposed approach also pushes for a new paradigm of globalisation, which one may call "**glocalisation**". It would be based on

---

[76] See https://en.wikipedia.org/wiki/Open_source.

[77] See https://en.wikipedia.org/wiki/Creative_Commons.

[78] See https://makecity.berlin/en/.

- thinking global,
- acting local (and diverse),
- experimentation,
- learning from each other, and
- helping each other.

The approach would be scalable. It would be more diverse and less vulnerable to disruptions than today's attempted global governance approaches. It would, furthermore, promote innovation and collective intelligence, while being compatible with privacy, freedom, participation, democracy, and a high quality of life. If cities would open up and engage in co-creation and sharing, they would quickly become more innovative and efficient. This brings us to the next subject.

12.6 Open Everything, Making, and Citizen Science

In recent years, we have seen the spread of new ways of addressing problems—both local and global. We are seeing novel solutions to problems that politics and capitalism 1.0 could not fix. These solutions are often based on the engagement of citizens and on contributions by civil society. Such contributions are plentiful and diverse. They are often useful and effective, particularly in places, where problems have not been noticed and fixed by business or politics, because they were too particular or too remote.

These new solutions have often been based on open approaches, ranging from Open Access over Open Data and Open Source to Open Innovation. Frequently, such approaches have been able to catalyze massive public engagement. Think of the thousands of hackathons in the past years. In the "We versus Virus Hackathon", for example, 40.000 people have been mobilized to work on solutions of all kinds.[79] However, public institutions still need to learn, how to integrate the power of collective action into public policies.

This concerns crowd-sourced approaches of all kinds: crowdsourcing,[80] crowd-funding,[81] and crowdsensing,[82] for example. Citizen Science has been another remarkable recent development, which has been able to address complex problems that machine learning could not tackle alone.[83] In the meantime, Citizen Science

---

[79] See https://www.deutschland.de/en/topic/knowledge/hackathon-on-corona-wirvsvirus-brings-solutions.

[80] See https://en.wikipedia.org/wiki/Crowdsourcing.

[81] See https://en.wikipedia.org/wiki/Crowdfunding.

[82] See https://en.wikipedia.org/wiki/Crowdsensing and https://en.wikipedia.org/wiki/Participatory_sensing.

[83] Hand [37]; https://obamawhitehouse.archives.gov/blog/2015/09/30/accelerating-use-citizen-science-and-crowdsourcing-address-societal-and-scientific.

and machine learning are combined, thereby nicely illustrating **human-machine symbiosis**[84] and **augmented intelligence**.[85]

Another pillar of this movement are **fab labs**[86] and **maker spaces**[87] as well as **GovLabs**[88] (government labs). Fablabs and Maker Spaces provide technology and machines such as 3D printers, which allow ordinary citizens to generate their own tools and products. Such capacities can be extremely valuable during crises and disasters, particularly if they have autonomous energy supply (e.g. based on solar power). Fablabs can produce tools needed for survival, when supply chains are interrupted or if delivery would take too long. The United States was even so excited about these perspectives that it wanted to become "**A Nation of Makers**".[89]

In a sense, City Olympics[90] would build on all these exciting recent developments, and take them to the next level. Over a period of several months, such approaches would be used to craft innovative solutions that could benefit cities, regions, or even the entire world. I firmly believe the success principles of the information age will be **co-* principles** such as

- co-learning,
- co-creation,
- combinatorial innovation,
- co-ordination,
- co-operation,
- co-evolution, and
- collective intelligence.

This holds, in particular, as the use of digital goods and services is not as competitive and exclusive as it applies to material resources. Sharing information can have a benefit for many, while a material good can typically be used by one person only at a time. Why don't we use this particular benefit of the information age more, given that the existential pressure of unsustainable economies is so big that a lot of people are in a danger of dying early? For example, why don't we have a **public data set of the world's resources** and materials flows (i.e. logistics), such that everyone can make an effort to improve supply chains and promote a circular economy? We have access to all sorts of stock market data, but we don't have access to the data needed to organize our survival. This appears pretty irresponsible to me.

---

[84] See https://en.wikipedia.org/wiki/Man-Computer_Symbiosis.

[85] Sean Gourley: Big Data and the Rise of Augmented Intelligence, https://www.youtube.com/watch?v=mKZCa_ejbfg (December 6, 2012).

[86] See https://en.wikipedia.org/wiki/Fab_lab.

[87] See http://edutechwiki.unige.ch/en/Maker_space.

[88] See https://en.wikipedia.org/wiki/The_GovLab.

[89] See https://obamawhitehouse.archives.gov/nation-of-makers.

[90] http://futurict.blogspot.com/2017/06/city-olympics-to-improve-world.html, https://www.csh.ac.at/event/csh-workshop-city-games/, https://www.csh.ac.at/new-global-movement-city-olympics/. https://www.csh.ac.at/event/csh-eth-workshop-1st-city-olympics/, https://climatecitycup.org, https://www.youtube.com/watch?v=TaRghSuzBYM, https://www.youtube.com/watch?v=SEsga1ZKsw4.

Last but not least, I must stress that living in a thriving society, in an age of peace and prosperity, is not just about having access to material resources. There are also a lot of "immaterial things" that matter, for example, **social capital** such as trust, reputation, solidarity, etc. I would, therefore, like to promote again the idea of creating a "**Culturepedia**"[91] or, as some people would say, running a "**cultural genome project**". The idea is as follows: Every culture is made up of all sorts of traditions and success principles, many of which have been invented to cope with particular problems. We should collect and document these specific practices and success principles, how they work, and what they are good for—namely, in a special *Wikipedia*.

By making these mechanisms explicit and by describing also the side effects and interaction effects (e.g. with solution approaches of other cultures), it would be possible to use them more consciously. Such a Culturepedia could certainly help us to solve current problems, by applying and combining the best solutions and success principles for the local situation at hand.

A Culturepedia would also help us to build "**social guides**", i.e. personal digital assistants that make other cultures better understandable and help us to navigate them (imagine something like a "cultural adapter").[92] Altogether, social and cultural diversity can be a great asset. A similar observation has been made for biological diversity, which we have learned to protect. So, why fight against other cultures? We should rather learn to make better use of the social and cultural diversity the world is offering us!

12.7 Open Source Urbanism[93]

Cities are the places where the engagement of citizens can have the greatest impact. The most livable cities manage to create opportunities to unfold the talents of many different people and cultures and to catalyse fruitful interactions among them. Opportunities for participation and co-creation are key for success.

Alexandros Washburn[94] said about the design process of New York City that he could not control anything, but influence everything; successful urban design requires the right combination of top-down and bottom-up involvement. It is therefore essential that urban development involves all stakeholders, including citizens. Vauban,[95] a quarter of the city of Freiburg, Germany, is a good example for this. The city

---

[91] See "Democratic Platforms" in this article: https://www.scientificamerican.com/article/will-dem ocracy-survive-big-data-and-artificial-intelligence/.

[92] Interaction Support Processor (August 13, 2015) see https://patentscope.wipo.int/search/en/det ail.jsf?docId=WO2015118455.

[93] This section also contains materials from the joint manuscript "Open Source Urbanism: Beyond Smart Cities" that Sergei Zhilin and I have written some time ago, see https://futurict.blogspot.com/ 2018/11/open-source-urbanism-beyond-smart-cities.html.

[94] Washburn [38].

[95] See https://en.wikipedia.org/wiki/Vauban,_Freiburg.

council encouraged the citizens to actively **participate in land-use planning and city budgeting**.

Sustainability and new energy-saving technologies were a primary focus of the planning strategy. In two new districts of Freiburg (namely, Rieselfeld and Vauban), self-built and community architecture was created, which led to urban environments conceived and designed by future inhabitants according to their own vision. Now, Freiburg counts as a benchmark city. Its concepts of sustainable urban planning and community participation are widely used by other cities all over the world.

So far, most urban planning professionals do not pay much attention to a long-term involvement of citizens in urban development. With the ubiquity of information and communication technologies, our cities are getting smarter, but not automatically more inclusive, just, and democratic.

The application of Open Source principles to the co-creation of urban environments could overcome these problems by supporting active participation, technological pluralism, and diversity. Thereby, it would also avoid technological lock-ins and dead-ends. The Open Source movement, which started with opening software (see the example of GitHub[96]) now promotes the co-production of open content (*Wikipedia*,[97] *OpenStreetMap*[98]), Open Hardware[99] (3D-printer *RepRap*), and even open architecture[100] (*WikiHouse*). Open Source Urbanism would be the next logical step of this Open Source trend.

In 2011, Saskia Sassen (*1947) wrote:

> *"I see in Open Source a DNA that resonates strongly with how people make the city theirs or urbanize what might be an individual initiative. And yet, it stays so far away from the city. I think that it will require making. We need to push this urbanizing of technologies to strengthen horizontal practices and initiatives."*[101]

Yochai Benkler (*1964) argues that Open Source projects indicate the beginning of a social, technological, organizational and economic transformation of society towards a new mode of production.[102] This new mode, called **commons-based peer production**, is a collective activity of volunteers, usually coordinated via the Internet, producing free-to-use knowledge. Open Source Urbanism, as a new way of urban development, would therefore build on concepts such as Open Innovation and commons-based peer production.

---

[96]See https://en.wikipedia.org/wiki/GitHub.

[97]See https://en.wikipedia.org/wiki/Wikipedia.

[98]See https://en.wikipedia.org/wiki/OpenStreetMap.

[99]See https://en.wikipedia.org/wiki/Open-source_hardware.

[100]See https://en.wikipedia.org/wiki/Open_architecture.

[101]S. Sassen, Open Source Urbanism, Domus (2011) http://www.domusweb.it/en/op-ed/2011/06/29/open-source-urbanism.html. (accessed on November 16, 2016)

[102]Benkler [39].

In fact, citizens are keen to be not just consumers, but co-creators of their future and co-producers of their urban habitats. Some of them already experiment with open-sourcing urban design by collecting, improving, and sharing their Do-It-Yourself design blueprints and manuals on the Internet. The maker movement as well promotes community-driven design, prototyping, and fabrication to solve local and global challenges by improving lives in local communities around the planet.

Such examples are presently still dispersed, and, therefore, not yet able to shift cities effectively towards more inclusive urban development on a global scale. For this, one would need a socio-technical platform to consolidate and strengthen the nascent movement. Such a platform could promote the exchange of best practices and solutions to frequently occurring problems. The results would be a digital commons designed to satisfy the citizens' needs.[103]

All in all, Open Source Urbanism could take our cities and societies to a new level. In particular, the approach could help to create better living conditions in developing countries, refugee camps, and regions suffering from war and disaster. It could, however, also help to improve the quality of living in local city quarters around the world.[104]

Acknowledgements This work was partially supported by the European Research Council (ERC) under the European Union's Horizon 2020 research and innovation programme (grant agreement No 833168). The support is gratefully acknowledged.

References

1. D.H. Meadows, Limits to Growth (Signet, 1972).
2. A. King, B. Schneider, The First Global Revolution: A Report (Simon & Schuster Ltd., 1991), p. 85.
3. F. Pasquale, The Black Box Society (Harvard University Press, 2016).
4. D. Helbing, Big Data Society: Age of Reputation or Age of Discrimination? In: D. Helbing (ed.) Thinking Ahead (Springer, 2015).
5. T. Vigen, Spurious Correlations (Hachette, 2015).
6. D. Helbing, The dream of controlling the world – and why it often fails (February 2017) https://www.researchgate.net/publication/313895579.
7. D. Helbing, Interaction Support Processor – And Why the Patenting System Is Broken (June 2020) https://www.researchgate.net/publication/342040513.
8. H.W. Chesbrough, Open Innovation: The New Imperative for Creating and Profiting from Technology (Harvard Business Review Press, 2006).
9. S.E. Page, The Difference: How the Power of Diversity Creates Better Groups, Firms, Schools, and Societies (Princeton University, 2008).
10. E. Bonabeau, M. Dorigo and G. Theraulaz, Swarm Intelligence: From Natural to Artificial Systems (Oxford University Press, 1999).
11. A. Dussutour, V. Fourcassie, D. Helbing, and J.-L. Deneubourg, Optimal traffic organization in ants under crowded conditions, Nature 428, 70–73 (2004).

---

[103] Schrijver [40]

[104] It does not have to end with "Guerilla Gardening", which, in fact, has made many streets a lot more beautiful, which also many politicians and urban planners had to admit…

12. J. Surowiecki, The Wisdom of Crowds (Anchor, 2005).
13. A.W. Woolley, C.F. Chabris, A. Pentland, N. Hashmi, and T.W. Malone, Evidence for a collective intelligence factor in the performance of human groups, Science 330, 686–688 (2010).
14. D. Helbing and E. Pournaras, Build Digital Democracy, Nature 527, 33–34 (2015).
15. J. Marichal, Facebook Democracy (Routledge, 2016).
16. B. Kaiser, Targeted: The Cambridge Analytica Whistleblower's Inside Story … (Harper, 2019).
17. C. MacKay, Extraordinary Popular Delusions and the Madness of Crowds (lulu.com, 2018).
18. J. Lorenz, H. Rauhut, F. Schweitzer and D. Helbing, How social influence can undermine the wisdom of crowd effect, PNAS 108, 9020–9025 (2011) https://www.pnas.org/content/108/22/9020.
19. T.H. Davenport and J.C. Beck, The Attention Economy (Harvard Business Review Press, 2002).
20. C. Wylie, Mindf*ck: Inside Cambridge Analytica's Plot to Break the World (Profile Books, 2019).
21. R.P. Mann and D. Helbing, Optimal incentives for collective intelligence, PNAS 114, 5077–5082 (2017).
22. N.N. Taleb, Skin in the Game (Random House, 2020).
23. E.A. Posner and E.G. Weyl, Voting Squared: Quadratic Voting in Democratic Politics, Vand. L. Rev. (2015) pp. 441ff, https://heinonline.org/HOL/LandingPage?handle=hein.journals/vanlr68&div=15.
24. E.A Posner and E.G. Weyl, Quadratic Voting as Efficient Corporate Governance, The University of Chicago Law Review 81(1), 251–272 (2014) https://www.jstor.org/stable/23646377.
25. G. Metcalf, Democratic by Design (St. Martin's Press, 2015).
26. N.L. Thomas, Democracy by Design, Journal of Public Deliberation 10 (1), 17 https://delibdemjournal.org/articles/abstract/10.16997/jdd.187/.
27. N.N. Talib, Antifragile (Random House, 2014).
28. D. Helbing, Globally networked risks and how to respond, Nature 497, 51–59 (2013).
29. I. Banerjee, M. Warnier, F.M.T. Brazier, and D. Helbing (2020) SOS—Self-Organization for Survival: Introducing fairness in emergency communication to save lives, preprint https://arxiv.org/abs/2006.02825.
30. D. Helbing, H. Ammoser, and C. Kühnert, Disasters as Extreme Events and the Importance of Network Interactions for Disaster Response Management. In: S. Albeverio, V. Jentsch, and H. Kantz (eds.) Extreme Events in Nature and Society (Springer, 2006).
31. R. Sennett, Building and Dwelling: Ethics for the City (Farrar, Straus and Giroux, 2018).
32. S. Hugel and T. Hoare, Disrupting cities through technology, Wilton Park (March 17–19, 2016) https://www.wiltonpark.org.uk/event/wp1449/.
33. G. West, Scale: The Universal Laws of Growth, Innovation, Sustainability, and the Pace of Lifein Organisms, Cities, Economies, and Companies (Penguin, 2017).
34. S. Sassen, Open Source Urbanism. Domus (2011) http://www.domusweb.it/en/op-ed/2011/06/29/open-source-urbanism.html (accessed on November 16, 2016).
35. L.M.A. Bettencourt and G. West, A unified theory of urban living. Nature 467, 912–913 (2010).
36. M. Batty, Cities and Complexity: Understanding Cities with Cellular Automata, Agent-Based Models, and Fractals. (MIT Press, 2007).
37. E. Hand, People Power, Nature 466, 685–687 (2010).
38. A. Washburn, The Nature of Urban Design: A New York Perspective on Resilience (Island Press, 2013).
39. Y. Benkler, Freedom in the Commons: Towards a Political Economy of Information. Duke Law J. 52, 1245–1276 (2003).
40. L. Schrijver, in: J. van den Hoven, P.E. Vermaas, and I. van de Poel (eds.) Handbook of Ethics, Values, and Technological Design: Sources, Theory, Values and Application Domains (Springer Netherlands, 2015), pp. 589–611.

Chapter 13
Democratic Capitalism

Why Not Give It a Try?

Democracy and capitalism have been struggling with each other for some time. Wouldn't it be possible to invent a new, digitally upgraded kind of capitalism that is aligned with the values and foundations of democracies? How would such a capitalism look like? What would be its elements? In this connection, I will discuss the monetary, financial, and taxation system. I will also propose investment premiums, to allow for bottom-up projects. Last but not least, I will introduce a socio-ecological finance system, which would combine measurements of environmental impacts with new incentive systems. What I have in mind is a multi-dimensional real-time feedback system, which would allow one to manage complex systems more successfully, or even enable revolutionary self-organizing and self-regulating systems.

13.1 The Failing Financial System

In 2007/08, a shot was heard around the world. A real estate crisis in the USA had triggered a financial and banking crisis,[1] which eventually turned into a world-wide economic and political crisis. More than 12 years later, we can still hear the echo of these events. Even worse, the crisis has not been resolved. Now, there are many countries with debt levels of the order of 100% of the Gross Domestic Product (GDP) or even higher.[2]

In order to allow countries to pay at least the interest rates on their debt, the central banks had to reduce them to almost zero. There have been even negative interest rates,[3] such that those who make debts will have to pay less money back. This sounds like paradise, but an economy cannot work like this for long. It will not anymore reward the companies, which do a good job and perform well. Instead, it will reward business models that do not work. In many cases, these would even harm our planet.

---

[1] See https://en.wikipedia.org/wiki/Financial_crisis_of_2007–2008.

[2] See https://en.wikipedia.org/wiki/List_of_countries_by_public_debt (accessed on August 7, 2020).

[3] Negative Interest Rates, Bloomberg (August 4, 2020) https://www.bloomberg.com/quicktake/negative-interest-rates.

© Springer Nature Switzerland AG 2021
D. Helbing, *Next Civilization*,
https://doi.org/10.1007/978-3-030-62330-2_13

Such developments did not come as a surprise. As interest rates for debts have been higher than interest rates for savings, debts increased more quickly over time. To compensate for this, new money had to be created, which—in the fiat currency system[4] of today—happens literally at the push of a button. As the volume of money increases while the material resources on this planet are limited, products and services tend to become ever more expensive. The more money exists in the system, the lower becomes the value of the money in the savings account. This phenomenon is called inflation,[5] and it requires us to work against it. As a consequence, economic growth is needed, but it makes the world economy less and less sustainable. It drives our planet closer and closer to an „abyss", where the current financial system will ultimately fail.

There are at least three factors, which make things worse. First, the dollar is currently a „petrodollar".[6] It is not backed by gold, but by oil. Hence, when new money is created, new oil has to be extracted and consumed, which will eventually turn into CO_2 and potentially add to climate change. Hence, the current monetary system makes a major contribution to the so-called "climate emergency" that we are now faced with.

Second, the dollar is the world's lead currency,[7] so that all the other countries have to buy dollars in order to buy certain kinds of goods (such as oil). Therefore, all countries that bought dollars contributed to the so-called "climate emergency," even if they did not raise CO_2 levels themselves.

Moreover, the lead currency system put the USA in a privileged position. They could import much more than they exported, and automatically enjoyed a higher standard of living. They could also spend more money on arms. In fact, for a long time the US military spending of the USA was as high as the military spending of all other countries on this planet together[8] (according to other sources, that applies to the USA and China together[9]). This circumstance made the USA the "world police" and global superpower. The power was needed to keep and expand the privileged position, and it resulted in many wars.

Third, some central banks are private in part.[10] This means that, when a country makes debts at the central bank, a few private individuals will make incredible amounts of money. Money creation works to their benefit, while tax payers will have to pay many years for the billions lent, maybe many generations.

---

[4]See https://en.wikipedia.org/wiki/Fiat_money.

[5]See https://en.wikipedia.org/wiki/Inflation.

[6]Petrodollars and the System that Created It, The Balance (April 8, 2020) https://www.thebalance.com/what-is-a-petrodollar-3306358.

[7]See https://en.wikipedia.org/wiki/Reserve_currency, https://de.wikipedia.org/wiki/Leitwährung.

[8]U.S. Defense Spending Compared to Other Countries (May 13, 2020) https://www.pgpf.org/chart-archive/0053_defense-comparison.

[9]See https://en.wikipedia.org/wiki/Military_budget_of_the_United_States.

[10]Who Owns the Worlds' Central Banks?, Zerohedge (October 22, 2019) https://www.zerohedge.com/markets/who-owns-worlds-central-banks; Wem gehören die Notenbanken wirklich? GodmodeTrader (August 24, 2016) https://www.godmode-trader.de/artikel/wem-gehoeren-die-notenbanken-wirklich,4843326.

The private "share-holders" of (central) banks are interested in a permanent increase of the debt level, because this generates new money for them. But their private interest is not aligned with public interest. If they want to get money, it is important that the government needs money, i.e. the situation in the country must be sufficiently bad. In fact, over the years the situation became so bad that it was claimed "quantitative easing"[11] was needed to keep the economy afloat, thereby, channeling trillions of public property (such as government bonds[12]) into private hands. Some would say, with raising debt levels, companies, countries, and indebted people were increasingly "owned" by a small banking elite.

It is hard to imagine how much wealth and power could be accumulated in this way over hundred years or more. However, no matter whether you believe in private shareholders of central banks and in the above money creation mechanism or not, it is certainly correct to say that banks control the world to a much larger extent than most people know. But how much longer would this system work?

In the end of 2019, the REPO markets got in trouble.[13] Apparently, some banks did not trust each other anymore, and they did not lend each other money as they used to. The central banks had to jump in.[14] In 2020, they created insane amounts of new money.[15] Market indicators reached levels comparable to those before the financial crisis back in 2007/08.[16] The oil market, too, became unstable. Shortly, oil prices even dropped below zero.[17] It became increasingly clear that the financial system was about to fail again, and there was no possibility to save it this time, given the accumulated levels of public debts. An entirely new system would be needed.

The Chinese Credit Score system could potentially have replaced this monetary system. It would certainly be possible to run an economy on its basis. Say, there were a hundred thousand cars produced. Who would get one? The principle could be "just (virtually) raise your hand (on an Internet platform)", and we will give cars to those with the highest Credit Scores, until we run out of cars. In other words,

---

[11] See https://en.wikipedia.org/wiki/Quantitative_easing.

[12] European Central Bank takes its pandemic bond buying to 1.35 trillion euro to try to prop up economy, CNBC (June 4, 2020) https://www.cnbc.com/2020/06/04/european-central-bank-ramps-up-its-pandemic-bond-buying-to-1point35-trillion-euros.html; Christine Lagarde says ECB is 'undeterred' by German court challenge, Financial Times (May 7, 2020) https://www.ft.com/content/d93008c5-2b3c-4b2e-9499-5eabaaa959db.

[13] Why the US Repo Market Blew Up and How to Fix It, Bloomberg (January 6, 2020) https://www.bloomberg.com/news/articles/2020-01-06/why-the-u-s-repo-market-blew-up-and-how-to-fix-it-quicktake.

[14] The Fed seems to have halted a potential crisis in the overnight lending market – for now, CNBC (December 30, 2019) https://www.ft.com/content/d93008c5-2b3c-4b2e-9499-5eabaaa959db.

[15] The Crash of 2020, QE and the Federal Reserve's Market, Forbes (April 20, 2020) https://www.forbes.com/sites/investor/2020/04/20/the-crash-of-2020-qe-and-the-federal-reserves-market-corner/.

[16] Market and Macro Data Signal COVID-10 Economic Crisis, Forbes (March 31, 2020) https://www.forbes.com/sites/mayrarodriguezvalladares/2020/03/31/market-and-macro-data-signal--covid-19-economic-crisis-will-be-worse-than-in-2008/.

[17] Over a barrel: how oil prices dropped below zero, The Guardian (April 20, 2020) https://www.theguardian.com/business/2020/apr/20/over-a-barrel-how-oil-prices-dropped-below-zero.

if you had a high Credit Score, you would get all sorts of goods and services—basically everything you wish. If you had a lower score, you would get *some* goods and services, but if you had a low Credit Score, you would have to live with what is left over. This is, what we call triage: some people would be "chosen" and some of them would be tolerated (they may be lucky sometimes), but some people would be doomed. There would possibly be a certain score below which one could not survive.

The Chinese Credit Score is largely behavior-based. He or she who does exactly what the government demands may hope to become a "chosen one" some day. However, as this is a neo-feudal pyramidal concept, there would not be a lot of them.

It is clear that the Chinese Credit Score is more a control system than a reward system. And it would be even more so, when the production capacity of the system was re-adjusted to meet sustainability goals.[18] This would mean to reduce production by one third (at least). You can imagine that a lot of people would then be doomed, and the Citizen Score could very well decide about life and death. Given this perspective, it is high time to think about an alternative, fair monetary and economic system, which would work for all of us, or at least for most of us, and which would also make the planet more sustainable.

13.2 Democratic Capitalism

In the past, our society was run by several different operating systems: One was culture, one was law, and one was capitalism. Now, we have also algorithms ("code is law").[19] Unfortunately, these operating systems are not only mutually inconsistent many times—they are also often contradictory. It is therefore expected that these systems would not coexist for long. They would potentially interfere with and damage each other. Eventually, one of them may dominate or even destroy all others, and it may not necessarily be consistent with the legitimate foundation of our society: the constitution.

Figure 13.1 illustrates on the left the society we used to have. Democracy was the framework, and the economy was one institution that contributed to its well-being. Capitalism was a "subroutine" that was running the economy. Other spheres of society—politics, law, media, science, health, religion etc.—were run by different subroutines. While the goal function in the economy was profit, the goal function of politics should have been human dignity. Law should have established justice, science should have revealed the truth, the media should have exposed it, the health sector should have tried to ensure human well-being, and religion should have cared about spiritual matters. However, the disruptions by the digital revolution have largely dismantled the old fabric of our society and its organization into separate spheres, which once ensured that we achieved the different goals that mattered for society.

---

[18] See https://en.wikipedia.org/wiki/Sustainable_Development_Goals.
[19] Lessig [1].

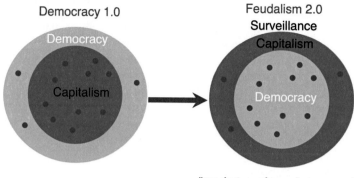

Fig. 13.1 In many countries, the digital revolution has silently transformed democracies into a new kind of system that is often called "Surveillance Capitalism" ("Feudalism 2.0")

With the invention and spread of "Surveillance Capitalism",[20] digital technology became ubiquitous, as is illustrated on the right of Fig. 13.1. Surveillance Capitalism was not restricted to new business models, products and services. It invaded our jobs, our consumption, our leisure time, our private homes, our friendship network, our thinking and our believes. It re-organized everything according to the principle of profit maximization, influence, or power (which are often largely aligned).

Utility maximization[21] took over. Basically, everything got a price tag, and human dignity was lost on the way. Democracy was hollowed out, and turned into a subroutine. Surveillance Capitalism became the new framework of society. A silent coup had happened. Most people had not noticed the coup, as it was gradual and slow. Politicians and courts were late—and basically let it happen. Democracy became increasingly an illusionary façade. Human rights were in retreat, as they did not produce profits in obvious ways. As they were not represented by numbers, optimization routines ignored them. Given this, it was no wonder that hate speech and conflict took over.

Such a collapse of a once thriving society should not surprise us. One should have demanded from the very beginning that algorithms follow a "Design for Values" approach.[22] Constitutional and cultural values should have been built into the algorithms behind the platforms organizing our lives—those that were used in business, financial markets, social media etc. Then, a conflict between the different societal operating systems would have been avoided, and damage of the constitutional operating system underlying democracies would have been prevented. It did not have to happen.

It happened, nevertheless, because some people did not want to care about environment, people, and values, if it would reduce their profit. And so, the digital revolution

---

[20]Zuboff [2].

[21]Aleskerov et al. [3].

[22]Friedman and Hendry [4].

undermined the constitution. It cannot be denied that disruptive innovators have tried to replace our previous societal operating system by another, digital one, and that they usually did not care much about democratic legitimacy.

In the meantime, we have seen where this can end. Some countries have implemented totalitarian police states, based on mass surveillance, behavioral control, censorship and propaganda. Moreover, the underlying algorithms can be easily transferred to other countries. As time passes, it becomes increasingly clear that similar principles are taking over in Western democracies as well. Sometimes, these principles are operated in grey zones, so most of the time, we do not know who is pulling the (digital) strings. Clearly, there is a lack of transparency.

In the meantime, people are increasingly worried about the dystopian potentials of digital technologies. They are losing trust. A new approach is needed. In the following, I suggest to upgrade capitalism democratically. Capitalism' success is largely based on the idea of fair competition, while democracy's success is based on cooperation and collective intelligence. Marrying both systems would promote "co-epetition", a synergetic socio-economic system that would work like a symbiotic ecosystem. In this connection, I would like to prose a number of measures:

– a better tax system,
– participatory steering boards,
– an information ecosystem,
– participatory budgeting,
– universal basic income,
– investment premiums,
– asset-backed money, and
– a socio-ecological finance system.

13.3 A Better Tax System

We are entering an age of Artificial Intelligence, where intelligent computer programs and robots increasingly do our work.[23] However, while we have to pay massive income taxes, robots are getting away for free. This is not fair, and it cannot work for societies in the long run. It is a serious discrimination of humans, many of whom may get unemployed and then be considered a burden of the social insurance system. Moreover, they would probably have to be alimented by fewer and fewer people. It is likely that the social insurance system would break down at some point in time.

However, what would replace income taxes? Besides value added tax (VAT), there would be transaction fees (for financial transactions at stock markets). If more money were needed, one could also tax the storage and transaction of information. This would certainly lead to more environment-friendly IT business models. The current approach is not sustainable at all. The electricity share spent on digital technology

---

[23] Brynjolfsson and McAfee [5].

is growing exponentially. A short time ago, it amounted to 3% of the overall energy consumption, but by the year 2030, the percentage is expected to be above 20%.[24]

Most importantly, however, we should realize how much less taxes we would have to pay

- if there was no inflation,
- if we did not have to pay interest rates on government debts,
- if we would not have to spend a lot of money on defense,
- if the tax system would be simple and effective,
- if administration would be digital and slim.

In such a system, the taxation level would drop dramatically. In fact, if we managed to establish world peace, military spending would drop and prosperity levels would raise significantly. A Tobin tax[25] on financial transactions is already being discussed for a long time, and in some European countries, it seems to be on the way.[26] When the next financial crises looms, a reset of the financial system would be unavoidable. Debts would have to be forgiven. And a simplification of administration would be surely possible, if we had a new social contract, including basic income.

13.4 Universal Basic Income

With the next wave of automation, based on robots and AI, many expect dramatic changes in the job market. Some even expect mass unemployment. This is often framed as dystopia. However, it could also be a chance to liberate humanity. Robots could do dirty, dull and dangerous work for us, and provide us with the basic goods and resources needed for our every-day life. This would allow humans to focus on social, environmental and cultural issues.

Many have proposed basic income as a means to take away the existential threats connected with the digital age. To some extent, it would be the "helicopter money" that the central banks have been talking about so many times, but never had the courage to introduce. Everyone would get it, so these payments would not have to be administered and approved.

Universal basic income would have many benefits:[27]

---

[24] How to stop data centres from gobbling up the world's electricity, Nature (September 12, 2018) https://www.nature.com/articles/d41586-018-06610-y.

[25] See https://en.wikipedia.org/wiki/Tobin_tax.

[26] Watered-down Tobin tax could enter into force in 2021, Euractiv (June 11, 2019) https://www.euractiv.com/section/economy-jobs/news/watered-down-tobin-tax-could-enter-into-force-in-2021/.

[27] From unthinkable to universal: Universal basic income gains momentum in America, The Economist (August 8, 2020) https://www.economist.com/united-states/2020/08/08/universal-basic-income-gains-momentum-in-america.

- The benefit for producers would be that there would be stable and predictable consumption patterns, while mass unemployment would mean serious recessions and bankruptcies of many companies.
- The benefit for workers would be that they would have their backs free to re-orient and educate themselves for the new jobs of the digital age.
- The benefit for politicians would be that the social situation would be stable, while mass unemployment would trigger protests and unrests, perhaps even a revolution.

Of course, a flat income structure would be somewhat like communism, which is known to work badly for people. For an innovative and thriving economy, incentive systems are needed. Hence, while the universal basic income would be just enough to cover food, shelter, and basic needs, a job would add income for a comfortable life. For sure, there would be well-paid jobs for specialists, but ordinary people would like to contribute to society as well. So, how can this be organized?

13.5   From Participatory Budgeting to Crowd Funding for All

For some time already, various cities have been experimenting with participatory budgeting,[28] which apparently was a response to the financial crisis. In these cities, citizens take part in the decision process how to spend tax payers' money. The most well-known example is perhaps the city of Barcelona.[29] Over there, with more than 600 citizen-based proposals, experiences with participatory budgeting have been quite positive. Accordingly, other cities have started to copy this approach. The innovative measure has increased collective action, the effectiveness of public spending, transparency, and trust. It helps people to perceive the city to be "their city", which comes with identity, care, and commitment.

However, we could even go a step further, by allowing citizens to run their own projects. This would have at least two benefits: First of all, citizens could locally contribute to the improvement of their city quarter and living conditions, in places, which political attention and economic investments did not reach. Second, citizens could earn some money for the projects they are running and engaging in.

It is important to realize that the capacity of any centralized structure is limited. Neither politics nor bosses of companies nor the military knows the exact local living conditions everywhere—and even less so can they take locally fitting action

---

[28] See https://en.wikipedia.org/wiki/Participatory_budgeting.

[29] Over six hundred proposals from citizens for the participatory budgets, Info Barcelona (March 1, 2020) https://www.barcelona.cat/infobarcelona/en/tema/participation/over-six-hundred-pro posals-from-citizens-for-the-participatory-budgets_923661.html; Democratic Innovation, The experience of participatory budgeting in the province of Barcelona, Universitat Autonoma de Barcelona, https://ajuntament.barcelona.cat/innovaciodemocratica/en/projects; http://cpa.uab.cat/ index.php/en/research-results-blog/92-the-experience-of-participatory-budgeting-in-the-province-of-barcelona.

everywhere. Accordingly, in a country, there are lots of "forgotten places", which could be best upgraded by locals, if they had access to the necessary means.

"Guerilla Gardening",[30] i.e. the distribution of seeds of colorful plants in public spaces, is a little example showing how easy it can sometimes be to turn a depressing city landscape into a colorful paradise.

As described in the chapter on Digital Democracy, it does not have to stay there. Hackathons, Make City Festivals, and Urban Source Urbanism have taken things big steps forward. Why shouldn't it be possible to lift this to a professional level?

In the future, I imagine that people would engage in economic, social, environmental, and cultural projects. In some of these projects, they would take a coordinating role, in others they would contribute. Digital assistants,[31] using augmented reality and other digital technologies, would provide supportive tools for organizing such projects (from recruiting project staff over budget planning and organization to detailed instructions for a state-of-the-art implementation). In this way, citizen-based projects could be run on a professional or semi-professional level. Of course, it is conceivable to include experts with a special education and city representatives as well, to achieve high standards and a coordination with other projects.

The citizen-run projects would be financed by means of crowdfunding.[32] If there was just enough money for crowd funding, in perspective one could even imagine to build a sports arena or develop a new medical drug using community-based funding.

So, here is the idea: What if everyone got a special budget every month, which would be reserved for crowd funding. Such an "investment premium"[33] would allow ordinary people to chip money into local projects that they consider important. This could be official projects or citizen-run projects. People could donate their investment premiums (or parts of them) to whatever seems to be good economic, social, environmental or cultural engagements. Projects that manage to attract enough money for their realization would go forward. Other projects, which lack maturity or popularity, would be dropped. The competition for funding should make sure that primarily high-quality projects would be realized.

13.6   A New Monetary System

How should we pay for all of this? In principle, in the same way as in the past: Let money work! However, in the future, it should work for everyone, not just for a few people or private entities. For this, we should have the equality principle in place.

---

[30] See https://en.wikipedia.org/wiki/Guerrilla_gardening.

[31] See https://en.wikipedia.org/wiki/Personal_digital_assistant; also: Interaction Support Processor, https://patents.google.com/patent/US20160350685A1/en.

[32] See https://en.wikipedia.org/wiki/Crowdfunding.

[33] Decomplexing the System, Roland Berger (June 4, 2018) https://www.rolandberger.com/fr/Point-of-View/Decomplexing-the-system.html, see also http://futurict.blogspot.com/2019/08/a-utopian-society-that-would-be-feasible.html and https://werteundwandel.de/inhalte/dirk-helbing-wir-muessen-wirtschaft-und-politik-neu-erfinden/.

Everyone should have the same fair chance to benefit from the money system. In other words, we should democratize capitalism. Then, it would probably work much better for the world.

So, suppose:

1. all money would be gold-or asset-backed,
2. all money would be property of The People,
3. one would pay a regular basic income for everyone,
4. one would pay a regular investment premium to enable "crowd funding" for all,
5. the basic income, investment premium and fundamental public services would be paid from a flat tax (VAT), transaction fees, a proper inheritance tax, and, if necessary, a flat tax on company revenues. Then, we would have a circular flow of money.

Should all of this not be enough, one could think about creating money with a time stamp. This would allow one to introduce "artificial ageing" of money. In other words, such money would be most valuable when handed out, but it would lose its value exponentially over time. However, the same amount of money that would be lost in this way would be newly generated for the payment of basic income and investment premiums.

In a sense, such a system would infuse fresh money at the bottom of society, on everyone's bank account. By spending it, it would "evaporate" and rise to the top. Today's money system instead creates money on the top and claims to benefit everyone through a "trickle-down effect".[34] Unfortunately, the trickle-down effect does not work well, as the growing inequality shows. Therefore, money concentrates in less and less hands, which eventually makes the monetary system dysfunctional (as one immediately understands by imagining that, one day in the future, all money would be in the hands of one person or company—as we know it from the game of "Monopoly"[35]).

In principle, it should not be too difficult to create a system that works for the environment and us. "Vested interests" seem to be the main obstacle, i.e. the people who profit over-proportionally from today's system and do not want to give up their exceptional wealth, power and privileges. But are "vested interests" really legitimate in times of "over-population", where millions or even billions of lives are at stake? I don't think so.

From an ethical point of view, it is not acceptable to sacrifice one life for another one, not even for two or more. Mathematically, this means that the value of a life is considered to be infinite. It cannot be bought for money. That would be immoral.

From an insurance point of view, the value of a "statistical life"[36] is not infinite, but it is officially still of the order of several million dollars, at least in various

---

[34] See https://en.wikipedia.org/wiki/Trickle-down_effect and https://en.wikipedia.org/wiki/Trickle-down_economics.

[35] See https://en.wikipedia.org/wiki/Monopoly_(game).

[36] See https://en.wikipedia.org/wiki/Value_of_life; The Value of Statistical Life: A Meta-Analysis (January 30, 2012) https://www.oecd.org/officialdocuments/publicdisplaydocumentpdf/?cote=ENV/EPOC/WPNEP(2010)9/FINAL&doclanguage=en.

industrialized countries.[37] From a capitalistic point of view, the value of a billionaire would correspond to the lives of about 150 ordinary people. So, why are we paying so much attention to what billionaires want, when having to solve the existential problems of this planet? We should just do the right thing! Otherwise we would all be responsible for the deaths of millions or even billions of people. Who could take such responsibility?

13.7 Participatory Steering Boards

It is worth asking how we got into this huge existential trouble in the first place? Why could things get so bad? The following seems to be among the main contributing factors:

- An economic system, where companies could largely do what they wanted and where profit maximization and competition were driving the economy into an unsustainable state.
- A system, in which over-consumption was encouraged, as it increased the main performance indicator "Gross Domestic Product per capita" (GDP per person).
- A system, in which non-sustainable business models were exported to developing countries in the context of "globalization".
- A system, in which the maximization of profit and power got more attention than solving existential problems, particularly if this did not translate into highly profitable business models.
- A system, in which one could earn more money on harm, wars and disasters than on avoiding them.
- A system, which was largely built on the exploitation of others and the environment.

It would not have to be this way. For example, China has decided to introduce a measurement-based company score,[38] which would reflect to what extent a company meets the expectations of the government. If the value would be too low, a company could simply be closed down.

Perhaps, less drastic measures would be effective as well. Assume, for example, that companies would be judged, to what extent they contribute to achieving societal goals (not just economic, but also environmental, social, and cultural ones). Based on this, the one third best performing companies would get a large reduction in taxation, the next 33% quantile would benefit from a moderate reduction, and the remaining

---

[37] https://www.tagesanzeiger.ch/sonntagszeitung/wie-viel-ist-unserer-gesellschaftein-menschenl eben-wert/story/18010226. In other countries, the value could be even higher: https://en.wikipedia. org/wiki/Value_of_life. Around 100.000 Dollars for a quality year of life in industrialized countries is quite normal.

[38] China's Corporate Social Credit System: What Businesses Need to Know, China Briefing (November 5, 2019) https://www.china-briefing.com/news/chinas-corporate-social-credit-system-how-it-works/.

33% would get no reduction in taxation. This would be a more liberal way of guiding economic activities towards satisfying the needs of humanity and the planet.

There is also a participatory, democratic approach. Specifically, I would like to argue for steering boards of companies and institutions that are composed not only of managers, but also of representatives of various groups that bring in views of people who are affected by their decisions. In other words, the steering boards should contain representatives of owners and shareholders, but also of workers, suppliers, customers, users or patients, and of local citizens as well. People representing environmental and ethical issues should be included, too. If every company would (have to) do this, products and services would fit consumer interests better, supply chains would be improved, and the environmental footprint reduced.

Would it really work? The answer is: Yes! The company *Caterpillar*, for example, has experimented with the above participatory approach.[39] In this way, it managed to increase not only customer satisfaction and the approval of the citizens, but also the company's revenues.

The question is: Why is it working so well? In a sense, the answer is "collective intelligence". By adding representatives of other affected groups, the overall business approach becomes better adjusted to the various expectations and needs. Interfaces are created with other businesses and institutions, and with citizens and the environment. In this way, the activities of different companies and institutions become more coordinated, integrated, context-sensitive and balanced.

Note that a very similar approach has been successfully applied to the self-control of traffic lights.[40] This produces astonishing performance gains, if local traffic flow optimization is combined with a coordination between neighboring intersections.[41]

Bringing the needs of all beings into a sustainable balance may be less difficult than one may think. To allow it to happen, one needs to give sufficient weight to the needs of others. In fact, game-theoretical studies have shown that giving a 40–50% weight to the utility of others will establish a high level of cooperation[42] in many social dilemma situations, in which otherwise a "tragedy of the commons"[43] would result. Such an approach can dramatically increase the expected success or, as game theorists would say, the "average payoff".[44] It does so by weighting different goals in a fair way, thereby getting different interests aligned. This is actually the reason why the principle "Love your neighbor as yourself" is so incredibly powerful.

The above principle could be the basis of a thriving, symbiotic information, innovation, production and service "ecosystem" or, as some people would say: prosperity and peace. Additional success principles will be described in the following section.

---

[39] https://drive.google.com/file/d/0B7AGP10l35ksWkEyemxQMV9nb3c/view.

[40] https://iopscience.iop.org/article/10.1088/1742-5468/2008/04/P04019/.

[41] https://link.springer.com/article/10.14441/eier.D2013002.

[42] See https://en.wikipedia.org/wiki/Co-operation_(evolution).

[43] See https://en.wikipedia.org/wiki/Tragedy_of_the_commons.

[44] https://www.nature.com/articles/srep01480.

13.8 Socio-Ecological Finance System

As we talk about ecosystems: have you ever wondered what makes them so surprisingly sustainable as compared to our own economy? After all, ecological systems are based on an almost perfect "**circular economy**":[45] there is basically no resource that is not being recycled[46] and reused. This is partly due to the modular, organic organization of nature. All elements are decomposable, and there is a shared genetic code with the same basic components.

Companies, in contrast, don't favor such compatibility of their products, and they often do not pay attention to building easily decomposable and repairable products. Legal regulations to promote recycling and a circular economy have been frustratingly inefficient so far. It's about time for an alternative approach: empowerment instead of regulation! What would this require?

I think, we would need a **multi-dimensional real-time coordination system**. Why multi-dimensional? Because nature is not controlled by a one-dimensional quantity such as money. Instead, the self-organization of an ecosystem (or of our body) is based on multiple feedback loops regulating the use and distribution of water, proteins, carbon, vitamins, minerals etc.

In many cases, there are **symbiotic relationships**.[47] For example, plants may exchange nutrients with each other. The situation for our own body is even more spectacular, because we are actually not an individual, but an ecosystem![48] In our gut, there are thousands of different kinds of bacteria. This is called the "microbiome",[49] and it seems to contain more cells than our body! The bacteria in it are responsible for the decomposition of the food we eat, i.e. the powering and regeneration of our body. In addition, however, they are also an important part of our immune system.[50] All of this would not work, if there was not a symbiotic relationship and reasonable balance in the microbiome.

What can we learn from this? How to build a **bio-inspired economy**![51] I believe, for this we would need multiple incentive systems, not just one kind of money. Today's economy is based on optimization[52] and utilitarian thinking,[53] while nature is based on (co-)evolution.[54] Utilitarian thinking means that everything would be compared (or made comparable) with each other, e.g. measured in money. This also calls for a seamless convertibility of different kinds of currencies and values.

---

[45]See https://en.wikipedia.org/wiki/Circular_economy.

[46]See https://en.wikipedia.org/wiki/Recycling.

[47]See https://en.wikipedia.org/wiki/Symbiosis.

[48]See https://en.wikipedia.org/wiki/Ecosystem.

[49]See https://en.wikipedia.org/wiki/Human_microbiome.

[50]See https://en.wikipedia.org/wiki/Immune_system.

[51]Helbing et al. [6].

[52]See https://en.wikipedia.org/wiki/Mathematical_optimization.

[53]See https://en.wikipedia.org/wiki/Utilitarianism, https://de.wikipedia.org/wiki/Utilitarismus.

[54]See https://en.wikipedia.org/wiki/Evolution, https://en.wikipedia.org/wiki/Coevolution.

Optimization is based on a one-dimensional goal function that allows one to apply ">" or "<" operations to determine which solution is better. However, a precondition for this is that the system under consideration must be mapped onto a one-dimensional function, which is often a gross over-simplification, as I have pointed out before. Consequently, one can only move up or down, while a society needs to be able to do more than profit maximization. For example, it wants to simultaneously improve education, health, and environmental conditions as well. This calls for a multi-dimensional control or coordination system.

In a complex non-linear system, whatever goal function we choose, we would often find one optimum solution, while there are often multiple solutions that reach 95% of the maximum performance. Among these solutions, there will typically be solutions that also perform well from the perspective of other goal functions. We are interested in these solutions with multiple high performance, which are usually eliminated by classical optimization.

Therefore, I propose to introduce multiple currencies, which are not seamlessly exchangeable against each other. These currencies would, in a sense, be administered through separate bank accounts. Each currency would be defined by a particular measurement procedure. In other words, a measurement would define a new kind of currency, which could be used to establish a new kind of incentive, or just an additional feedback effect.

The "Finance 4.0" system (or shorter: "Fin4+" system) we have developed does exactly this.[55] It combines Internet of Things and Blockchain technology in order to turn measurements into currencies, which can then be used to create feedback effects. Such multiple feedback effects enable the control of complex systems, or even the design of self-organizing or self-regulating complex systems, as I have discussed them before.

In this way, it would be possible, for example, to separately incentivize the reuse of different kinds of resources. One could separately account for CO_2, noise, and various other environmental impacts, but also positive effects such as health and education. One could furthermore reward environmental-friendly production, socially responsible management, and cultural engagement.

While today, those companies often grow the fastest that neglect good working conditions and the protection of the environment, a socio-ecological finance system such as "Fin4+" could promote profitable production that would also generate social, environmental, and cultural benefits for our society.

By the way, having multiple currencies would dramatically expand the space of possibilities. Say, we are faced with a win-lose situation, where one person would make a lot of profit, if the interaction took place, but the other person would suffer from a loss. Such an exploitative situation could be turned into a symbiotic win-win situation, if the first person would make a compensation payment.[56] Indeed, a currency transfer would create a profitable situation for both.

---

[55] http://ebook.finfour.net; see also [7].

[56] Interaction Support Processor, WiPo (August 13, 2015) https://patentscope.wipo.int/search/en/detail.jsf?docId=WO2015118455.

Having multiple currencies, there would be a lot more situations in which such interactions between multiple parties, which today would be lossful for some, could be turned into a symbiotic, beneficial situation for all in the future.[57] Hence, a multi-currency system could act like a catalyst, creating new opportunities, where they did not exist before. In this way, our current economy, which is based on the exploitation of nature and people, could be turned into a system that is based on the creation of new opportunities for all. It could very well be the underlying principle for a new economy, characterized by more sustainability and new participatory opportunities. It could be the formula for future prosperity and peace. Why not give it a try?

Acknowledgements This work was partially supported by the European Research Council (ERC) under the European Union's Horizon 2020 research and innovation programme (grant agreement No. 833168). The support is gratefully acknowledged.

References

1. L. Lessig, Code Is Law: On Liberty in Cyberspace, Harvard Magazine (1.1.2000) https://harvardmagazine.com/2000/01/code-is-law-html.
2. S. Zuboff, The Age of Surveillance Capitalism: The Fight for a Human Future at the New Frontier of Power (PublicAffairs, 2019).
3. F. Aleskerov, D. Bouyssou, and B. Monjardet, Utility Maximization, Choice and Preferences (Springer, 2007).
4. B. Friedman and.G. Hendry, Value Sensitive Design: Shaping Technology with Moral Imagination (MIT Press, 2019); https://www.delftdesignforvalues.nl.
5. E. Brynjolfsson and A. McAfee, The Second Machine Age (W.W. Norton, 2016).
6. D. Helbing et al. Biologistics and the Struggle for Efficiency: Concepts and Perspectives, Advances in Complex Systems 12(6), 533–548 (2009).
7. M.M. Dapp, D. Helbing and S. Klauser (eds.) Finance 4.0: Towards A Socio-Ecological Finance System (Springer, 2021).

---

[57] Interaction Support Processor, in Helbing, [7].

Chapter 14
Summary: What's Wrong with AI?

Humanistic Technology Needed!

The end is near... We have arrived at the end of the book, and it is time to summarize our findings and to wrap up. The great potentials of the digital revolution cannot be denied. However, we need to think twice how to use this technology, so it will be at our service and not endanger what humanity has built over centuries. The lessons of wars and revolutions should not be forgotten. We need technology that empowers people, while helping us to coordinate our actions, so conflict is avoided. In this book, I have presented quite a lot of ideas how this can be done, and how to avoid the dark sides of the digital revolution. Therefore, I hope we will soon see humanistic digital technology that helps us live in harmony with nature.

14.1 AI on the Rise

There are probably not many people who would doubt that we have arrived in an age of Big Data and Artificial Intelligence (AI). Of course, this opens up many previously untapped opportunities, ranging from production to automated driving and everyday applications. In fact, not only digital assistants such as *Google* Home, Siri or Alexa, but also many Web services and apps, smartphones and home appliances, cleaning robots and even toys already use AI or at least some kind of machine learning.

Technology visionaries such as Ray Kurzweil (*1948) predicted that AI would have the power of an insect brain in 2000, the power of a mouse brain around 2010, human-like brainpower around 2020 and the power of *all* human brains on Earth before the middle of this century. Many do not share this extremely techno-optimistic view, but it cannot be denied that back in 1997 *IBM*'s Deep Blue computer beat the Chess genius Garry Kasparov (*1963), that *IBM*'s Watson computer won the knowledge game Jeopardy back in 1997, and that *Google*'s AlphaGo system beat the world champion Lee Sedol (*1983) in the highly complex strategy game "Go" in 2016—about 10–20 years before many experts had expected this to happen. When

This chapter appeared before as FuturICT Blog (September 1, 2019) http://futurict.blogspot.com/2019/09/whats-wrong-with-ai.html and in the SI Magazine—the SwissInformatics Society's New Online Platform (March 2020), see https://magazine.swissinformatics.org/en/whats-wrong-with-ai/.

AlphaZero managed to outperform AlphaGo without human training shortly later, just by playing Go a lot of times against itself, the German news journal *Der Spiegel* wrote on October 29, 2017: "Gott braucht keine Lehrmeister"[1] ["God does not need teachers."].

14.1.1 AI as God?

This was around the time when Anthony Levandovski (*1980), a former head of *Google*'s self-driving car project, founded a religion that worships an AI God.[2] By that time, many considered *Google* to be almost all-knowing. With the *Google* Loon project, they were also working on omni-presence. Omnipotence was still a bit of a challenge, but as the world learned by the end of 2015, it was possible to manipulate people's attention, opinions, emotions, decisions, and behaviors with personalized information.[3] In a sense, our brains had been hacked. However, only in summer 2017 did a previous member of a *Google* control room, Tristan Harris (*1984), reveal in his *TED* talk "How a handful of tech companies control billions of minds every day".[4] At the same time, *Google* was trying to build superintelligent systems and to become something like an emperor over life and death, namely with its *Calico* project.[5]

---

[1] Ein Gott braucht keine Lehrmeister, SPIEGEL (October 29. 2017) https://www.spiegel.de/wissen schaft/technik/kuenstliche-intelligenz-gott-braucht-keine-lehrmeister-kolumne-a-1175130.html.

[2] The engineer at the center of a bombshell Uber lawsuit has founded a religion that warships an AI god, Business Insider (September 27, 2017) https://www.businessinsider.com/anthony-levand owski-religion-worships-ai-god-report-2017-9; Inside the First Church of Artificial Intelligence (November 15, 2017). https://www.wired.com/story/anthony-levandowski-artificial-intelligence-religion/.

[3] Das Digital-Manifest: Digitale Demokratie statt Datendiktatur, Spektrum der Wissenschaft (November 12, 2015) https://www.spektrum.de/thema/das-digital-manifest/1375924, in English: Will Democracy Survive Big Data and Artificial Intelligence? Scientific American (February 25, 2017) https://www.scientificamerican.com/article/will-democracy-survive-big-data-and-artificial-intelligence/.

[4] Tristan Harris, How a handful of tech companies controls billions of minds every day, (July 28, 2017) https://www.youtube.com/watch?v=C74amJRp730.

[5] Google's project to 'cure death,' Calico, announces $1.5 billion research center, The Verge (September 3, 2014) https://www.theverge.com/2014/9/3/6102377/google-calico-cure-death-1-5-billion-research-abbvie.

14.1.2 Singularity

So, was *Google* about to give birth to a digital God—or had done so already?[6] Those believing in the "singularity"[7] and AI as "our final invention"[8] already saw the days of humans counted. "Humans, who are limited by slow biological evolution, couldn't compete and would be superseded," said the world-famous physicist Stephen Hawking (1942–2018).[9] Elon Musk (*1971) warned: "We should be very careful about Artificial Intelligence. If I had to guess at what our biggest existential threat is, it's probably that."[10] Bill Gates (*1955) stated he was "in the camp that is concerned about super intelligence".[11] And *Apple* co-founder Steve Wozniak (*1950) asked: "Will we be the Gods? Will we be the family pets? Or will we be ants that get stepped on?"[12] AI pioneer Jürgen Schmidhuber (*1963) wants to be the father of the first superintelligent robot and appears to believe that we would be like cats, shortly after the singularity happens.[13] Sofia, the female humanoid robot, who is now counted as a citizen of Saudi Arabia, seems to see things similarly.[14]

Therefore, it is time to ask the question, what will happen with humans and humanity after the singularity? There are different views on this. Schmidhuber claims superintelligent robots will be as little interested in humans as we are in ants, but this does, of course, not mean we would not be in a competition for material resources with them. Others believe the next wave of automation will make millions or even billions unemployed. Combined with the world's expected sustainability crisis (i.e. predicted shortage of certain resources), this is not good news. Billions of humans

---

[6]Google as God? Opportunities and Risks of the Information Age (April 11, 2013) https://arxiv.org/abs/1304.3271; Google and the Birth of a Digital God? The Globalist (December 25, 2017) https://www.theglobalist.com/google-artificial-intelligence-big-data-technology-future/; Dirk Helbing, The Birth of a Digital God? (October 21, 2018) https://www.youtube.com/watch?v=BgoU3koKmTg.

[7]See https://en.wikipedia.org/wiki/Technological_singularity.

[8]Barrat [1].

[9]Stephen Hawking Warns Artificial Intelligence Could End Mankind (December 2, 2014) https://en.wikipedia.org/wiki/Technological_singularity.

[10]Elon Musk: Artificial Intelligence Is Our Biggest Existential Threat, The Guardian (October 27, 2014) https://www.theguardian.com/technology/2014/oct/27/elon-musk-artificial-intelligence-ai-biggest-existential-threat.

[11]Microsoft's Bill Gates Insists AI Is A Threat, BBC (January 29, 2015) https://www.bbc.com/news/31047780.

[12]Steve Wozniak on AI: Will we be pets or mere ants to be squashed [by] our robot over-loards? Computerworld (March 25, 2015) https://www.computerworld.com/article/2901679/steve-wozniak-on-ai-will-we-be-pets-or-mere-ants-to-be-squashed-our-robot-overlords.html.

[13]Künstliche Intelligenz: Werden wir für sie wie Katzen sein? Frankfurter Allgemeine (November 30, 2017) https://www.faz.net/aktuell/feuilleton/debatten/kuenstliche-intelligenz-maschinen-ueberwinden-die-menschheit-15309705.html; Jürgen Schmidhuber: „Die Geschichte wird nicht mehr von Menschen dominiert" NZZ am Sonntag (October 21, 2017) https://nzzas.nzz.ch/gesellschaft/juergen-schmidhuber-geschichte-wird-nicht-mehr-von-menschen-dominiert-ld.1322558.

[14]When Asked With Sophia Robot „Are Humans Necessary" (March 9, 2019) https://www.youtube.com/watch?v=ntMy2wk4aFg.

might (have to) die early, if the predictions of the Club of Rome's "Limits to Growth" and other studies were right.[15] This makes the heated debate on ethical dilemmas, algorithms deciding about life and death, as well as killer robots and autonomous weapons[16] understandable[17] (see discussion below).

Will we face something like a "digital holocaust", where autonomous systems decide about our life based on a Citizen Score or some other approach? Not necessarily so. Others believe that, in the "Second Machine Age",[18] humans will experience unprecedented prosperity, i.e. we would enter some kind of technologically enabled "paradise", where we would finally have time for friends, hobbies, culture and nature rather than being exploited for work. But even those optimists have often issued warnings that societies would need a new framework in the age of AI, such as a universal basic income.[19] We would certainly need a new societal contract.

14.1.3 Transhumanism

Opinions are not only divided about the future of humanity, but also on the future of humans. Some, like Elon Musk, believe that humans would have to upgrade themselves with implanted computer chips in order to stay competitive with AI (and actually merge with it).[20] In perspective, we would become cyborgs and replace organs (degraded by aging, handicap, or disease) by technological solutions. Over time, a bigger and bigger part of our body would be technologically upgraded, and it would be increasingly impossible to tell humans and machines apart.[21] Eventually, some people believe, it might even become possible to upload the memories of humans into a computer cloud and thereby allow humans to live there forever, potentially connected to several robot bodies in various places.[22]

---

[15] Meadows [2].

[16] Autonomous Weapons: An Open Letter from AI/ Robotics Researchers (July 28, 2015) https://futureoflife.org/open-letter-autonomous-weapons/.

[17] Autonome Intelligenz ist nicht nur in Kriegsrobotern riskant, NZZ (February 20, 2018) https://www.nzz.ch/meinung/autonome-intelligenz-ist-nicht-nur-in-kriegsrobotern-riskant-ld.1351011.

[18] Brynjolfsson and McAfee [3].

[19] We Must Remake Society in the Coming Age of AI, Wired (October 12, 2016) https://www.wired.com/2016/10/obama-aims-rewrite-social-contract-age-ai/.

[20] Elon Musk unveils plan to build mind-reading implants: 'The monkey is out of the bag', The Guardian (July 17, 2019) https://www.theguardian.com/technology/2019/jul/17/elon-musk-neural-ink-brain-implants-mind-reading-artificial-intelligence.

[21] „In Zukunft werden wir Mensch und Maschine wohl nicht mehr unterscheiden können", NZZ (August 22, 2019) https://www.nzz.ch/zuerich/mensch-oder-maschine-interview-mit-neuropsychologe-lutz-jaencke-ld.1502927.

[22] Geraci [4].

Others think we would genetically modify humans and upgrade them biologically to stay competitive.[23] Genetic manipulation might also extend life spans considerably (at least for those who can afford it). However, in a so-called "over-populated world" this would increase the existential pressure on others, which brings us back to the life and death decisions mentioned before and, actually, to the highly problematic subject of eugenics.[24]

Overall, it appears to many technology visionaries that humans as we know them today would not continue to exist much longer. All those arguments, however, are based on the extrapolation of technological trends of the past into the future, while we may also experience unexpected developments, for example, something like "networked thinking"—or even an ability to perceive the world beyond our own body, thanks to the increasingly networked nature of our world. ("Links" are expected to become more important than the system components they connect, thereby also changing our perception of our world.)

14.1.4 Is AI Really Intelligent?

The above expectations may also be wrong for another reason: perhaps the extrapolations are based on wrong assumptions. Are humans and robots comparable at all, or did we fall prey to our hopes and expectations, to our definitions and interpretations, to our approximations and imitations? Computers process information, while humans think and feel. But is it actually the same? Or do we compare apples and oranges?

Are todays robots and AI systems really autonomous to the degree humans are autonomous? I would say "no", at least the systems that are publicly known still depend a lot on human maintenance and external resources provided by humans.

Are today's AI systems capable of emotions? I would say "no". Changing color when gently touching the head of a robot, as "Pepper" does, has nothing to do with emotions. Being able smile or look surprised or angry, or being able to read our mimics is also not the same as "feeling" emotions. And sex robots, I would say, are neither able to feel love nor to love humans or other living beings. They can just "make love", i.e. make sex and talk as if they would have emotions. But this is not the same as having emotions, e.g. feeling pain.

Are today's AI systems creative? I would say "hmmm". Yes, we know AI systems that can mix cocktails and can generate music that sounds similar to Bach or any other composer, or create "paintings" that look similar to a particular artist's body of work. However, these creations are, in a sense, variations of a lot of inputs that have been fed into the system before. Without these inputs, I do not expect those systems

---

[23] The Power to Upgrade Your Own Biology Is in Sight—But Is Society Ready for Human Enhancement? Singularity Hub (February 15, 2018) https://singularityhub.com/2018/02/15/the-power-to-upgrade-our-biology-and-the-ethics-of-human-enhancement/.

[24] See https://en.wikipedia.org/wiki/Transhumanism.

to be creative by themselves. Did we really see some AI-created piece of art that blew our minds, something entirely new, not seen or heard before? I am not sure.

Are today's AI systems conscious? I would say "no". We do not even know exactly what consciousness is and how it works. Some people think consciousness emerges when many neurons interact, in a similar way as there are waves, if many water molecules interact in an environment exposed to wind. Other people, among them some physicists, think that consciousness is related to perception—a measurement process rooted in quantum physics. If this were the case, as the famous hypothetical "cat experiment" of Erwin Schrödinger (1887–1961) suggests, reality would be *created* by consciousness, not the other way round. Then, the "brain" would be something like a projector producing our perceived reality.

So far, we do not even know whether AI systems understand the texts they process. When they generate text or translations, they are typically combining existing elements of a massive database of human-generated texts. Without this massive database of texts produced by humans, AI-based text and translations would probably not sound like human language at all. It is kind of obvious that our brain works in a very different way. While we don't have nearly as many texts stored in our brain, we can nevertheless speak fluently. Suggesting that we approximately understand the brain because we can now build deep learning neural networks that communicate with humans seems to be misleading.

Are today's AI systems intelligent at least? I would say "no". The systems that are publicly known are "weak" AI systems, that are very powerful in particular tasks and often super-human in specific aspects. However, so far, "strong" AI systems that can flexibly adjust to all sorts of environments and tasks as humans can do, are not publicly known. I am also not aware of any AI systems that have invented something like the physical laws of electrodynamics, quantum mechanics, etc.

In conclusion, I would say that today, even humanoid robots are not like humans. They are a simulation, an emulation, an imitation, or approximation, but it is hard to tell how similar they really are. Recent reports even suggest that humans are often involved in generating "AI" services—and it would then be more appropriate to speak of "pseudo-AI".[25] I would not be surprised at all, if our attempts to build human-like beings in Silico would finally make us aware of how different humans are, and what makes us special. In fact, this is the lesson I expect to be learned from technological progress. What is consciousness? What is love? We might get a better understanding of who we really are and what is our role in the universe, if we can't just build it the way we have tried it so far.

Of course, I could be wrong with my judgment that true Artificial Intelligence does not currently exist. However, I have been waiting for years that someone would allow me to put their most advanced AI system to the test. They might, of course, be hiding it from me. To convince me of the existence of true Artificial Intelligence, I would have to see more than a chat bot that is capable of self-diagnosis (simulated

---

[25]The rise of 'pseudo-AI': how tech firms quietly use humans to do bot's work, The Guardian (July 6, 2018)_https://www.theguardian.com/technology/2018/jul/06/artificial-intelligence-ai-humans-bots-tech-companies.

"self-awareness"). I would need to have scientific and philosophical conversations with the system, and it should not just reply to my questions and statements by trying to find the best response in a huge database of human statements.

14.1.5 What Is Consciousness?

I sometimes speculate that our (3+1)-dimensional world might actually be an interpretation of higher-dimensional data (as elementary particle physics suggests as well). Then, there could be different kinds of interpretations, i.e. different ways of perceiving the world, depending on how we learn to perceive it. Imagine the brain to work like a filter—and think of Plato's allegory of the cave,[26] where people just see two-dimensional shadows of a three-dimensional world and, hence, would find very different "natural laws" governing what they see. What if our world was not three-dimensional with changes in time, but if we would see only a three-dimensional projection of a higher-dimensional world? Then, one day, we may start seeing the world in an entirely different way,[27] for example, from the perspective of quantum logic rather than binary logic. For such a transition in consciousness to take place, we would probably have to learn to interpret weaker signals than what our five senses send to our brain. Then, the permanent distractions by the attention economy would just be the opposite of what would be needed to advance humanity.

Let me make this thought model a bit more plausible: Have you ever wondered why Egyptian and ancient paintings used to look flat, i.e. two-dimensional, for hundreds of years, while suddenly three-dimensionally looking perspective was invented and became the new standard in art? What if ancient people have really seen the world "with different eyes"? And what if this would happen again? Remember that the invention of photography was not the end of paintings. Instead, this invention freed arts from naturalistic representation, and entirely new painting styles were invented, such as impressionistic and expressionistic ones. Hence, will the creation of humanoid robots finally free us from our mechanistic, materialistic view of the world, as we learn how different we really are?

If we were not just "biological robots", other fields besides science, engineering, and logic would be a lot more important than some may think, such as psychology, sociology, history, philosophy, the humanities, ethics, and maybe even religion. Trying to reinvent society and humanity without the proper consideration of such fields of knowledge could easily end in major mistakes of historical proportions.

---

[26]https://en.wikipedia.org/wiki/Allegory_of_the_Cave.
[27]Mueller [5], Phil Ball [6].

14.2   Can We Trust It?

14.2.1   Big Data Analytics

In an article of the "Wired" magazine 2008, Chris Anderson (*1961) claimed that we would soon see the end of theory, and the data deluge would make the scientific method obsolete.[28] If one just had enough data, certain people started to believe, data quantity could be turned into data quality, and, therefore, Big Data would be able to reveal the truth by itself. In the past few years, however, this paradigm has been seriously questioned. As data volume increases much faster than processing power, there is the phenomenon of "dark data", which will never be processed and, hence, it will take scientists to decide which data should be processed and how.[29] In fact, it is not trivial at all to distill raw data into useful information, knowledge and wisdom. In the following, I will describe some of the problems related to the question "how to connect the dots".

It is frequently assumed that more data and more model parameters are better to get an accurate picture of the world. However, it often happens that people "can't see the forest for the trees". "Overfitting", where one happens to fit models to random fluctuations or otherwise meaningless data, can easily happen. "Sensitivity", where outcomes of data analyses change significantly, when some data points are added or subtracted, or another algorithm or computer hardware is used, is another problem. A third problem are errors of first and second kind, i.e. "false positives" ("false alarms") and cases, where alarms should go off, but fail to do so. A typical example is "predictive policing", where false positives are overwhelming (often above 99 percent), and dozens of people are needed to clean the suspect lists.[30] And these are by far not all the problems …

14.2.2   Correlation Versus Causality

In Big Data, it is easy to find patterns and correlations. But what do they actually mean? Say, one finds a correlation between two variables A and B. Then, does A cause B or B cause A? Or is there a third factor C, which causes A and B? For example, consider the correlation between the number of ice-cream-eating children and the number of forest fires. Forbidding children to eat ice cream will obviously not reduce the number of forest fires at all—despite the strong correlation. It is, of

---

[28]The End of Theory: The Data Deluge Makes the Scientific Method Obsolete, Wired (June 24, 2008) https://www.wired.com/2008/06/pb-theory/.

[29]Das Digital-Manifest: Digitale Demokratie statt Datendiktatur, Spektrum der Wissenschaft (November 12, 2015) https://www.spektrum.de/thema/das-digital-manifest/1375924.

[30]BKA: Überwachung von Flugpassagieren liefert Fehler über Fehler, Süddeutsche Zeitung (April 24, 2019) https://www.sueddeutsche.de/digital/fluggastdaten-bka-falschtreffer-1.4419760.

course, a third factor, namely outside heat, which causes both, increased ice cream consumption and forest fires.

Finally, there may be no causal relationship between A and B at all. The bigger a data set, the more patterns will be found just by coincidence, and this could be wrongly interpreted as meaningful or, as some people would say, as a signal rather than noise.[31] In fact, spurious patterns and correlations are quite frequent.[32]

Nevertheless, it is, of course, possible to run a society based on correlations. The application of predictive policing may be seen as example. However, the question is, whether this would really serve society well. I don't think so. Correlations are frequent, while causal relationships are not. Therefore, using correlations as basis of certain kinds of actions unnecessarily restrains our freedom (effectively introducing new laws through code).

14.2.3 Trustable AI

There has been the dream that Big Data is the "new oil" and Artificial Intelligence something like a "digital motor" running on it. So, if it is difficult for humans to make sense of Big Data, AI might be able to handle it "better than us". Would AI be able to automate Big Data analytics? The answer is, partly.

In recent years, it was discovered that AI systems would often discriminate against women, non-white people, or minorities. This is because these systems are typically trained with data of the past. That is problematic, since learning from the past may stabilize old societal paradigms we should actually better replace by something else, given that today's world is not sustainable.

Lack of explanation is another important issue. For example, you may get into the situation that your application for a loan or life insurance is turned down, but nobody can explain you why. The salespeople would just be able to tell you that their AI system has recommended them to do so. The reason may be that two of your neighbors had difficulties paying back their loans. But this is again messing up correlations and causal relationships. Why should you suffer from this? Hence, experts have recently pushed for explainable results under labels such as "trustable AI". So far, however, one may say that we are still living in a "blackbox society".[33]

---

[31] Silver [7].

[32] Vigen [8].

[33] Pasquale [9], The Dark Secret at the Heart of AI, MIT Technology Review (April 11, 2017) https://www.technologyreview.com/s/604087/the-dark-secret-at-the-heart-of-ai/.

14.2.4 Profiling, Targeting, and Digital Twins

Since the revelations of Edward Snowden (*1983), we know that we have all been targets of mass surveillance, and based on our personal data, "profiles" of us have been created—no matter whether you wanted this or not.[34] In some cases, such profiles have not only been data bases or unstructured data about us. Instead, "digital doubles", i.e. computer agents emulating us, have been created of all people in the world (as much as this was possible). You may imagine this like a black box that has been created for everyone, which is continuously fed with surveillance data and learns to behave like the humans they are imitating. Such platforms may be used to simulate countries or even the entire world. One of these systems is known under the name of "Sentient World".[35] It contains highly detailed profiles of individuals. Services such as "Crystal Knows" may give you an idea.[36] Detailed personal information can be used to personalize information and to manipulate our attention, opinions, emotions, decisions and behaviors.[37] Such "targeting" has been used for "neuromarketing"[38] and also to manipulate elections.[39] Furthermore, it plays a major role for today's information wars and the current fake news epidemics.[40]

14.2.5 Data Protection?

In seems, the EU General Data Protection Regulation (GDPR) should have protected us from mass surveillance, profiling and targeting. In fact, the European Court of Human Rights has ruled that past mass surveillance as performed by the GCHQ was unlawful.[41] However, it appears that lawmakers have come up with new reasons for

---

[34] Sentient World Simulation and NSA Surveillance—Exploiting Privacy to Predict the Future? (May 19, 2019) https://emerj.com/ai-future-outlook/nsa-surveillance-and-sentient-world-simulation-exploiting-privacy-to-predict-the-future/.

[35] Sentient world: war games on the grandest scale, The Register (June 23, 2007) https://www.theregister.co.uk/2007/06/23/sentient_worlds/.

[36] Crystal, https://www.crystalknows.com, "The app that tells you anyone's personality" https://web.archive.org/web/20181205032704/, https://www.crystalknows.com/.

[37] How Covert Agents Infiltrate the Internet to Manipulate, Deceive, and Destroy Reputations, The Intercept (February 25, 2014) https://theintercept.com/2014/02/24/jtrig-manipulation/.

[38] Morin and Renvoise [10].

[39] Cambridge Analytica: how did it turn clicks into votes? The Guardian (May 6, 2018) https://www.theguardian.com/news/2018/may/06/cambridge-analytica-how-turn-clicks-into-votes-christopher-wylie.

[40] Meet the weaponized AI system that know you better than you know yourself, Extreme Tech (March 1, 2017) https://www.extremetech.com/extreme/245014-meet-sneaky-facebook-powered-propaganda-ai-might-just-know-better-know.

[41] GCHQ data collection regime violated human rights, court rules, The Guardian (September 13, 2018) https://www.theguardian.com/uk-news/2018/sep/13/gchq-data-collection-violated-human-rights-strasbourg-court-rules.

surveillance such as being the friend of a friend of a friend of a suspected criminal, where riding a train without a ticket might be enough to consider someone a criminal or suspect.

Moreover, it is now basically impossible to use the Internet without agreeing to Terms of Use beforehand, which typically forces you to agree with the collection of personal data, even if you don't like this—otherwise you will usually not get a service. The personal data collected by companies, however, will often be aggregated by secret services, as Edward Snowden's revelations about the NSA have shown. In other words, it seems that the GDPR, which claims to protect us from unwanted collection and use of personal data, has actually enabled it. Consequently, there are huge amounts of data about everyone, which can be used to create digital doubles.

Are our personal profiles reliable at least? How similar to us are our digital twins really? Some skepticism is in place. We actually don't know exactly how well the learning algorithms, which are fed with our personal data, converge. Social networks often have features similar to power laws. As a result, the convergence of learning algorithms may not be guaranteed. Moreover, when measurements are noisy (which is typically the case), chances that digital twins behave identical to us are not very high. Hence, we may be easily misjudged. This does, of course, not necessarily exclude that averages or distributions of behaviors may be rather accurate (but there is no guarantee).

14.2.6 Scoring, Citizen Scores, Superscores

The approach of "scoring" goes a step further. It assesses people based on personal (e.g. surveillance) data and attributes a certain economic or societal value to them. People would be treated according to their score. Their lives would be "curated". Only people with a high enough score would get access to certain products or services, while others would not even see them on their digital devices at all, or see a downgraded offer. Personalized prizing is just one example for the personalization of our digital world.

According to my assessment, scoring is not compatible with human rights, particularly human dignity (see below).[42] However, you can imagine that there are currently quite a lot of scores about you. Each company working with personal data may have several of them.[43] You may have a consumer score, a health score, an environmental footprint, a social media score, a Tinder score, and many more. These scores may then be used to create a superscore, by aggregating different scores into an index.[44] In

---

[42] Superscoring: Wie wertvoll sind Sie für die Gesellschaft? PC Welt (February 5, 2020) https://www.pcwelt.de/ratgeber/Superscoring-Wie-wertvoll-sind-Sie-fuer-die-Gesellschaft-10633488.html.

[43] Silicon Valley is building a Chinese-style social credit system, Fast Company (August 26, 2019) https://www.fastcompany.com/90394048/uh-oh-silicon-valley-is-building-a-chinese-style-social-credit-system.

[44] Super-Scoring? https://www.superscoring.de.

other words, individuals would be represented by one number, which is often referred to as "Citizen Score".[45] This appears to be done, for example, by the "Karma Police Program",[46] which was revealed on September 25, 2015—the day on which Pope Francis demanded the Agenda 2030, a set of 17 Sustainable Development Goals, in front of the UN General Assembly.[47] China is currently testing a Citizen Score approach named "Social Credit Score".[48] The program may be seen as an attempt to make citizens obedient to the government's wishes. This has been criticized as data dictatorship[49] or technological totalitarianism.[50]

A Citizen Score establishes, in principle, a Big-Data-driven neo-feudalistic order.[51] Those with a high score will get access to good offers, products, and services, and basically get anything they desire. Those with a low score may lose their human rights and may be deprived from certain opportunities. This may include such things as travel visa to other countries, job opportunities, the allowance to use a plane or fast train, the Internet speed, or certain kinds of medical treatment. In other words, high-score citizens will live kind of "in heaven", while a large number of low-score citizens may experience something like "hell on Earth". It's something like a digital judgment day scenario, where you get negative points, when you cross a red pedestrian traffic light, if you read critical political news, or if you have friends who read critical political news, to give some realistic examples.

The idea behind this is to establish "total justice" (in the language of the Agenda 2030: "strong institutions"), particularly for situations when societies are faced with scarce resources. However, a Citizen Score will not do justice to people at all, as these are different by nature. All the arbitrariness of a superscore lies in the weights of the different measurements that go into the underlying index.[52] These weights will be to

---

[45] Big data meets Big Brother as China moves to rate ist citizens, Wired (October 21. 2017) https://www.wired.co.uk/article/chinese-government-social-credit-score-privacy-invasion.

[46] Profiled: From Radio to Porn, British Spies Track Web Users' Online Identities, The Intercept (September 25, 2015) https://theintercept.com/2015/09/25/gchq-radio-porn-spies-track-web-users-online-identities/.

[47] Sustainable development summit 2015: World leaders agree sustainable development goals—as it happened, The Guardian (September 25, 2015) https://www.theguardian.com/global-development/live/2015/sep/25/un-sustainable-development-summit-2015-goals-sdgs-united-nations-general-assembly-70th-session-new-york-live.

[48] China has started ranking citizens with a creepy 'social credit' system … Business Insider (October 29, 2018) https://www.businessinsider.com/china-social-credit-system-punishments-and-rewards-explained-2018-4.

[49] Das Digital-Manifest: Digitale Demokratie statt Datendiktatur, Spektrum der Wissenschaft (November 12, 2015) https://www.spektrum.de/thema/das-digital-manifest/1375924, English translation: Will Democracy Survive Big Data and Artificial Intelligence? (February 25, 2017) https://www.scientificamerican.com/article/will-democracy-survive-big-data-and-artificial-intelligence/.

[50] Schirrmacher [11].

[51] Superscoring: Wie wertvoll sind Sie für die Gesellschaft? PC Welt (February 5, 2020) https://www.scientificamerican.com/article/will-democracy-survive-big-data-and-artificial-intelligence/.

[52] https://www.pcwelt.de/ratgeber/Superscoring-Wie-wertvoll-sind-Sie-fuer-die-Gesellschaft-10633488.html.

Fig. 14.1 Full automation
scenario

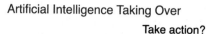

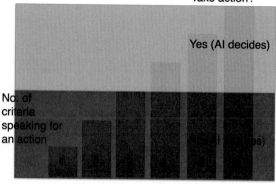

the advantage of some people and to the disadvantage of others.[53] A one-dimensional index will brutally oversimplify the nature and complexity of people, and tries to make everyone the same, while societies thrive on differentiation and diversity. It is clear that scoring will treat some people very badly, in many cases without good reasons (remember my notes on predictive policing above). This would affect societies in a negative way. Creativity and innovation (which challenge established ways of doing things) are expected to suffer. This will be detrimental to changing the way our economy and society work and, thereby, obstruct the implementation of a better, sustainable system.

14.2.7 Automation Versus Freedom

In the age of AI, there is a great temptation to automate processes of all kinds. This includes the automation of decisions. Within the framework of Smart Cities and Smart Nations concepts, there are even attempts to automate societies, i.e. to run them almost like a machine.[54] Such automation of decisions is often (either explicitly or implicitly) based on some index or one-dimensional decision function with one or several decision thresholds. Let us assume you are measuring 6 different quantities to figure out whether you should decide "yes" or "no" in a certain situation. Then, of course, you could specify that a measurement-based algorithm should decide for "no", if between 0 and 3 measurements speak for "yes", while it should decide for "yes", if between 4 and 6 measurements recommend to choose "yes" (see Fig. 14.1).

---

[53] See Appendix A in Helbing and Balietti [12].

[54] Politik der Algorithmen: Google will den Staat neu programmieren, FAZ (October 14, 2015) http://www.faz.net/aktuell/feuilleton/medien/google-gruendet-in-den-usa-government-innovaton-lab-13852715.html.

Fig. 14.2 Creating room for
free human choice in an age
of automation

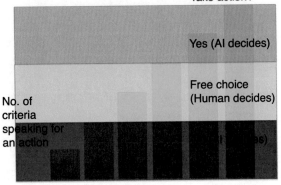

This may sound plausible, but it eliminates the human freedom of decision-making completely.

Instead, it would be more appropriate to automate only the "sure" cases (e.g. "no", if a clear minority of measurements recommend "yes", but "yes" if a clear majority of measurements recommend "yes"). In contrast, the "controversial" cases should prompt human deliberation (also if some additional fact comes to light that makes it necessary to revisit everything). Such a procedure is justified by the fact that there is no scientific method to fix the weights of different measurements exactly (in particular, given the problem of finite confidence intervals). Figure 14.2 shows the clear "yes" cases in green, the clear "no" cases in red, and the controversial cases in yellow.

The suggested semi-automated approach would reduce human decision workload by automation where it does not make sense to bother people with time-consuming decisions that are clear. Moreover, it would give people more time to decide important affairs. I believe this is important to choose a future that works well for humans, who derive happiness, in particular, from exercising autonomy and maintaining good relationships.[55]

---

[55]The No. 1 Contributor to Happiness, Psychology Today (June 30, 2011) https://www.psycholog ytoday.com/intl/blog/bouncing-back/201106/the-no-1-contributor-happiness; The Secret of Happiness Revealed by Harvard Study, Forbes (May 27, 2015) https://www.forbes.com/sites/georgebradt/2015/05/27/the-secret-of-happiness-revealed-by-harvard-study/.

14.2.8 *Learning to Die?*

Indices based on multiple measurements or data sets are also being used to recommend prison sentences,[56] to advise on whether a certain medical operation is economic, given the overall health and age of a person, or even to make life-or-death decisions.[57] One scenario that has recently been obsessively discussed is the so-called "trolley problem".[58] Here, an extremely rare situation is assumed, where an autonomous vehicle cannot brake quickly enough, and one person or another will die (e.g. one pedestrian or another one, or the car driver). Given that, thanks to camera technology, pattern recognition and Big Data, a future car might distinguish between age, race, gender, social status etc., controversial questions arise. For example, should the car save a grandmother, while sacrificing an unemployed person, or vice versa? Note that such considerations are not in accordance with the equality principle of democratic constitutions[59] and also not compatible with classical medical ethics.[60] Nevertheless, questions like these have been recently raised in what was framed as the "Moral Machine Experiment",[61] and recent organ transplant practices already apply new policies considering previous lifestyle.[62]

Now, assume that we would put such data-based life-and-death situations into laws. Can you imagine what will happen if such principles originally intended to save human lives would be applied in possible future scenarios characterized by scarce resources? Then, the algorithm would turn into a killer algorithm bringing a "digital holocaust" on the way, which might sort out old and ill people, and probably also people of low social status, including certain minorities. This is the true danger of creating autonomous systems that may seriously interfere with our lives.[63] It does not take killer robots for autonomous systems to become a threat to our lives, if the supply of resources falls short.

---

[56] AI is convicting criminals and determining jail time, but is it fair? World Economic Forum (November 19, 2018) https://www.weforum.org/agenda/2018/11/algorithms-court-criminals-jail-time-fair/; AI is sending people to jail—and getting it wrong, MIT Technology Review (January 21, 2019) https://www.technologyreview.com/s/612775/algorithms-criminal-justice-ai/.

[57] Wenn Software über Leben und Tod entscheidet, ZDF (December 20, 2017) https://web.archive.org/web/20171222135350/, https://www.zdf.de/nachrichten/heute/software-soll-ueber-leben-und-tod-entscheiden-100.html.

[58] Nagler et al. [13].

[59] Automatisiertes und vernetztes Fahren, Bericht der Ethikkommission, Bundesministerium für Verkehr und digitale Infrastruktur, https://www.bmvi.de/SharedDocs/DE/Publikationen/DG/bericht-der-ethik-kommission.pdf.

[60] WMA Declaration of Geneva: The Physician's Pledge, World Medical Association, https://www.wma.net/policies-post/wma-declaration-of-geneva/.

[61] Awad et al. [14].

[62] Wer erhält die Organspende? Swiss Transplant https://www.swisstransplant.org/de/organspende-transplantation/rund-umsspenden/wer-erhaelt-die-spende/ (accessed August 8, 2020).

[63] Artificial Intelligence: Lethal Autonomous Weapons Systems and Peace Time Threats (November 28, 2017) https://ict4peace.org/activities/artificial-intelligence-lethal-autonomous-weapons-systems-and-peace-time-threats/.

14.2.9   A Revolution from Above?

The above scenario is, unfortunately, not just a phantasy scenario. It has been seriously discussed by certain circles already for some time. One book even proposes "Learning to Die in the Anthropocene".[64] In this connection it should be remembered that the Club of Rome's "Limit to Growth" study suggests we will see a serious economic and population collapse in the twenty-first century.[65] Accordingly, billions of people would die early. I have heard similar assessments from various experts, so we should take these forecasts easy. Some PhD theses have even discussed AI-based euthanasia.[66] There is also a possible connection with the eugenics agenda.[67] Sadly, it seems that the argument to "save the planet" is now increasingly being used to justify the worst violations of human rights. China, for example, has recently declined to sign a human rights declaration in a treaty with Switzerland,[68] and has even declined to receive an official German human rights delegation in China.[69] Certain political forces in other countries (such as Turkey, Japan, UK and Switzerland) have also started to propose human rights restrictions.[70] In the meantime, many cities and even some countries (including Canada and France) have declared a state of "climate emergency", which may eventually end with emergence laws and human rights restrictions.

Such political trends remind one of the proposed "revolution from above" that has been demanded by the Club of Rome already decades ago.[71] In the meantime, such a system—based on technological totalitarianism such as mass surveillance and

---

[64]Scranton [15].

[65]Meadows et al. [16].

[66]Hamburg [17].

[67]Kühl [18]; see also https://en.wikipedia.org/wiki/William_Shockley; Private jets, parties and eugenics: Jeffrey Epstein's bizarre world of scientists, The Guardian (August 19, 2019) https://www.theguardian.com/us-news/2019/aug/18/private-jets-parties-and-eugenics-jeffrey-epsteins-bizarre-world-of-scientists.

[68]Der Bundesrat plante offenbar eine Menschenrechtsvereinbarung mit China – und blitzte damit ab, NZZ (May 4, 2019) https://www.nzz.ch/schweiz/seidenstrasse-bundesrat-wollte-vereinbarung-zu-menschenrechten-ld.1479438.

[69]China verweigert Einreise von Delegationen des Deutschen Bundestages, Handelsblatt (August 3, 2019) https://www.handelsblatt.com/politik/international/menschenrechtsausschuss-china-verweigert-einreise-von-delegationen-des-deutschen-bundestages/24868826.html.

[70]Theresa May: Human rights laws could change for terror fight, BBC (June 7, 2017) https://www.bbc.com/news/election-2017-40181444; Die SVP ist bereit Menschenrechte zu opfern, Tagesanzeiger (August 13, 2014) https://www.tagesanzeiger.ch/schweiz/standard/Die-SVP-ist-bereit-die-Menschenrechte-zu-opfern/story/12731796; Erdogan nennt Hitler-Deutschland als Beispiel, SPIEGEL (January 1, 2016) https://www.spiegel.de/politik/ausland/tuerkei-recep-tayyip-erdogan-nennt-hitler-deutschland-als-beispiel-fuer-praesidialsystem-a-1070162.html; Japans Vizepremier nennt Hitlers Absichten „richtig", SPIEGEL (August 30, 2017) https://www.spiegel.de/politik/ausland/japan-vizepremier-taro-aso-nennt-adolf-hitlers-absichten-richtig-a-1165232.html.

[71]King and Schneider [19].

citizen scores—has been created.[72] Note, however, that the negative climate impact of the oil industry was already known about 40 years ago,[73] and the problem of limited resources as well.[74] Nevertheless, it was decided to export the non-sustainable economic model of industrialized countries to basically all areas of the world, and the energy consumption has (been) dramatically increased since then.[75] I would consider this reckless behavior and say that politics has clearly failed to ensure responsible and accountable business practices around the world, so that human rights now seem to be at stake.

14.3 Design for Values

14.3.1 Human Rights

Let us now recall what was the reason to establish human rights in the first place. They actually resulted from terrible experiences such as totalitarian regimes, horrific wars, and the holocaust. The establishment of human rights as the foundation of modern civilization, reflected also by the UN's Universal Declaration of Human Rights,[76] was an attempt to prevent the repetition of such horrors and evil. However, the promotion of materialistic consumption-driven societies by multi-national corporations led to the current (non-)sustainability crisis, which has become an existential threat, which might cause the early death of hundreds of millions, if not billions of people.

Had industrialized countries reduced their resource consumption just by 3% annually since the early 70ies, our world would be sustainable by now, and discussions about overpopulation, climate crises and ethical dilemmas would be baseless. Would we have engaged in the creation of a circular economy and would we have used alternative energy production schemes early on rather than following the wishes of big business, our planet could easily manage our world's population.

---

[72]Log of the Rightwing Power Grab of Society, The Connectivist Blog (August 6, 2019) https://the connectivist.wordpress.com/2019/08/06/log-of-the-rightwing-power-grab-of-society/.

[73]Oil industry knew of 'serious' climate concerns more than 45 years ago, The Guardian (April 13, 2016) https://www.theguardian.com/business/2016/apr/13/climate-change-oil-industry-enviro nment-warning-1968.

[74]Global 2000 Report to the [US] President, http://www.geraldbarney.com/G2000Page.html; the report was published in 1980.

[75]See https://en.wikipedia.org/wiki/World_energy_consumption, https://www.e-education.psu. edu/earth104/node/1347 (accessed August 20, 2020).

[76]Universal Declaration of Human Rights, United Nations, https://www.un.org/en/universal-declar ation-human-rights/.

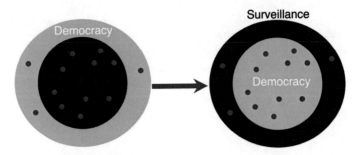

Fig. 14.3 Illustration of the transition from democracy to surveillance capitalism, which appears to quickly become an all-encompassing "operating system" of many societies

14.3.2 Happiness Versus Capitalism

Psychology knows that autonomy and good relationships—one might also call it "love"—are the main factors promoting happiness.[77] It is also known that (most) people have social preferences.[78] I believe, if our society would be designed and managed in a way that supports the happiness of people, it would also be more sustainable, because happy people do not consume that much. From the perspective of business and banks, of course, less consumption is a problem and, therefore, consumption is being pushed in various ways: by promoting individualism through education, which causes competition rather than cooperation between people and makes them frustrated; they will then often consume to compensate their frustration. Furthermore, using personalized methods such as "big nudging" and neuromarketing, advertisements have become extremely effective in driving us into more consumption. These conditions could, of course, be changed.

I would, therefore, say the average human is not the enemy of humanity (and nature), in contrast to what the book "The First Global Revolution" by the Club of Rome claimed.[79] It is rather the calculus of selfishness followed by companies, known as profit maximization or utilitarian thinking, often neglecting "externalities", which has driven the world to the edge. In the framework of surveillance capitalism (see Fig. 14.3), this system has been further perfected. Now, human personality and social capital have become resources that are being exploited. People's lives are being downloaded for free by means of mass surveillance—and our personal data are then being sold to companies that we don't know and probably do a lot of things we would not approve of. We have become the victims of this system, and we have to pay for it by means of data and by paying for the products we buy. It is often stated, if services

---

[77] The No. 1 Contributor to Happiness, Psychology Today (June 30, 2011) https://www.psychologytoday.com/intl/blog/bouncing-back/201106/the-no-1-contributor-happiness; The Secret of Happiness Revealed by Harvard Study, Forbes (May 27, 2015) https://www.forbes.com/sites/georgebradt/2015/05/27/the-secret-of-happiness-revealed-by-harvard-study/.

[78] Murphy et al. [20].

[79] King and Schneider [21].

are free, we are the product. I have serious doubts that such a system is compatible with human dignity, and it will probably cause a lot of problems.

For such reasons, I am also critical about utilitarian ethics, which makes numerical optimization, i.e. business-like thinking, the foundation of "ethical" decision-making. Such an approach gives everyone and everything (including humans and their organs) a certain value or price and implies that it would be even justified to kill people in order to save others, as the "trolley problem" seems to suggest. In the end, this would probably turn peaceful societies into a state of war or "unpeace", in which only one thing would count (namely, money, or whatever is chosen as utility function—it could also be power or security, for example, or any kind of Citizen Score).

This is the main problem of the utilitarian approach. While it aims to optimize society, it will actually destroy it little by little. The reason is that a one-dimensional optimization and control approach is not suited to handle the complexity of today's world. Societies cannot be steered like a car, where everyone moves right or left, as desired.[80] For societies to thrive, one needs to be able to steer into different directions at the same time, such as better education, improved health, reduced consumption of non-renewable energy, more sustainability, and increased happiness. This requires a multi-dimensional approach rather than one-dimensional optimization, and therefore, cities and countries are managed differently from companies (and live much longer!). The before mentioned requirement of pluralism has so far been best fulfilled by democratic forms of organization. However, a digital upgrade of democracies is certainly overdue. For this reason, it has been proposed to build "Digital Democracies" (or Democracies x.0, where x is a natural number greater than 1).[81]

14.3.3 Human Dignity

It is for the above reasons that human dignity has been put first in many constitutions. Human dignity is considered to be a right that's given to us by birth, and it is considered to be the very foundation of many societies. It is the human right that stands above everything else. Politics and other institutions must take action to protect human dignity from violations by public institutions and private actors, also abroad. Societal institutions lose their legitimacy if they do not engage effectively in this protection.[82] Improving human dignity on short, medium and long time scales should, therefore, be the main goal of political and human action. This is obviously not just about protecting or creating jobs, but it also calls for many other things such as sustainability, for example.

The question is, what is human dignity really about? It means, in particular, that humans are not supposed to be treated like animals, objects or data. They have

---

[80] Society Is Not A Machine, Optimization Not the Right Paradigm! Edge Question (2016) https://www.edge.org/response-detail/26795.

[81] Helbing and Pournaras [22], Helbing [23].

[82] See https://de.wikipedia.org/wiki/Menschenwürde, https://en.wikipedia.org/wiki/Dignity.

the right to be involved in decisions and affairs concerning them, including the right of informational self-determination. Exposing people to surveillance, while not giving them the possibility to easily control the use of their personal data is a significant violation of human dignity. Moreover, human dignity implies certain kinds of freedoms that serve to protect not only individual interests, but also to avoid the abuse of power and to ensure the functioning and well-being of society altogether.

Of course, freedoms should be exercised with responsibility and accountability, and this is where our current economic system fails. It diffuses, reduces and dissolves responsibility, in particular with regard to the environment. So-called externalities, i.e. external (often unwanted, negative) [side] effects are not sufficiently considered in current pricing and production schemes. This is what has driven our world to the edge, as the current environmental crisis shows in many areas of the world.

Furthermore, human dignity, like consciousness, creativity and love, cannot be quantified well. In other words: what makes us human is not well reflected by Big Data accumulated about our world. Therefore, a mainly data-driven and AI-controlled world largely ignores what is particularly important for us and our well-being. This is not expected to lead to a society that serves humans well.

While many people are talking about human-centric AI and a human-centric society, they often mean personalized information, products and services, based on profiling and targeting. This is getting things wrong. The justification of such targeting is often to induce behavioral change towards better environmental and health conditions, but it is highly manipulative, often abused, and violates privacy and informational self-determination. In the ASSET project [EU Grant No. 688364], instead, we have shown that behavioral change (e.g. towards the consumption of more sustainable and healthy products) is possible also based on informational self-determination and on respect of privacy.[83] I do, therefore, urge public and private actors to push informational self-determination forward quickly.

14.3.4   Informational Self-determination

Informational self-determination is a human right that follows directly from human dignity, and this cannot be given away under any circumstances (in particularly not by accepting certain Terms of Use). Nevertheless, in times of Big Data and AI, we have largely lost self-determination, little by little. This must be corrected quickly.

Figure 14.4 summarizes a proposed platform for informational self-determination, which would give control over our personal data and our digital doubles back to us. With this, all personalized services and products would be possible, but companies would have to convince us to share some of our data with them for a specific purpose, period of time, and perhaps price. The resulting competition for consumer trust would eventually promote a trustable digital society.

---

[83] Asikis et al. [24].

Informational Self-Determination

1. Send (a copy of) all personal data to a personal data mailbox
2. Require legally that personal data can only be used with the informed consent of individuals
3. Create a public platform that allows individuals to determine who is allowed to use what kind of data for what period of time and purpose (and what amount of money)
4. Build AI-based digital assistants that help people to easily administer personal data according to their preferences
5. Allow governments and scientists to run statistics on data
6. Report data use transparently to data mailbox

Fig. 14.4 Main suggested features of a platform for informational self-determination

The platform would also create a level playing field: not only big businesses, but also small and medium-sized enterprises (SMEs), spinoffs, non-government organizations (NGOs), scientific institutions and civil society could work with the data treasure, if they would get data access approved by the people (but many people may actually select this by default). Overall, such a platform for informational self-determination would promote a thriving information ecosystem that catalyses combinatorial innovation.

Government agencies and scientific institutions would be allowed to run statistics. Even a benevolent superintelligent system that helps desirable activities (such as social and environmental projects and the production of public goods) to succeed more easily while not interfering with the free will of people would be possible. Such a system should be designed for values such as human dignity, sustainability, fairness, as well as further constitutional and cultural values that support the evolvement of creativity and human potential with societal and global benefits in mind.

Data management would be done by means of a personalised AI system running on our own devices, i.e. digital assistants that learn our privacy preferences and the companies and institutions we trust or don't trust. Our digital assistants would comfortably preconfigure personal data access, and we could always adapt it.

Over time, if implemented well, such an approach would establish a thriving, trustable digital age that empowers people, companies and governments alike, while making quick progress towards a sustainable and peaceful world.[84]

---

[84]How to stop surveillance capitalism, The Globalist (April 26, 2018) https://www.theglobalist.com/capitalism-democracy-technology-surveillance-privacy and The Japan Times (April 30, 2018) https://www.japantimes.co.jp/opinion/2018/04/30/commentary/world-commentary/stop-surveillance-capitalism/; Facebook-Skandal: Experte rät zu digitalem Datenassistenten, Berliner Morgenpost (March 28, 2018) https://www.morgenpost.de/web-wissen/web-technik/article21386 8509/Facebook-Skandal-Experte-raet-zu-digitalem-Datenassistenten.html.

14.3.5 Design for Values

As our current world is challenged by various existential threats, innovation seems to be more important than ever. But how to guide innovation in a way that creates large societal benefits while keeping undesirable side effects at bay? Regulation often does not appear to work well. It is often quite restrictive and typically comes late, which is problematic for businesses and society alike.

However, it is clear that we need responsible innovation.[85] This requires pro-actively addressing relevant constitutional and ethical, social and cultural values already in the design phase of new technologies, products, services, spaces, systems, and institutions.

There are several reasons for adopting a design for values approach:[86] (1) the avoidance of technology rejection due to a mismatch with the values of users or society, (2) the improvement of technologies by better embodying these values, and (3) the stimulation of value-oriented behavior by design.

Value-sensitive design[87]—and ethically aligned design[88]—have quickly become quite popular. It is important to note, however, that one should not focus on a single value. Value pluralism is important.[89] Moreover, the kinds of values chosen may depend on the functionality or purpose of a system. For example, considering findings in game theory and computational social science, one could design next-generation social media platforms in ways that promote cooperation, fairness, trust and truth.[90] Also note that a list of 12 values to support flourishing information societies has recently been proposed (see Appendix 14.1).[91]

14.3.6 Democracy by Design

Among the social engineers of the digital age, it seems that democracy has often been framed as "outdated technology".[92] Larry Page (*1973) once said that *Google* wanted to carry out experiments, but could not do so, because laws were preventing

---

[85]How we can engineer a more responsible digital future, World Economic Forum (March 16, 2018). https://www.weforum.org/agenda/2018/03/engineering-a-more-responsible-digital-future.

[86]Engineering Social Technologies for a Responsible Digital Future, TU Delft, https://www.tudelft.nl/en/tpm/research/projects/engineering-social-technologies-for-a-responsible-digital-future/; see also the next references.

[87]TU Delft Design for Values, http://designforvalues.tudelft.nl (accessed on August 8, 2020).

[88]Ethically Aligned Design, Version 2, IEEE https://standards.ieee.org/content/dam/ieee-standards/standards/web/documents/other/ead_v2.pdf.

[89]Jeroen van den Hoven, Design for Values and Democracy (June 11, 2019) https://www.youtube.com/watch?v=sZYH53j5dFc.

[90]Helbing [25].

[91]Helbing [26].

[92]Randoph Hencken in: Mikrogesellschaften. Hat die Demokratie ausgedient? Video veröffentlicht am 15.5.2014. Joachim Gaertner, München, Bayerischer Rundfunk (2014).

What Does Democracy Mean?

- Human rights, human dignity
- Freedom
- Self-determination
- Pluralism
- Protection of minorities
- Division of power
- Checks and Balances
- Participation
- Transparency
- Fairness
- Justice
- Legitimacy
- Anonymous, equal votes

Design for values,
value-sensitive
design

- Privacy
  ✓ Protection from
  misuse/exposure
  ✓ Right to be left
  alone

Fig. 14.5 Values that matter for democracies, in particular

this.[93] Later, in fact, *Google* experimented in Toronto, for example, but people did not like it much.[94] Peter Thiel (*1967), on the other hand, claimed there was a deadly race between politics and technology, and one had to make the world save for capitalism.[95] In other words, it seems that, among many leading tech entrepreneurs, there has been little love and respect for democracy (not to talk about their interference with democratic elections by trying to manipulate voters).

However, democratic institutions have not occurred by accident. They have been lessons from wars, revolutions and other trouble in human history. Can we really afford to ignore them? At least, we should have a broad public debate about this beforehand. Personally, I consider the following democratic values to be relevant: human dignity and human rights, (informational) self-determination, freedom (combined with accountability), pluralism, protection of minorities, division of power, checks and balances, participation, transparency, fairness, justice, legitimacy, anonymous and equal votes, and privacy (in the sense of protection from unfair exposure and the right to be left alone) (see Fig. 14.5). I am not sure it would benefit

---

[93]Google CEO Larry Page Wants a Totally Separate World Where Tech Companies Can Conduct Experiments on People, Business Insider (May 16, 2013) https://www.businessinsider.com/google-ceo-larry-page-wants-a-place-for-experiments-2013-5.

[94]'Surveillance capitalism': critic urges Toronto to abandon smart city project, The Guardian (June 6, 2019) https://www.theguardian.com/cities/2019/jun/06/toronto-smart-city-google-project-privacy-concerns.

[95]Peter Thiel is trying to save the world: The apocalyptic theory behind his actions, Business Insider (December 8, 2016) https://www.businessinsider.com/peter-thiel-is-trying-to-save-the-world-2016-12.

society to drop any of these values. Maybe there are even more values to consider. For example, I have been thinking about possible ways to upgrade democracy, by unleashing collective intelligence and constructive collective action, particularly on a local level.[96] New formats such as City Olympics or a Climate City Cup[97] could establish a new paradigm of innovation and change in a way that can engage all stakeholders in a bottom-up way, including civil society. In fact, networks of cities and the regions around them could become a third pillar of transformation besides nations and global corporations.

14.3.7 Fairness

I want to end this contribution with a discussion of the power of fairness. People have often asked: Is it good that everyone has one vote? Shouldn't we have a system in which smart people have more weight? Shouldn't we replace the democratic "one man one vote" principle by a "one dollar one vote" system? (Btw, today we probably have the latter, because of voter manipulation by "big nudging".) Experimental evidence about the "wisdom of crowds" surprisingly suggests that, giving people different weights, whatever the criteria are, does not improve results.[98] On the contrary, studies in collective intelligence show that largely unequal influence on a debate will reduce social intelligence.[99] Diversity of information sources, opinions, and solution approaches is what makes collective intelligence work.

In conclusion, it seems a fair system based on the principle of equality is the best. In fact, it can be mathematically shown for many complex model systems that they will evolve towards an optimal state if and only if interactions are symmetrical.[100] If symmetry is broken, all sorts of things can happen. However, one can surely say that a hierarchical system or one controlled by utilitarian principles will very unlikely achieve the best systemic performance. For example, replacing tree-like supply systems by a circular economy could potentially improve the quality of living for everyone while making our economy more sustainable.

---

[96]How to Make Democracy Work in the Digital Age, Huffington Post (August 4, 2016) http://www.huffingtonpost.com/entry/how-to-make-democracy-work-in-the-digital-age_us_57a2f488e4b0456cb7e17e0f, https://www.researchgate.net/publication/305571691.

[97]The Climate City Cup, http://climatecitycup.org, Dirk Helbing, Introducing Climate City Cup (July 22, 2019) https://www.youtube.com/watch?v=lX4OQ1mDEA4.

[98]Page [27].

[99]Woolley et al. [28].

[100]Helbing and Vicsek [29], Helbing [30].

14.3.8 Network Effects for Prosperity, Peace and Sustainability

In our increasingly networked world, we currently experience a transition from component-dominated to interaction-dominated systems. The resulting network effects can change everything. Combinatorial innovation (i.e. innovation ecosystems), which would be enabled by a platform for informational self-determination, could boost our economy. Supporting collective intelligence (which should be the foundational principle of digital democracies) would benefit society. And a multidimensional real-time feedback system (a novel, socio-ecological finance system, which we sometimes call Finance 4.0,[101] FIN4, FIN 4+ or just FIN+), would be able to promote a sustainable circular and sharing economy and, thereby, help improve the state of nature. In this way, harmony between humans and nature could be reached, and everyone could benefit.

Finally, I would like to draw the attention to a computer simulation we have performed in order to understand the evolution of "homo economicus", the utility-maximizing selfish man assumed in economics. To our surprise we found that, after dozens of generations, people would develop other-regarding preferences and cooperative behavior, if parents raise children in their close neighborhood, as humans do.[102] In fact, compared to other species, it is quite exceptional how many years of their lives children spend with their parents. This makes a big difference, as it makes (most) people social.

There is a huge benefit in this. If people are friendly to each other, they can accomplish more and reach better outcomes. If it happens frequently enough that "I help someone and this person helps someone else", then also somebody will help me. In such a way, "networked thinking" will emerge—and with it a powerful support network (so-called "social capital"). Such cooperation will cause much higher average success (see Fig. 14.6).

Interestingly, such networked thinking and benevolent behavior can now be supported with digital technologies.[103] I believe, however, technological telepathy[104] is not needed for this—it might have even negative effects. But altogether, it is entirely in our hands to create a better world. "Love your neighbor as yourself" (i.e. be fair and give the concerns of others [and nature] as much weight as yours) is the simple

---

[101]FuturICT 2.0: Towards a Sustainable Digital Society with a Socio-Ecological Finance System (Finance 4.0) http://ebook.finfour.net; Falling Walls Lab 2020: The FIN4 project: Towards a sustainable digital Society with A Socio-Ecological finance system https://bit.ly/2VJEqEE; Dapp et al. [31], Helbing [32].

[102]Grund et al. [33].

[103]Interaction Support Processor (2015) https://patents.google.com/patent/US20160350685A1/en, Helbing [34].

[104]Is Tech-Boosted Technology on Its Way, Forbes (December 4, 2018) https://www.forbes.com/sites/forbestechcouncil/2018/12/04/is-tech-boosted-telepathy-on-its-way-nine-tech-experts-weigh-in/, Telepathic Technology is Here, But Are We Ready? Singularity (July 23, 2015) https://www.singularityweblog.com/telepathic-technology-is-here-but-are-we-ready/.

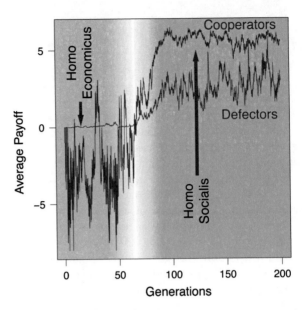

Fig. 14.6 Results of computer simulations showing the evolution of "homo socialis" with other-regarding preferences and "networked thinking", starting with a selfish type of human, called "homo economicus". The transition is expected to take dozens of generations, but it may just be happening… (Fig. 1A of Grund et al. [33]. Reproduction with kind permission of the Springer Nature Publishing Group.)

success principle, which will eventually be able to create prosperity and peace. We could have known this before, but now we have the scientific evidence for it…

14.4   Appendix 1: Success Principles for Our Future

In this book, I have argued that we need to allow diverse sets of rules to create a socio-economic system that serves a large variety of functions, but also to allow companies and people to experiment in order to find better rules for the future. Nevertheless, it would be favorable to have a number of globally shared fundamental principles—a guiding set of rules small enough for everyone to remember, which would support interoperability and peaceful co-existence.

As I have demonstrated before, in a strongly connected world, maximizing individual payoffs does not produce the best results. To avoid undesirable systemic insta-bilities and tragedies of the commons, superior principles are needed. The following set of fundamental rules is the result of extensive discussions I have had with many people. The similarity of these principles with those advocated by philosophers and world religions is probably not by chance. It is clear that these principles have been the foundation on which the success of many societies has been based for thousands of years. As I pointed out before, these cultural principles are more persistent than steel and more powerful than wars. They also create social capital, which is one of the preconditions for economic well-being. However, the rules below are particularly attuned to the problems implied by complex interdependencies, strong interactions, and the increasing importance of information, which are characteristic of our current and future world.

1. **Respect:** Treat all forms of life respectfully; protect and promote their (mental, psychological and physical) well-being.
2. **Diversity and non-discrimination:** Support socio-economic diversity and pluralism (also by the ways in which Information and Communications Technologies are designed and operated). Counter discrimination and repression, prioritize signaling and rewards over punishment.
3. **Freedom:** Support the principle of informational self-determination; respect creative freedom (opportunities for individual development) and the freedom of non-intimidating expression; abstain from mass surveillance.
4. **Participatory opportunities:** Enable self-determined decisions, offer participatory opportunities and a choice of good options. Ensure to properly balance the interests of all relevant (affected) stakeholders, particularly political and business interests, and those of citizens.
5. **Self-organization:** Create a framework to support flexible, decentralized, self-organized adaptation, e.g. by using suitable reputation systems.
6. **Responsibility:** Commit yourself to timely, responsible and sustainable actions, by considering their externalities.
7. **Quality and awareness:** Commit yourself to honest, high-quality information and good practices and standards; support transparency and awareness.
8. **Fairness:** Reduce negative externalities that are directly or indirectly caused by your own decisions and actions, and fully compensate the disadvantaged parties (in other words: "pay your bill"); reward others in a fair way for positive externalities.
9. **Protection:** Protect others from harm, damage, and exploitation; refrain from aggressive or war-like activities (including cybercrime, cyberwar, and misuse of information).
10. **Resilience:** Reduce the vulnerability of systems and increase their resilience.
11. **Sustainability:** Promote sustainable systems and long-term societal benefits; increase systemic benefits.
12. **Compliance:** Engage in protecting and complying with these fundamental principles.

To summarize the above even more briefly, the most important rule is to **increase positive externalities, reduce negative ones, and ensure fair compensation.** Some would just say "**Love nature and your neighbor as yourself!**"

This fundamental principle takes care of the implications of our interactions, and is probably enough to create a better world that will benefit everyone! Mastering our future isn't that complicated, after all!

Acknowledgements This work was partially funded by the European Community's H2020 Program under the funding scheme "FETFLAG-01-2018 (CSA)", grant agreement #820437, "Toward AI Systems that Augment and Empower Humans by Understanding us, our Society and the World Around Us—Humane AI" (https://www.humane-ai.eu).

References

1. J. Barrat, Our Final Invention: Artificial Intelligence and the End of the Human Era (Griffin, 2015).
2. D.H. Meadows, Limits to Growth (Signet, 1972).
3. E. Brynjolfsson and A. McAfee, The Second Machine Age (W.W. Norton, 2016).
4. R.M. Geraci, Apocalyptic AI: Visions of Heaven in Robotics, Artificial Intelligence, and Virtual Reality (Oxford University Press, 2012).
5. M.P. Mueller, Law without law: from observer states to physics via algorithmic information theory, https://arxiv.org/abs/1712.01826.
6. Phil Ball, Reality? It's What You Make It, New Scientist (November 11, 2017), pp. 29–32.
7. N. Silver, The Signal and the Noise: Why So Many Predictions Fail – But Some Don't (Penguin, 2015).
8. T. Vigen, Spurious Correlations (Hachette Books, 2015).
9. F. Pasquale, The Black Box Society (Harvard University Press, 2016).
10. C. Morin and P. Renvoise, The Persuasion Code: How Neuromarketing Can Help You Persuade Anyone, Anywhere, Anytime (Wiley, 2018).
11. F. Schirrmacher (ed.) Technologischer Totalitarismus (Suhrkamp, 2015).
12. D. Helbing and S. Balietti, How to Create an Innovation Accelerator, EPJ ST 195, 101–136 (2011) https://link.springer.com/article/10.1140/epjst/e2011-01403-6.
13. J. Nagler, J. van den Hoven, and D. Helbing, An Extension of Asimov's Robotics Laws, in: D. Helbing (ed.) Towards Digital Enlightenment (Springer, 2018) https://link.springer.com/chapter/10.1007/978-3-319-90869-4_5, also available as SSRN preprint (February 7, 2018) https://papers.ssrn.com/sol3/papers.cfm?abstract_id=3110582.
14. E. Awad et al. The Moral Machine experiment, Nature 563, 59–64 (2018) https://www.nature.com/articles/s41586-018-0637-6.
15. R. Scranton, Learning to Die in the Anthropocene: Reflections on the End of a Civilization (City Lights Publishers, 2015).
16. D.H. Meadows, J. Randers, and D.L. Meadows, Limits to Growth: The 30-Year Update (Chelsea Green Publishing, 2004).
17. F. Hamburg, Een Computermodel voor het Ondersteunen van Euthanasiebeslissingen (Maklu, 2005) https://books.google.ch/books/about/Een_computermodel_voor_het_onders teunen.html?id=eXqX0Ls4wGQC.
18. S. Kühl, The Nazi Connection: Eugenics, American Racism, and German National Socialism (Oxford University Press, 1994).
19. A. King and B. Schneider, The First Global Revolution: A Report by the Council of the Club of Rome, https://ia800202.us.archive.org/33/items/TheFirstGlobalRevolution/TheFirstG lobalRevolution.pdf.

20. R.O. Murphy, K.A. Ackermann and M. Handgraaf, Measuring Social Value Orientation, Judgment and Decision Making 6(8), 771–781 (2011) http://journal.sjdm.org/11/m25/m25. pdf.

21. A. King and B. Schneider, The First Global Revolution (Simon & Schuster Ltd, 1991).

22. D. Helbing and E. Pournaras, Build Digital Democracy, Nature 527, 33–34 (2015): http://bit. ly/1WCSzi4.

23. D. Helbing, Why We Need Democracy 2.0 and Capitalism 2.0 to Survive, Jusletter IT (May 25, 2016) https://jusletter-it.weblaw.ch/en/issues/2016/25-Mai-2016/why-we-need-dem ocrac_72434ad162.html, also available as preprint at https://www.researchgate.net/public ation/303684254.

24. T. Asikis, J. Klinglmayer, D. Helbing, and E. Pournaras (2020) How Value-Sensitive Design Can Empower Sustainable Consumption, preprint, https://arxiv.org/abs/2004.09180.

25. D. Helbing, The Automation of Society Is Next: How to Survive the Digital Revolution (CreateSpace, 2015) (this Springer book is an extended version).

26. D. Helbing, Why We Need Democracy 2.0 and Capitalism 2.0 to Survive, in: D. Helbing (ed.) Towards Digital Enlightenment (Springer, 2018) https://link.springer.com/chapter/10. 1007/978-3-319-90869-4_11, previously appeared as FuturICT Blog (April 21, 2016) http:// futurict.blogspot.com/2016/04/why-we-need-democracy-20-and-capitalism.html.

27. S.E. Page, The Difference: How the Power of Diversity Creates Better Groups, Firms, Schools, and Societies (Princeton University Press, 2008).

28. A.W. Woolley et al. Evidence for a Collective Intelligence Factor in the Performance of Human Groups, Science 330, 686–688 (2010) https://science.sciencemag.org/content/330/6004/686. short.

29. D. Helbing and T. Vicsek, Optimal self-organization, New Journal of Physics 1, 13 (1999) https://iopscience.iop.org/article/10.1088/1367-2630/1/1/313/meta.

30. D. Helbing, Quantitative Sociodynamics (Springer, 2010).

31. M.M. Dapp, S. Klauser, and M. Ballandies, Finance 4.0 Concept (February 2018) https:// www.researchcollection.ethz.ch/bitstream/handle/20.500.11850/286469/D.3.2ReportonFin ance4.0Concept(M12report)-PUBLICv2.pdf.

32. D. Helbing, Economics 2.0: The Natural Step towards a Self-Regulating, Participatory Market Society, Evolutionary and Institutional Economics Review 10, 3–41 (2013) https://link.spr inger.com/article/10.14441/eier.D2013002.

33. T. Grund, C. Waloszek, and D. Helbing, How Natural Selection Can Create Both Self- and Other-Regarding Preferences and Networked Minds, Scientific Reports 3, 1480 (2013) https:// www.nature.com/articles/srep01480.

34. D. Helbing, Social Mirror: More Success Through Awarenss and Coordination, In: D. Helbing (ed.) Towards Digital Enlightenment (Springer, 2018) https://link.springer.com/chapter/10. 1007/978-3-319-90869-4_17.

Epilogue

Why did I write this book and how did it come about? My research activities, on which this book is based, had three essential triggers. While I studied statistical physics in Göttingen, Germany, I specialized in modeling complex systems and became increasingly interested in addressing real-life problems, such as the conflicts between neo-Nazis and left-wing students that were happening at that time.

Later, when I was Managing Director of the Institute of Transport & Economics at Dresden University of Technology, I was shocked by September 11, 2001. Just a year later, the beautiful city of Dresden was terribly flooded. This led me to study disasters and how to respond to them.

Finally, when I worked as a professor of sociology at ETH Zurich in Switzerland, the financial crisis in 2008 clearly revealed that we needed entirely new ways of thinking about socio-economic systems. As a result of ongoing cascading effects, the emerging economic crisis would eventually result in political extremism and increasing levels of societal conflict—a scenario that had to be avoided.

To address these concerns, I launched the FuturICT initiative back in 2010.[1] This was a response to a European call for proposals for two €1 billion "flagship" projects to boost innovation in the information and technology sector—so we are talking about the perspective of significant funding, here. The FuturICT project aimed to develop new scientific and technological systems to manage our future in an increasingly complex world. In particular, FuturICT wanted to develop a "Living Earth Simulator" that would allow us to explore and understand the opportunities and risks implied by possible decisions we might take. This simulator was intended to be an open and participatory platform, committed to the protection of personal privacy.[2]

FuturICT was (and—in various aspects—still is) recognized as a highly innovative project, bringing the best researchers in the social, natural, and engineering sciences together. Eventually, we established a global network of interdisciplinary

---

[1] See http://www.futurict.eu.

[2] Helbing et al. [1], http://link.springer.com/article/10.1140/epjst/e2012-01686-y#page-1, http://arxiv.org/abs/1211.2313.

research communities, including more than 25 European countries, the USA, Japan, Singapore, Australia, and many other states. Far more than 100 academic institutions and a similar number of companies wanted to be partners of the project. As a result, about €90 million of co-funding had been promised for the first 2.5 years.

Leading the FuturICT project, however, also turned out to be an adventure. Big players from all over the world became interested in the project. The USA quickly launched Big Data research programs amounting to $150 million or more. China broadcast a film about the project on national TV, watched by hundreds of millions of people. Russia sent big TV teams to cover the project. Moreover, on the title page of the Christmas edition of the Scientific American in 2011, FuturICT was featured as No. 1 world-changing idea.[3]

Months later, the FuturICT project was in the final round and performed well, but contrary to everyone's expectations, it wasn't funded. At this point in time, I started to worry that governments might enter a digital arms race rather than building the global, participatory information and communication system that FuturICT had proposed. I also worried that a powerful surveillance-based digital society may be brought on the way, ignoring privacy, informational self-determination, democracy and human rights. Therefore, I wrote an article entitled "*Google* as God?" to make the public aware of the potential dangers of information and communication systems that lack transparency. In fact, the digital revolution created the scary possibility to control entire societies and every one of us.

When I later searched the Internet for "*Google* as God?", I was surprised to find results such as the "Church of *Google*". While these sites are perhaps not meant to be serious, they provide interesting food for thought nevertheless. At some webpages[4] one can find "proofs" that *Google* is God, for example: "Google is the closest thing to an Omniscient (all-knowing) entity in existence" or "*Google* is everywhere at once (Omnipresent)" or "*Google* can 'do no evil' (Omnibenevolent)". My book explored these ideas further and asks how realistic it is to realize such ambitions.

When I talked about a "crystal ball", "magic wand", "wise king" or "benevolent dictator", I used these terms to represent some abstract concepts. Even when I used names for illustration, I didn't mean any particular company or institution, such as *Google*, the *NSA*, or a particular computer network with Artificial Intelligence. Nevertheless, we must ask ourselves where we are heading. Could we unintentionally create a digital nightmare, even if all the Big Data companies and institutions had the very best intentions? And if so, what can we do to minimize the risk of such a scenario? In other words, what institutions and technological solutions will the digital society need? How much decentralization and encryption are required to ensure sufficient Internet security? How much transparency and informational self-determination are needed in order to enter an age of digital enlightenment, i.e. to avoid "digital slavery" and ensure "digital freedom"?

---

[3]See https://www.scientificamerican.com/article/the-machine-that-would-predict/; in the meantime, several FuturICT-like projects are running or being prepared in the world, but most of them don't seem to be participatory and public, in contrast to what we proposed.

[4]Such as http://www.thechurchofgoogle.org/Scripture/Proof_Google_Is_God.html.

Without any doubt, we must develop a better understanding of the new world we are currently creating—a world characterized by more data, more processing speed and more connectivity. We are at a crossroads, where we might mistakenly take the wrong turn, which would lead to more instability and, potentially,global disaster.

Remember that, even with the best technology ever, huge amounts of information, and the very best intentions, our world might become impossible to control. A good analogy of the hyper-connected world we are living in is perhaps an atomic bomb, which may explode as a result of chain reactions that are triggered when a certain "critical mass" (a critcal density) is reached. It turns out that similar kinds of "explosions" happen in socio-economic systems, too. They are much slower, but similarly destructive—think, for example, of a political revolution, a war, or the collapse of a civilization. Has our global system unintentionally become a "complexity time bomb"? If so, was it already ticking?

I certainly don't want to worry you, but history tells what the results of such "explosive" socio-economic processes can be. This includes political instabilities and regimes that might be unjust and cruel. It is important to avoid these scenarios by developing a better understanding of the causes of societal problems such as economic crises, crime, and war. In this book, I tried to provide a new and integrated perspective of how society works, and how we can use this knowledge to master our future. In fact, I believe we shouldn't be too pessimistic. We should rather take the future into our own hands, because we can make our society more resilient to crises and change the world for the better.

Further Reading

David Bornstein and Susan Davis, Social Entrepreneurship: What Everyone Needs to Know (Oxford University Press, Oxford, 2010).

Nick Bostrom and Milan M. Cirkovic (eds.) Global Catastrophic Risks (Oxford University Press, Oxford, 2008).

Joel Luc Cachelin, Baustellen der Digitalen Wissensgesellschaft (Wissensfabrik, 2013), see https://www.wissensfabrik.ch/pdfs/baustellen.pdf.

Joel Luc Cachelin, Schattenzeitalter (Stämpfli Verlag, Bern, 2014).

Joel Luc Cachelin, Offliner (Stämpfli Verlag, Bern, 2015).

John L. Casti, X-Events: The Collapse of Everything (William Morrow, 2012).

David Colander and Roland Kupers, Complexity and the Art of Public Policy: Solving Society's Problems from the Bottom Up (Princeton University Princeton NJ, 2014).

W.H. Dutton, G. Law, G. Bolsover, and S. Dutta, The Internet Trust Bubble: Global Values, Beliefs and Practices (The World Economic Forum, 2013), see http://www3.weforum.org/docs/WEF_InternetTrustBubble_Report2_2014.pdf.

Richard Florida, The Rise of the Creative Class … and how it's transforming work, leisure, community & everyday life (Basic Books, New York, 2002).

Karin Frick and Bettina Höchli, Die Zukunft der vernetzten Gesellschaft: Neue Spielregeln, neue Spielmacher (Gottlieb Duttweiler Institute, Zürich, 2014), s. https://www.gdi.ch/de/neue-gdi-studie-die-zukunft-der-vernetzten-gesellschaft-neue-spielregeln-neue-spielmacher, https://www.researchgate.net/publication/305659563.

Francis Fukuyama, Trust: The Social Virtues and the Creation of Prosperity (Free Press, New York, 1995).

Glenn Greenwald, No Place to Hide (Metropolitan Books, New York, 2014).

© Springer Nature Switzerland AG 2021
D. Helbing, *Next Civilization*,
https://doi.org/10.1007/978-3-030-62330-2

Christopher Hadnagy, Social Engineering: The Art of Human Hacking (Wiley, Indianapolis, 2011).

David Halpern, Social Capital (Polity, Malden MA, 2005).

Robert Hassan, The Information Society (Polity, Cambridge, 2008).

Dirk Helbing (ed.) Managing Complexity: Insights, Concepts, Applications (Springer, Berlin, 2008).

Dirk Helbing (ed.) Social Self-Organization: Agent-Based Simulations and Experiments to Study Emergent Social Behavior (Springer, Berlin, 2012).

D. Helbing, Thinking Ahead: Essays on Big Data, Digital Revolution, and Participatory Market Society (Springer, Berlin, 2015).

D. Helbing (ed.) Towards Digital Enlightenment: Essays on the Dark and Light Sides of the Digital Revolution (Springer, Berlin, 2019).

Silke Helfrich und Heinrich-Böll-Stiftung (eds.) Commons: Für eine neue Politik jenseits von Markt und Staat (Transcript, Bielefeld, 2012).

Yvonne Hofstetter, Sie wissen alles: Wie intelligente Maschinen in unser Leben eindringen und warum wir für unsere Freiheit kämpfen müssen (Bertelsmann, München, 2014).

Steven Johnson, Future Perfect: The Case for Progress in a Networked Age (Riverhead, New York, 2012).

McKinsey Global Institute, Disruptive Technologies: Advances that will transform life, business, and the global economy (May 2013), see https://www.mckinsey.com/business-functions/mckinsey-digital/our-insights/disruptive-technologies.

Rudi Klausnitzer, Das Ende des Zufalls (Ecowin, Salzburg, 2013).

Thomas R. Köhler, Der programmierte Mensch: Wie uns Internet und Smartphone manipulieren (Frankfurter Allgemeine, 2012).

Loet Leydesdorff, The Knowledge-Based Economy: Modeled, Measured, Simulated (Universal Publishers, Boca Raton, 2006).

Klaus Mainzer, Die Berechnung der Welt: Von der Weltformel zu Big Data (C.H. Beck, München, 2014).

Viktor Mayer-Schönberger and Kenneth Cukier, Big Data: A Revolution That Will Transform How We Live, Work, and Think (Eamon Dolan/Mariner Books, 2014).

Miriam Meckel, Wir verschwinden (Kein & Aber, 2013).

Enrico Moretti, The New Geography of Jobs (Houghton Mifflin Harcourt, Boston, 2012).

Risk Nexus, Beyond Data Breaches: Global Interconnections of Cyber Risk (Atlantik Council/Zurich Insurances, April 2014), see https://www.atlanticcoun cil.org/in-depth-research-reports/report/beyond-data-breaches-global-interconnect ions-of-cyber-risk/.

Peter Schaar, Überwachung total: Wie wir in Zukunft unsere Daten schützen (Aufbau Verlag, Berlin, 2014).

H. Schaffers et al. Smart cities and the future Internet: Towards cooperation frameworks for open innovation (Springer, 2011).

Kai Schlieter, Die Herrschaftsformel: Wie Künstliche Intelligenz uns berechnet, steuert und unser Leben verändert (Westend, 2015).

Rupert Scofield, The Social Entrepreneur's Handbook (McGraw Hill, New York, 2011).

Edition Unseld (ed.): Big Data - Das neue Versprechen der Allwissenheit (Suhrkamp, Berlin, 2013).

The White House, Consumer Data Privacy in a Networked World: A Framework for Protecting Privacy and Promoting Innovation in the Global Digital Economy (2012), see https://web.archive.org/web/20120304042853/ https://web.archive.org/web/ 20120304042853/https://www.whitehouse.gov/sites/default/files/privacy-final.pdf.

The World Economic Forum, Delivering Digital Infrastructure: Advancing the Internet Economy (April 2014), see http://www3.weforum.org/docs/WEF_TC_Del iveringDigitalInfrastructure_InternetEconomy_Report_2014.pdf.

The World Economic Forum, Global Risks 2015, see http://www3.weforum.org/ docs/WEF_Global_Risks_2015_Report15.pdf.

The World Economic Forum, Rethinking Personal Data: A New Lens for Strength-ening Trust (May 2014), see http://www3.weforum.org/docs/WEF_RethinkingPe rsonalData_ANewLens_Report_2014.pdf.

Muhammad Yunus, Building Social Business: The New Kind of Capitalism that Serves Humanity's Most Pressing Needs (Public Affairs, New York, 2010).

About the Author

Dirk Helbing is perhaps one of the most imaginative experts in the world when it comes to envisioning the opportunities and risks of the digital revolution. He is an advocate of value-sensitive design and strongly contributed to the public debate around Big Data and Artificial Intelligence. He also coordinated the FuturICT initiative, which built a global interdisciplinary community of experts at the interface of complexity, computer and data science. These activities, aimed at confronting global problems and crises were featured by Scientific American as the number one world-changing idea and earned him an honorary doctorate from TU Delft, where he later coordinated the PhD school on "Engineering Social Technologies for a Responsible Digital Future".

Dirk Helbing is Professor of Computational Social Science at the Department of Humanities, Social and Political Sciences and also affiliated to the Computer Science Department at ETH Zurich. He has a Ph.D. in physics, was Managing Director of the Institute of Transport & Economics at Dresden University of Technology in Germany, and Professor of Sociology at ETH Zurich.

Helbing is an elected member of the German Academy of Sciences "Leopoldina" and worked for the World Economic Forum's Global Agenda Council on Complex Systems for some time. He was also a co-founder of the Physics of Socio-Economic Systems Division of the German Physical Society and ETH Zurich's Risk Center. Furthermore, he is a member of the International Centre for Earth Simulation in Geneva and various high-level committees to assess the implications of the digital revolution. He was also a board member of the Global Brain Institute in Brussels.

The motivation for his research may be summarized by "What can complexity science and information systems contribute to saving human lives?" This ranges from avoiding crowd disasters over reducing crime and conflict to the reduction of epidemic spreading. His work brings theoretical studies, data science, and lab experiments together with agent-based computer models, where agents may have cognitive features. Furthermore, his publication on globally networked risks called for a Global Systems Science.

Using the "Internet of Things", his team was also engaged in establishing the core of a decentralized Digital Nervous System as a Citizen Web (see nervousnet.info).

© Springer Nature Switzerland AG 2021
D. Helbing, *Next Civilization*,
https://doi.org/10.1007/978-3-030-62330-2

This was intended to be an open, transparent and participatory information platform to support real-time measurements of our world, situational awareness, successful decision-making, and self-organization. The goal of this system was to open up the new opportunities of the digital age for everyone, but the project was heavily obstructed.

Photo: Jannick Timm

Reference

1. D. Helbing, S. Bishop, R. Conte, P. Lukowicz, and J.B. McCarthy (2012) FuturICT: Participatory computing to understand and manage our complex world in a more sustainable and resilient way, EPJ Special Topics 214, 11–39.

Printed in the United States
By Bookmasters